Bioimpedance in Biomedical Applications and Research

Franco Simini · Pedro Bertemes-Filho
Editors

Bioimpedance in Biomedical Applications and Research

Springer

Editors
Franco Simini
Núcleo de Ingeniería Biomédica de las
Facultades de Medicina e Ingeniería
Universidad de la Republica
Montevideo, Uruguay

Pedro Bertemes-Filho
Departamento de Engenharia Elétrica
Universidade do Estado de Santa Catarina
Joinville, Santa Catarina, Brazil

ISBN 978-3-030-08978-8 ISBN 978-3-319-74388-2 (eBook)
https://doi.org/10.1007/978-3-319-74388-2

Softcover re-print of the Hardcover 1st edition 2018

Printed on acid-free paper

This Springer imprint is published by the registered company Springer International Publishing AG part of Springer Nature.
The registered company address is: Gewerbestrasse 11, 6330 Cham, Switzerland

Dedicated to people of Latin America and Caribbean and within each of them to the medical communities, which include biomedical engineering groups, companies, researchers, and students related to or studying bioimpedance applications. The effort of writing this book is intended to match the ongoing efforts of all Countries struggling for a better life, materially and spiritually speaking.

Preface

A book is both an end point and a starting point. The present collection of chapters on bioimpedance and its applications goes to press after a collective work was started during academic meetings in Latin America and Europe. The elegance and relatively low financial requirements of our research are part of the enthusiasm the reader will find in between lines of text and behind the figures of the book. The initial meetings during which researchers shared doubts and achievements were followed by the idea of putting the experience together under a form that could be disseminated with fellow staff and students, as well as with industry. It took 2 years of work to get to this end point: publishing the book.

But a book is also a starting point, and there will be questions as readers come to the results of research, always ongoing and partial by definition. A great number of new instruments will be designed in the near future based on bioimpedance, an intrinsically innocuous measurement. Imagine ions being just temporarily disturbed at, say, 50 kHz to characterize tissue using spectroscopy or tomography. Too few of them are present in our hospitals and clinics, especially if compared to instruments based on other physical principles, either inaccurate or dangerous. We hope that as many professional designers, clinical users, and academics will adopt the present book as a starting point for their original ideas to be tested under investigation by trial and error. We are therefore asking for more research, more clinical trials, more basic physics applied to the use of electrical currents in biological tissues, and more mathematics to ease the computation of bioimpedance raw data.

The only part of the history of a book that can be included in its pages is its gestation, leaving the battles, limitations, and successes to be added later and disseminated by others. This book is no exception, and it is interesting to put in words how it came about. It all started during a transfer on the Paraná River in the Province of Entre Ríos, Argentina, under a clear sky, with an open landscape, huge amounts of slow water, and only the mind boggling extensive Pampa prairies to be seen. It was during the equally magnificent 2014 Latin American Congress on Biomedical Engineering CLAIB. The editors were discussing papers written by colleagues, discovering complementarities of different research groups. A first decision to unite valuable material into a useful tool was made, with the hope of finding support that would match their enthusiasm. During the 2015 Toronto World Congress on Medical Physics and Biomedical Engineering, the editors were discussing publication matters at the SPRINGER booth where a general agreement was reached and work started during the following months.

Now that the book is in print, the editors are happy to hand it to the readers with the hope that it may help foster research and clinical use as well as biomedical research both using bioimpedance and developing knowledge on bioimpedance itself.

Joinville, Brazil — Pedro Bertemes-Filho
Montevideo, Uruguay — Franco Simini

Acknowledgments

We thank the **Universidad de la República** (Uruguay) and the **Universidade do Estado de** Santa **Catarina** (Brazil), in which the editors are full-time professors, for supporting in several indirect ways by allowing the long and continuous dedication put into the writing, supervision, and compilation of this book.

We also thank our families, both near and extended, for their unconditional support, especially in difficult moments along the months the book took to see the day.

This bioimpedance volume would not have existed without the professional encouragement and trust offered by Marta Moldvai and Christian Baumann and their co-workers at Springer. Ms. Moldvai, upon getting word from CLABIO (Congresso Latino Americano de \mpedancia), supported our idea of a completely new book, extrapolated from the best contributions. Consequently, the authors thank the whole CLABIO community for their competitive work toward the present editorial milestone, including those who did not match the high standard of the reviewing process.

We should not forget to thank the professors of the International Society of Bioelectrical Impedance (ISEBI) in helping us in the reviewing process.

The editors thank, last but certainly not the least, the authors of the chapters who have taken a considerable part of their time to put their knowledge in a way that could be understood by students and scholars alike, sharing their experience to be used for the improvement of bioimpedance applications.

Contents

1 **Introduction** 1
Franco Simini

2 **Electrical Impedance Spectroscopy** 5
Pedro Bertemes-Filho

3 **Basic Electrical Impedance Tomography** 29
Richard Bayford

4 **Electrical Impedance Tomography to Detect Trends in Pulmonary Oedema** 45
Franco Simini, Eduardo Santos, and Martín Arregui

5 **Electrical Impedance Signal Analysis for Medical Diagnosis** 65
Francisco Miguel Vargas Luna, Marco Balleza-Ordaz, María Raquel Huerta Franco, and Pere Riu

6 **Tissue Engineering Instrumentation Based on Electrical Impedance Measurements** 87
Marcio Nogueira de Souza

7 **Basics of Numerical Simulations of Bioimpedance Phenomena** 101
Dejan Križaj

8 **Numerical Basics of Bioimpedance Measurements** 117
Alexander Danilov, Sergey Rudnev, and Yuri Vassilevski

9 **Focused Impedance Method: Basics and Applications** 137
Khondkar Siddique-e Rabbani

10 **Clinical Applications of Electrical Impedance Spectroscopy** 187
Carlos-Augusto González-Correa

11 **Body Composition by Bioelectrical Impedance Analysis** 219
Clara Helena González-Correa

12 **Bioimpedance for Analysis of Body Composition in Sports** 243
Lucia Malá, František Zahálka, and Tomas Maly

13 **Wavelet Analysis in Impedance Rheocardiography** 257
Rodion Stepanov, Andrey Dumler, Sergey Podtaev, and Peter Frick

Index 271

List of Contributors

Martín Arregui Núcleo de Ingeniería Biomédica de las Facultades de Medicina e Ingeniería, Universidad de la República, Montevideo, Uruguay

Marco Balleza-Ordaz Physical Engineering Department, Universidad de Guanajuato, Guanajuato, Mexico

Richard Bayford Biophysics and Engineering, Middlesex University, London, UK

Pedro Bertemes-Filho Departamento de Engenharia Elétrica, Universidade do Estado de Santa Catarina, Joinville, Santa Catarina, Brazil

Alexander Danilov Institute of Numerical Mathematics, Russian Academy of Sciences, Moscow, Russia

Moscow Institute of Physics and Technology, Moscow, Russia

Sechenov University, Moscow, Russia

Andrey Dumler E.A. Vagner Perm State Medical University, Perm, Russia

Peter Frick Institute of Continuous Media Mechanics, Perm, Russia

Carlos-Augusto González-Correa Universidad de Caldas, Grupo de Bio-Impedancia Eléctrica (GruBIE), Manizales, Caldas, Colombia

Clara Helena González-Correa Research Group on Bioelectrical Impedance, Department of Basic Sciences for Health, Universidad de Caldas, Manizales, Colombia

María Raquel Huerta Franco Applied Science to Work, Universidad de Guanajuato, Guanajuato, Mexico

Dejan Križaj Laboratory for Bioelectromagnetics, Faculty of Electrical Engineering, University of Ljubljana, Ljubljana, Slovenia

Lucia Malá Faculty of Physical Education and Sport, Charles University, Prague, Czech Republic

Tomas Maly Faculty of Physical Education and Sport, Charles University, Prague, Czech Republic

Marcio Nogueira de Souza Biomedical Engineering Program – COPPE/Federal University of Rio de Janeiro, Rio de Janeiro, Brazil

Sergey Podtaev Science Technology Center "FM Diagnostics", Perm, Russia

Pere Riu Departament d'Enginyeria Electrònica, Universitat Politecnica de Catalunya, Barcelona, Spain

Sergey Rudnev Institute of Numerical Mathematics, Russian Academy of Sciences, Moscow, Russia

Lomonosov Moscow State University, Moscow, Russia

Federal Research Institute for Health Organization and Informatics, Moscow, Russia

Eduardo Santos Núcleo de Ingeniería Biomédica de las Facultades de Medicina e Ingeniería, Universidad de la República, Montevideo, Uruguay

Khondkar Siddique-e Rabbani Department of Biomedical Physics & Technology, University of Dhaka, Dhaka, Bangladesh

Franco Simini Núcleo de Ingeniería Biomédica de las Facultades de Medicina e Ingeniería, Universidad de la Republica, Montevideo, Uruguay

Rodion Stepanov Institute of Continuous Media Mechanics, Perm, Russia

Francisco Miguel Vargas Luna Physical Engineering Department, Universidad de Guanajuato, Guanajuato, Mexico

Yuri Vassilevski Institute of Numerical Mathematics, Russian Academy of Sciences, Moscow, Russia

Moscow Institute of Physics and Technology, Moscow, Russia

Sechenov University, Moscow, Russia

František Zahálka Faculty of Physical Education and Sport, Charles University, Prague, Czech Republic

List of Abbreviations

3D	Three dimensional
A/D	Analog/digital
AC	Alternate current
ACI	Acceleration index
ACT	Adaptive current tomography
ADC	Analog to digital converter
ALI	Acute lung injury
ALS	Amyotrophic lateral sclerosis
APT	Applied potential tomography
ARDS	Acute respiratory distress
ATP	Adenosine triphosphate
BC	Body composition
BCM	Body cell mass
BCMrel	Relative body cell mass
BCRL	Breast cancer-related lymphedema
BEM	Boundary element method
BF	Body fat
BH	Body height
BIA	Bioimpedance analysis
BIS	Bioelectrical impedance spectroscopy
BIVA	Bioelectrical impedance vector analysis
BM	Body mass
BMI	Body mass index
BNP	Brain natriuretic peptide
CAD	Computer-aided design
CC	Current carrying
CE	*Conformité Européene* (European Conformity)
CGAL	Computational Geometry Algorithms Library
CI	Cardiac index
CIMBIOS	Research Group on Semiconductor Material Science
CLDI	Chronic lung disease of infancy

CM	Common mode
CMRR	Common-mode rejection ratio
CO	Cardiac output
COPD	Chronic obstructive pulmonary disease
CPE	Constant phase element
CPU	Central processing unit
CQ	Cell quote
CT	Computed tomography
DAQ	Data acquisition
DC	Direct current
DDS	Direct digital synthesizer
DE	Doppler echocardiography
DSP	Digital signal processor
DXA	Dual-energy X-ray absorptiometry
EAV	Electroaccupuncture according to Voll
EBI	Electrical bioimpedance
EBIS	Electrical Bioimpedance spectroscopy
ECG	Electrocardiography
ECM	Extracellular mass
ECT	Electroconvulsive therapy
ECW	Extracellular water
EDA	Electrodermal activity
EDP	Electrodermal potentitals
EDR	Electrodermal response
EEG	Electroencepjalography
EI	Electrical impedance
EIDORS	Electrical Impedance Tomography and Diffuse Optical Tomography Reconstruction Software
EIM	Electrical impedance myography
EIS	Electrical impedance spectroscopy
EIT	Electrical impedance tomography
EITS	Electrical impedance tomography and spectroscopy
ELCM	Electrical caries monitor
EM	Electromagnetic
EMG	Electromyography
EOG	Electrooculography
EP	Evoked potentials
EPROM	Erasable programmable read-only memory
ES	Extracellular space
EVM	Evaluation module
EXC	Excitation coil
EXP	Extremely premature
FDA	Food and Drug Administration
FEM	Finite element model
FEMM	Finite element magnetics

FET	Field effect transistor
FFM	Fat-free mass
FFT	Fast Fourier transform
FIM	Focused impedance method
FM	Fat mass
FPGA	Field-programmable gate array
GA	Gestational age
GERD	Gastroesophageal reflux episodes
GMRES	Generalized minimal residual method
GREIT	Graz consensus reconstruction algorithm for EIT
GruBIE	Research Group on Electrical BioImpedance
GSR	Galvanic skin response
HD	Hydrodensitometry
HIV	Human immunodeficiency virus
Ht^2/R	$Stature^2$/resistance
ICG	Electric impedance cardiography
ICG	Impedance cardiography
ICM	Intracellular mass
ICU	Intensive care unit
ICW	Intracellular water
IIR	Infinite impulse response
ILU	Incomplete LU factorization
IME	Interdigitated microsensor electrodes
IMPETOM	Impedance tomography Montevideo
IPG	Impedance plethysmography
ISTI	Initial systolic time interval
LBM	Lean body mass
LF-MRCDI	Low-frequency magnetic resonance current density imaging
LVET	Left ventricular ejection time
McASP	Multichannel audio serial port
MeSH	Medical Subject Headings
MF-BIA	Multi-frequency bioelectrical impedance analysis
MFLOPs	Million floating point operations per second
MG	Mammography
MM	Muscle mass
MREIT	Magnetic resonance electrical impedance tomography
MRI	Magnetic resonance imaging
MSC	Mesenchymal stem cells
MUX	Multiplexer
NCO	Numerically controlled oscillator
NI	National instruments
NIB	Núcleo de Ingeniería Biomédica
NOSER	Newton's one-step error reconstructor
OA	Osteoarthritis
OpAmp	Operational amplifier

PCG	Phonocardiogram
PEEP	Positive end-expiratory pressure
PEP	Pre-ejection period
PET	Positron emission tomography
PGA	Programmable gain amplifier
PGRAD	Planar gradiometers
pH-MII	pH-multichannel intraluminal impedance
PU	Picking up
RDS	Respiratory distress syndrome
REBIS	Rectal electrical bioimpedance spectroscopy
RFBR	Russian Foundation for Basic Research
RISC	Reduced instruction set computer
ROI	Region of interest
RTCA	Real-time cell analyzer
SDRAM	Synchronous dynamic random-access memory
SEE	Standard estimated error
SF-BIA	Single-frequency bioelectrical impedance analysis
SNR	Signal-to-noise ratio
SoC	System on chip
STR	Systolic time ratio
SV	Stroke volume
SVR	Systemic vascular resistance
SVRI	Systemic vascular resistance index
TBW	Total body water
TCM	Traditional Chinese medicine
TDM	Time-division multiplexing
TEER	Transepithelial electrical resistance
TENS	Transcutaneous electrical nerve stimulation
TFC	Thoracic fluid content
TII	Total impedance index
TKA	Total knee arthroplasty
TNS	Transcutaneous nerve stimulation
TPIM	Tetrapolar impedance method
VCCS	Voltage-controlled current source
VCO	Voltage-controlled oscillator
VHP	Visible Human Project
VI	Velocity index
WebGL	Web-based Graphics Library

Chapter 1
Introduction

Franco Simini

Humanity is evolving rapidly from a biological entity to a mixture of biological and technical complex species, where the result of biological activity in the form of human intelligence adopts the shape and functions of technological objects. This is why our species could be called "human technical" rather than simply "homo sapiens." Medicine is one of the pretechnological characteristics of our species, a specificity by which some individuals develop a special empathy toward their peers and acquire anticipatory insight and knowledge of facts they apply to counsel consulting people in need. The other part of the medical dyad, i.e., the suffering person, goes to the healing man or doctor for relief, material solutions of psychological comfort. The essence of medicine has not changed since Hippocrates or even earlier times in its inclusion of somewhat magic expectations put by the patient into a transcendental entity and its revelation by some minister or a knowledgeable and empathic physician.

What has greatly changed over the centuries is the gradual accumulation of information, models, techniques, and experience to be used by the medical system with more or less empathy with people. Medical technology has benefited from research and industrial production of goods, instruments, and methods originally developed for other tasks. Mechanical, chemical, and logical tools have been applied for centuries by medical doctors to help diagnostic, therapeutic, or prosthesis matters. Electronics and information technology were also used since the end of the twentieth century, including an ever-increasing share of communication and artificial intelligence capacity.

F. Simini (✉)
Núcleo de Ingeniería Biomédica de las Facultades de Medicina e Ingeniería, Universidad de la República, Montevideo, Uruguay
e-mail: simini@fing.edu.uy

F. Simini, P. Bertemes-Filho (eds.), *Bioimpedance in Biomedical Applications and Research*, https://doi.org/10.1007/978-3-319-74388-2_1

Probably due to its near-magic and even spiritual nature, medicine has a tendency to include complicated, expensive, and sometimes dangerous elements to cure, let alone to diagnose, medical conditions. This is the case of ionizing radiations in medicine, as a result of available technology and the general sense of guilt derived from the first use of the atomic bomb in 1945. Modern hospitals use routinely very complex systems such as magnetic resonance imaging (MRI) and very complicated and potentially hazardous systems such as computed tomography (CT), positron emission tomography (PET), or linear accelerators. All have been evaluated and deemed reasonably safe to be used considering the benefits derived both from the diagnostic and treatment points of view. Nevertheless, they are the results of applied biomedical instrument research within the area of knowledge of ionizing radiation physics.

Other potential diagnostic or treatment methods based on less dangerous and expensive fundamentals are not so successful and widespread in medicine. Research on ultrasound and its complex interaction with biological matter, optic analysis of surfaces by spectroscopic illuminations, or even scent determinations have much weaker funding opportunities both by states and industry worldwide. The measurement of current and voltages in biological tissues or individuals is yet another neglected area of research, far behind expensive and "dangerous" technologies. A quick survey of research publications comparing such keywords as PET, CT, and linear accelerator with others such as "impedance spectroscopy," electrical impedance tomography (EIT), and body composition analyzer reveals two orders of magnitude difference of publications on ionizing applications with respect to the few on electrical impedance or bioimpedance.

This book is about bioimpedance and its clinical applications, including a general invitation to further research on the topic to unleash the great potential of simple physical principles such as the flow of ions in living tissues.

The first approach derives from the observation that electrical current is described as both voltage and current, giving rise to the concept of impedance, as well as frequency which allows to present spectroscopy in several biomedical domains. One third of the chapters can be associated with the differentiated behavior of tissues according to the current frequency. The second approach draws from signal processing and mathematical modeling of measurements such as tomographic reconstruction. Another third of the material in this book is of this kind. Finally, the information obtained from the elementary physical phenomena of flowing current is used to infer clinical conditions, a matching addressed in the remaining third of the book.

The book is indented to be used both as an introduction to research in the field of bioimpedance and as study material for those wishing to further innovate and investigate how electric currents can help diagnose, monitor, or compare medical conditions. Readers will find here many suggestions on how to mix wish lists of clinicians with physical theory, mathematical methods, and either existing or future technology.

The authors of the chapters of this book are active in almost a dozen countries, and all take active part in the triennial academic event started in 2012 in Joinville,

Brazil, followed by the Montevideo edition of 2015 and the present 2018 Latin American Conference on Bioimpedance (CLABIO) in Manizales, Colombia.

The reader has now the word, to inspect, study, and enjoy the result of simple and yet breathtaking research on bioimpedance.

Chapter 2
Electrical Impedance Spectroscopy

Pedro Bertemes-Filho

2.1 Electrical Impedance Spectroscopy Basics

The world we live in is surrounded by electromagnetic waves. When electromagnetic (EM) waves interact with matter, radiation may be transmitted (e.g. visible light is transmitted by water), scattered (e.g. small particles suspended in water scatter blue light preferentially) and absorbed (e.g. the red part of visible light is absorbed by copper sulphate solution which causes the remaining transmitted light to appear blue) (Hollas 1998).

The absorption of electromagnetic (EM) radiation can involve many processes, but they all depend upon frequency and hence energy. For example, very high frequency radiation is ionising. The spectrum of the nonionising EM waves is considered to be below the infrared waves, i.e. less than 3 THz. The equation $E = h\nu$ gives the relationship between the energy E of the radiation and the frequency ν, where $h = 6.626 \times 10^{-34}$ Js is the Planck's constant (Brown et al. 1999). Figure 2.1 shows some of the major components of the EM spectrum and the associated energies and frequencies. The energies are expressed in terms of electronvolt, i.e. $1 \text{ eV} = 1.6022 \times 10^{-22}$ kJ.

Any study of how absorption depends upon frequency may be considered as *spectroscopy*. They are in the range 100 Hz—10 MHz where absorption is mainly determined by the cellular structure of tissue. Thus we may define *electrical impedance spectroscopy* (EIS) as the study of the interaction of an alternating electrical field with biological tissue in the frequency range 100 Hz–10 MHz (see Fig. 2.1).

P. Bertemes-Filho (✉)
Departamento de Engenharia Elétrica, Universidade do Estado de Santa Catarina, Joinville, Santa Catarina, Brazil
e-mail: pedro.bertemes@udesc.br

F. Simini, P. Bertemes-Filho (eds.), *Bioimpedance in Biomedical Applications and Research*, https://doi.org/10.1007/978-3-319-74388-2_2

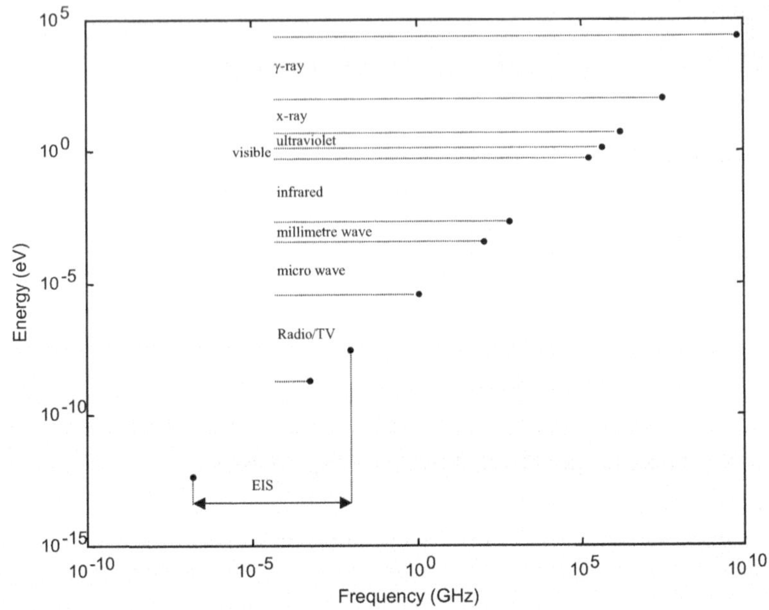

Fig. 2.1 Regions of the electromagnetic spectrum

EIS technique is related to the absorption phenomena of the EM waves, which is measured by varying the frequency of the applied "radiation" (e.g. electrical field) and plotting the "energy absorbed" (e.g. potential induced) by the sample at each frequency. The resultant plotting is called a *spectrum* which may contain the intensity of the absorption, i.e. electrical impedance.

The general approach in EIS is to apply an electrical stimulus (a known voltage or current) to the material and then to observe the resulting current or voltage. The stimulus can be applied in many forms. McAdams (1987) gives three mainly possible forms, as described in the following:

1. *Step function*: a step voltage $v(t)$ is applied at $t = 0$ to the material and a time-varying current $i(t)$ is measured. It is then Fourier-transformed into the frequency domain in order to calculate the frequency-dependent impedance.
2. *Noise signal*: a continuous voltage composed of random noise with energy over a known frequency range is applied to the material, and the resulting current is measured and then Fourier-transformed into the frequency domain.
3. *Sinusoidal signal*: a single-frequency voltage or current is applied to the material, and the resulting frequency-dependent current or voltage is measured. The response is measured in the frequency range of interest in terms of phase shift and amplitude or real and imaginary parts.

Other types of stimulus have been recently used in EIS (Min et al. 2007), such as chirp signal (Nahvi and Hoyle 2009) and binary multifrequency signals (Land et al.

2011; Yang et al. 2010). The first is based on short frequency-swept signals, whose advantage is that the characteristics of a biological object can be obtained in a wide frequency range during a very short measurement cycle (Paavle et al. 2012). The second is based on the excitation of waveforms with spectrally sparse distribution of frequency components, which is basically a technique for improving the Signal to Noise Ratio (SNR) of bioimpedance measurements.

Most EIS use current injection technique for stimulating the material under study, and then the resulting voltage is measured (Bertemes-Filho 2002). However, voltage stimulus may also be used and then the resulting current is measured instead. The current technique is a load-independent circuit, whereas the voltage one is more stable and easy to generate. It has been shown that non-linearities can arise in the electrodes used both to inject current and measure voltage (Pliquett et al. 2010). This effect is particularly important when using small electrodes with high impedance. Nevertheless, electrode-tissue interface issues have been investigated since Oliver Heaviside introduced the concept of electrical impedance in the 1880s.

EIS technique involves measuring electrical impedance Z, admittance Y, the impedance modulus $|Z|$, or the permittivity ε as a function of frequency in order to characterise the biological material under study. Cole and Cole (1941) showed that many tissues exhibit charges of complex permittivity ε^* which can be described by a relatively simple equation. This is considered later in this chapter.

EIS can be assisted by the use of electrical equivalent circuits to represent biological tissue, which is mostly studied over the last decades. Such equivalent circuits can be used to describe parameters which characterise a particular impedance spectrum. However, the representation of tissue is not easy as it must involve complex combination of resistors and capacitors to represent the conductivity and insulating components of tissue (Paterno et al. 2012). Some equivalent circuit models will be described in this chapter.

2.2 Tissue Impedance Basics

Tissues exhibit the properties of both conductors and dielectrics, i.e. they contain both free and bound (fixed) charges. As a result tissue impedance contains both conduction and dielectric terms. The conductivity term (σ) appears to account for the movement of free charges, and the relative permittivity term (ε_r) appears to account for movement of bound charges in the dielectric due to an applied electrical field of unit amplitude.

At the frequencies tissue can be considered as a collection of membranes separating intracellular and extracellular spaces. We can describe the electrical properties of tissue by considering that the current flows in the extracellular and in the intracellular fluids of tissue at low frequency and high frequency, respectively. The presence of current flow inside the cell is likely due to the cell membrane constituents. Cell membranes are composed mainly of proteins and water-insoluble lipids. Hence the membranes are highly resistive (conductivity less than 10^{-7} S/m)

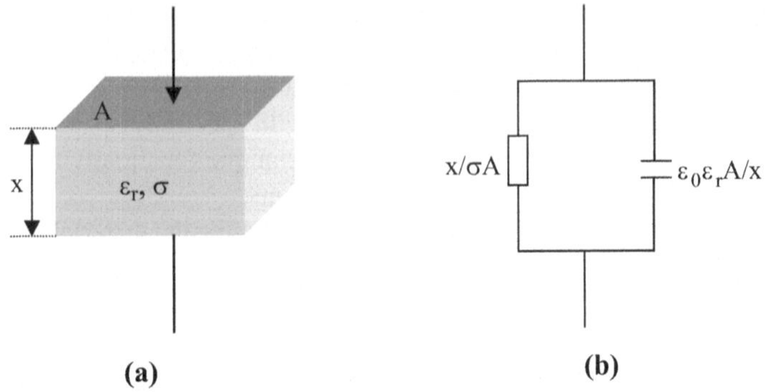

Fig. 2.2 (**a**) Idealised slab of tissue, where A is the area, x is the thickness, ε_r is the relative permittivity and σ is the conductivity. (**b**) Equivalent circuit of the slab of tissue represented by a resistance in parallel with a capacitance, where ε_0 is the permittivity of free space (8.85×10^{-12} F/m)

and mainly capacitive (typically 10^{-2} F/m^2) (Pethig 1984). Therefore, it is expected that impedance will drop with increasing frequency. Measurements have shown that intracellular resistivity is typically higher than 0.6 Ωm, whereas extracellular resistivity is within the range 0.5–0.6 Ωm (González-Correa et al. 1999).

A slab of tissue of area A and thickness x may have conductivity σ and relative permittivity ε_r, as shown in Fig. 2.2a. We can represent this tissue by a model consisting of a resistance in parallel with a capacitance, as shown in Fig. 2.2b. However, this model cannot explain the whole of tissue properties over a wide range of frequencies. The electrical properties may not be homogeneous within the slab and both σ and ε_r may be function of frequency.

Both relative permittivity ε_r and conductivity σ are found to be a function of frequency. The permittivity of a real dielectric, which is represented as a slab of tissue in figure, can be related to an admittance Y^*, where the symbol * represents a complex variable. Mathematically, the admittance Y^* can be considered either in terms of a complex conductivity σ^* (see Eq. 2.1) or in terms of a complex permittivity ε^* which is defined by a complex capacitance C^* (see Eq. 2.2).

$$Y^* = G + j\omega C = \frac{A}{x} \cdot (\sigma + j\omega\varepsilon_0\varepsilon_r) = \frac{A}{x} \cdot \sigma^* \tag{2.1}$$

$$C^* = \frac{Y^*}{j\omega} = \frac{A}{x} \cdot \left(-\frac{j\sigma}{\omega} + \varepsilon_0\varepsilon_r\right) = \frac{A}{x} \cdot \varepsilon_0\varepsilon^* \tag{2.2}$$

where ω is the angular frequency in radians and $j = (-1)^{1/2}$.

It can be noticed in Eq. (2.1) that as the frequency tends to zero, the conductivity becomes dominant, whereas permittivity is dominant at high frequencies (see Eq. 2.2).

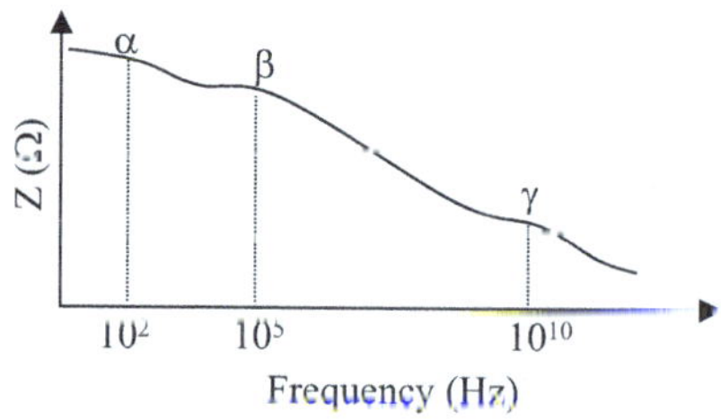

(a)

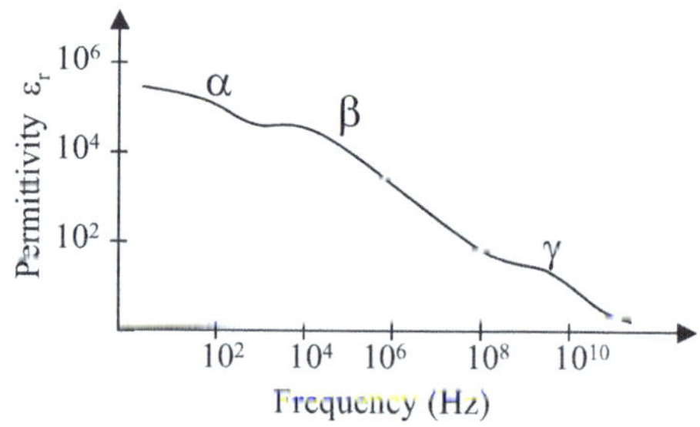

(b)

Fig. 2.3 (**a**) Theoretical impedance diagram showing the α, β and γ dispersions for muscle tissue; (**b**) an idealised plot of the relative permittivity as a function of frequency for a typical biological tissue (modified from Pethig 1987)

The frequency dependence of tissue is known as dispersion. Within the frequency range covered by this chapter, the cellular structure of tissue produces the dispersion. At higher frequencies, dispersion arises from the ability of molecules to reorient in an applied field. Dispersions also occur at lower frequencies but their origin is not clear. The dispersions within biological tissue are often separated into three regions: α (*alpha*), β (*beta*) and γ (*gamma*).

We would thus expect that the electrical impedance Z of biological materials would be frequency-dependent over a wide range of frequencies. For example, the major dispersions of the impedance for muscle tissue can be seen in Fig. 2.3a (Pethig 1987), which covers the range from 100 Hz to 20 GHz. The origin of the α dispersion is still not well understood. When collections of cells are considered, the α dispersion is important. The β dispersion results mainly from the capacitive charging of the cell membranes, and the γ dispersion results mainly from the dielectric relaxation of free water. Figure 2.3b shows an idealised representation of these dispersions according to the relative permittivity of most biological tissue against frequency. This shows that the effective capacitance of tissue falls with increasing frequency.

In practice, tissue is not totally homogeneous or isotropic. Many biological tissues are highly anisotropic; that is, the conductivity in different directions is not the same. Anisotropy can be represented by three principal conductivities in three mutually perpendicular directions as determined by the structure of the material. Foster and Schwan (1989) have concluded that muscle tissue exhibits an extreme anisotropy, mainly at *alpha* dispersion frequencies.

So far, we have seen that the electrical impedance of biological tissue decreases with increasing frequency and this dependence on frequency was due to the cell membrane, which behaves like a capacitor. By considering the extracellular and intracellular constituents of tissue, we can relate them to the electrical equivalent circuit, as shown in Fig. 2.4.

However, this cell model does not represent the behaviour of the membrane impedance over a wide frequency range. Debye produced a similar model by considering the impedance of a suspension of free dipoles. He produced Eq. 2.3 for the complex relative permittivity $\varepsilon_r{}^*$ of the suspension.

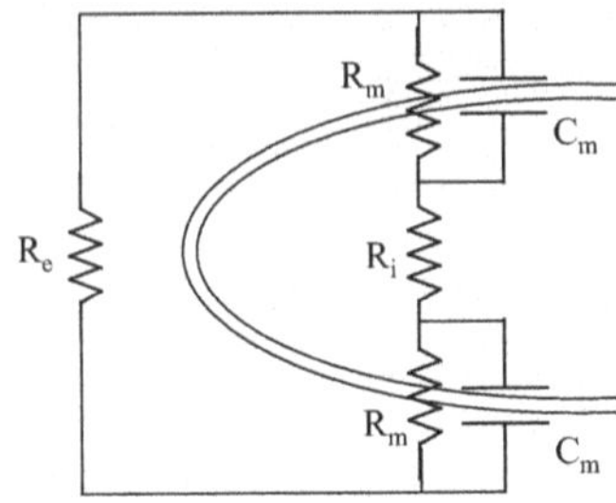

Fig. 2.4 Diagram schematic of the electrical cell model, where R_e is the extracellular fluid resistance, R_i is the intracellular fluid resistance, R_m is the membrane resistance and C_m is the membrane capacitance (redrawn from Webster 1990), which is here represented by the space between the ellipses

Fig. 2.5 Equivalent circuit for Eq. (2.3)

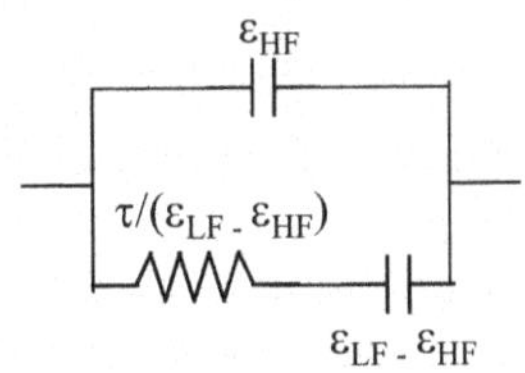

$$\varepsilon_r{}^* = \varepsilon_{\text{HF}} + \frac{\varepsilon_{\text{LF}} - \varepsilon_{\text{HF}}}{1 + j\omega\tau} \tag{2.3}$$

where $\varepsilon_r{}^*$ is the complex relative permittivity, ε_{HF} is the permittivity at high frequency, ε_{LF} is the permittivity at low frequency and τ is the relaxation time constant. The equivalent circuit representation of this equation is shown in Fig. 2.5.

However, neither the Debye model nor the cell model can predict values for the relative permittivity over a wide frequency range. Cole and Cole (1941) took into account dispersion by including a dimensionless distribution parameter α in Eq. (2.3). α can be chosen to produce a good fit between a measured spectrum and Equation 2.4. It should be pointed out that the Cole Equation (2.4) is not derived from a model but was an empirical modification to the Debye equation:

$$\varepsilon_r{}^* = \varepsilon_{\text{HF}} + \frac{\varepsilon_{\text{LF}} - \varepsilon_{\text{HF}}}{1 + (j\omega\tau)^{1-\alpha}} \tag{2.4}$$

where α is the dimensionless numerical constant to fit the data, which is not related to the *alpha* dispersion.

When $\alpha = 0$, Eqs. (2.3) and (2.4) are identical. The full analytic formulation for real and imaginary parts ($\varepsilon'$and ε'') of Eq. (2.4) can be seen in Cole and Cole (1941). As a result of Eq. (2.4), the membrane is a perfect capacitance (i.e. the membrane is fully ion impermeable) if α is equal to zero. On the other hand, the membrane is a pure resistance (i.e. the membrane is permeable for all ions) if α is equal to one.

2.3 Measuring Tissue Impedance

The ratio of the potential resulting between two electrodes in contact with tissue to the current injected between two other electrodes will be called "transfer impedance". All methods for measuring tissue transfer impedance use electrodes, which inject a known current into the tissue. The effect of electrodes on the measuring technique will be briefly described in this section.

Most tissue impedance spectroscopy systems involve the injection of a constant current into tissue at different frequencies and measuring the resultant voltages. The resultant voltage can be measured by using the same electrodes which were used for current injection (bipolar technique), or a separate pair of electrodes can be used for potential measurement (tetrapolar technique). Most tissue impedance measurements use the tetrapolar technique. However, both techniques will be described into this section. The accuracy of the measurements is dependent on the characteristics of the electronics. The key characteristics of these will also be described later in the chapter.

2.3.1 *The Bipolar Technique*

An impedance spectrum can be obtained by injecting current between two electrodes and measuring the resulting voltages between the same electrodes through which the current was injected. The impedance of skin and the electrode can be a problem for this kind of system due to unknown and varying contact impedance at each electrode site. Figure 2.6a shows the general concept of this system, where a voltage-controlled current source (VCCS) applies a constant current and a differential amplifier measures the voltage across the electrodes.

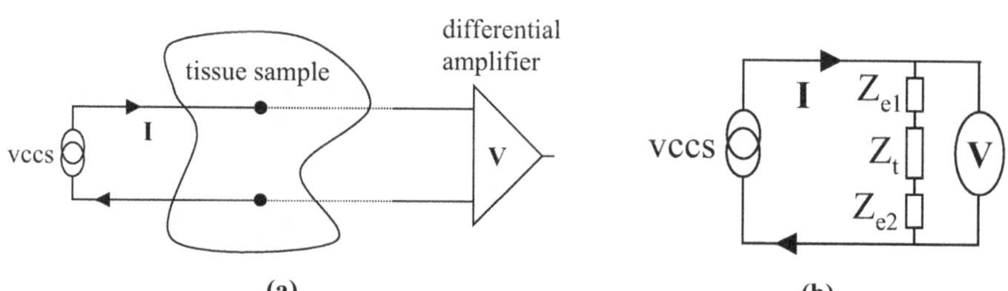

Fig. 2.6 (**a**) Bipolar constant current method to measure tissue impedance, where I is the drive current and V is the measured differential voltage; (**b**) equivalent circuit of two-electrode measurement, where Z_t is the tissue impedance and Z_{e1} and Z_{e2} are both the electrode impedances

It can be observed in Fig. 2.6b that there is only one possible measurement and there are three variables in the system (Z_{e1}, Z_{e2} and Z_t). Hence, the system is indeterminate. As a result, it is not possible to say if the change in the voltage V is due to tissue impedance Z_t changes or due to electrode impedance changes.

2.3.2 *The Tetrapolar Technique*

In the tetrapolar impedance method, a constant current is injected through one pair of electrodes, and the impedance-dependent voltage is measured with a second pair of electrodes. This technique was firstly introduced by Bouty in 1884. The method is relatively insensitive to changes in the electrode-subject impedance of both voltage-sensing and current-injecting electrodes (Raghed et al. 1992).

The insensitive conditions of the tetrapolar measurement are satisfied when the output impedance of the current generator Z_S and the input impedance of the differential amplifier Z_V are large as compared with the sum of the electrode and tissue impedances. In other words, the currents i_S and i_V can be assumed to be negligible. The tetrapolar measurement can be represented by an equivalent electrical circuit shown in Fig. 2.7.The equivalent circuit shown in Fig. 2.7 is a dual balance circuit (bipolar) for measuring the transfer impedance $T_Z = V/I$. By assuming equal electrode impedances, i.e. $Z_{e1} = Z_{e2} = Z_{e3} = Z_{e4} = Z$, the transfer impedance T_Z can be evaluated according to Eq. (2.5).

$$T_Z = \frac{Z_t}{1 + Z_t \cdot \left[\frac{4Z+Z_V+Z_S}{Z_V \cdot Z_S}\right] + 2Z \cdot \left[\frac{2Z+Z_V+Z_S}{Z_V \cdot Z_S}\right]} \tag{2.5}$$

In theory, the measured transfer impedance T_Z is assumed to be equal to the impedance Z_t measured across the tissue sample. However, it can be seen in

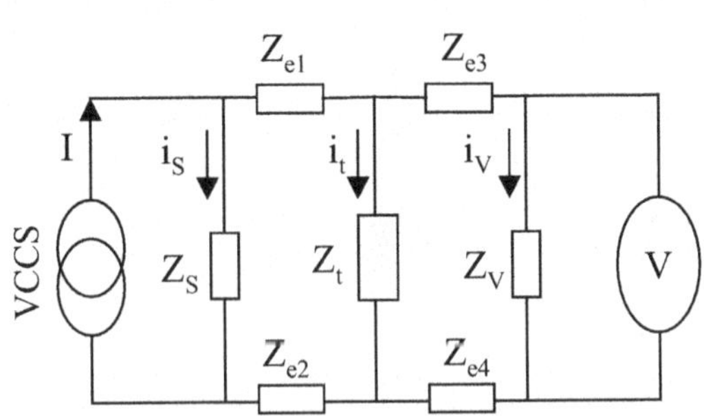

Fig. 2.7 Equivalent electrical circuit for the tetrapolar measurement, where I is the drive current, i_S is the equivalent leakage current because of the source impedance Z_S, i_t is the "real" current through the tissue, i_V is the current flowing through the input impedance Z_V of the differential amplifier and Z_{e1}, Z_{e2}, Z_{e3} and Z_{e4} are the electrode impedances

Eq. (2.5) that it will be the case only if Z_S and Z_V tend to infinity. In practice, stray capacitances connected to ground will decrease significantly Z_S and Z_V, especially at high frequencies. Hence, the non-idealised characteristics of the electronics will limit the measurements of tissue impedance.

2.3.3 *Basic Hardware Concepts*

Most transfer-impedance measurement systems use a constant multifrequency current for the driving system. Hence the most important characteristic in this system is the accuracy of the current generators, which should have a constant output current over a wide range of frequencies. Firstly, a voltage signal is generated in order to have it converted into a constant current. Most tissue impedance measurements use a sine wave or a pulse as the voltage signal in the drive system. The general characteristics of this will be described into this section.

As already shown here, there are mainly five types of methods to measure tissue impedance over a wide frequency range: (1) changing the frequency of the drive current and measuring the voltage (e.g. Griffiths and Ahmed 1987); (2) applying a current with a component containing multiple summed sine waves (e.g. Lozano et al. 1990); (3) applying a pulse, which contains a broad spectral energy, via current source (e.g. Record et al. 1992; Waterworth et al. 2000); (4) injecting either a current or voltage chirp signal (Nahvi and Hoyle 2009); and (5) applying current/voltage binary multifrequency signals (Land et al. 2011; Yang et al. 2010). The general principle of most EIS system for measuring the resulting voltage in order to calculate the transfer impedance is schematised in Fig. 2.8.

The sine wave is digitally generated and a VCCS circuit is used to convert it into a constant current. The voltage receive circuit measures the differential voltage between the electrodes. However, both circuits require a control circuit, for example, to select different electrode combinations from the probe and to set the voltage gain.

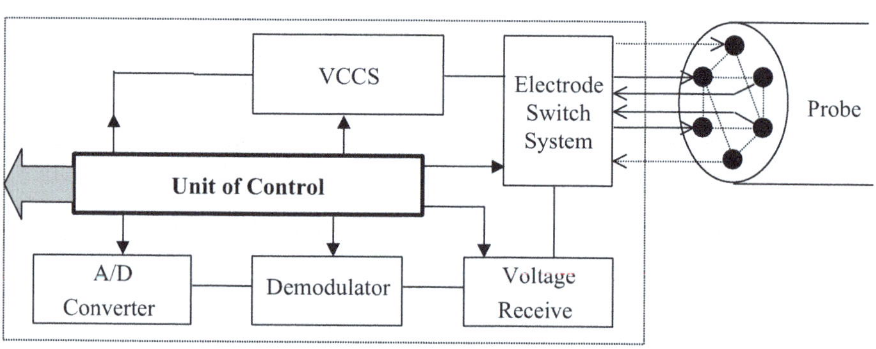

Fig. 2.8 General concepts of a system to measure transfer impedances by using a probe with six electrodes equally placed in a ring, where VCCS is the voltage-controlled current source

Other components required include a demodulator, low-pass filter and switches to measure the differential voltage from many different electrode combinations. Once the data are stored in memory, the measurements can be recovered and processed by computer, and then the transfer impedance can be calculated.

It is difficult to design a VCCS circuit to inject a constant current over a wide frequency range. The presence of parasitic capacitance to ground is a major factor which reduces the accuracy of the current injection circuitry. Furthermore, it is not always possible to get the desired performance over the whole frequency range because of other limiting factors, such as electrode impedance and nonideal characteristics of the instrumentation. The purpose of an EIS system is to measure the electrical impedance offered by electrodes placed on the biological material and then to interpret the results in terms of an equivalent resistivity or conductivity. In practice, the measured transfer impedance T_Z (the ratio of the measured voltage to the applied current) is a combination of the biological impedance, the electrode/tissue interface impedance and the impedance offered by the instrumentation involved in the measurements.

Generally, a VCCS circuit converts a sine wave voltage V_S into a current I_L whose magnitude is unaffected by load voltage V_L. Figure 2.9 shows a simple model of a monopolar current source, which means that both VCCS and load are grounded.

An analysis of the circuit will show that the voltage V_A at node A is equal to the load voltage V_L and that $I_S = I_Z + I_L$. The load current I_L is given by Eq. (2.6) and is ideally equal to I_S when assuming $Z_S >> Z_L$. In practice, stray capacitances decrease the magnitude of Z_S at higher frequencies. Hence the load current I_L decreases with increasing frequency.

$$I_L = \frac{1}{1 + Z_L/Z_S} \cdot I_S \tag{2.6}$$

The best possible performance from the current source is very important to the design of systems for electrical impedance tomography (EIT), as reported in Denyer et al. (1993, 1994, b), Blad et al. (1994), Li et al. (1994), Lu and Brown (1994) and Jossinet et al. (1994a). These show that it is very difficult to design a VCCS whose

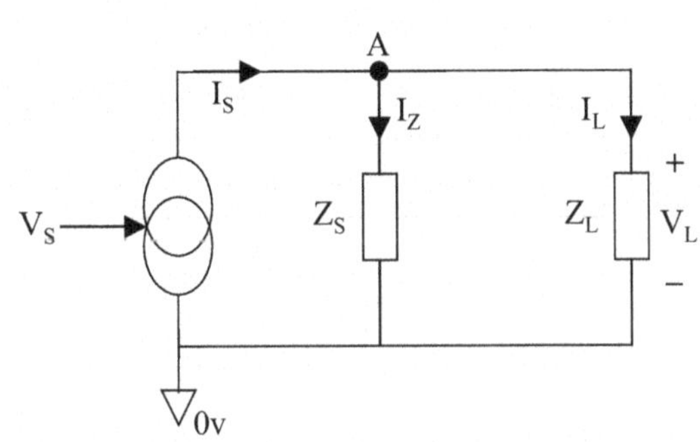

Fig. 2.9 Ideal model of a current source (modified from Webster 1990), where I_S is the output current of the current source controlled by V_S, Z_S is the output impedance of the current source and Z_L is the load

output impedance Z_S is greater than 1 MΩ at frequencies above 1 MHz, as reported in Lu (1995).

The simplest way to implement a VCCS for biological impedance measurements is to use an isolated negative-feedback current source (Smith 1990), where the patient is isolated from the circuitry by a transformer. This current source works well at a single frequency, but it is limited to low frequencies by the performance of the transformer. The majority of VCCS designs for working over a wide frequency range are of two categories: one uses positive feedback in a modified Howland circuit (Cusick et al. 1994; Jossinet et al. 1994b); the other uses current mirror architecture (Casas et al. 1996; Bragós et al. 1994; Riu et al. 1992).

The current mirror design is based on operational amplifier supply currents, as reported in Hart and Barker (1976) and Wilson (1981). Current mirrors for sensing the supply currents of the operational amplifier are used to obtain a VCCS without the use of positive feedback (Toumazou and Lidgey 1989; Denyer et al. 1993). Although high output impedance can be obtained in this type of current source, it was found in practice that there are mismatches between current mirrors which contribute to reduce the frequency response and produce distortions in the current injected into the load. Bragós et al. (1994) have reduced these mismatches by using the integrated current-feedback amplifier AD844 with a DC feedback unit. However, this circuit exhibits large changes in output impedance with consequent changes in load current, as demonstrated by Bertemes-Filho et al. (2000).

The Howland design uses a single operational amplifier with both negative and positive feedback loops, as shown in Fig. 2.10. The output current I_L can be calculated according to Eq. (2.7). It can be seen from Eq. (2.7) that for constant current I_L, it is the ratio R_1/R_2 and R_4/R_3 that is important. Also, the output current I_L will be defined by the resistance R_1 and the input voltage V_i. On the other hand, the output impedance will in theory be infinite if the two feedbacks are identical, i.e. $R_1 = R_2 = R_3 = R_4 = R$. However, mismatching between resistors, stray capacitances and the frequency limits of the op-amp will limit the output impedance.

$$I_L = \frac{V_i}{R_1 + R_L \cdot \left(1 - \frac{R_1}{R_2} \cdot \frac{R_4}{R_3}\right)} \tag{2.7}$$

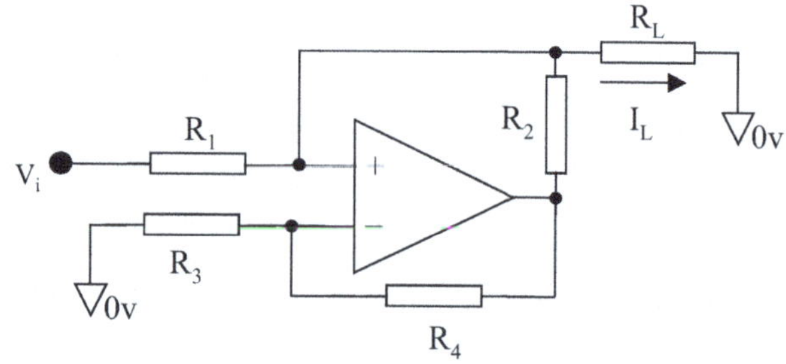

Fig. 2.10 Basic structure of the Howland generator, where V_i is the input voltage and I_L is the driven current in load R_L

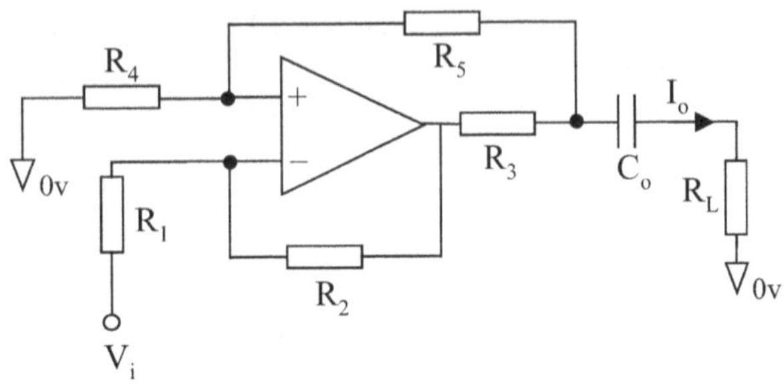

Fig. 2.11 Diagram schematic of the improved Howland current generator, where V_i is the input voltage, I_o is the output current and C_o is the DC-blocking capacitor

In addition to the limitation of the output impedance, the maximum voltage V_L which may develop across the load R_L is limited by the close-loop gain G and by the supply voltages V_{CC} of the operational amplifier, as shown in Eq. (2.8). Higher values of load will saturate the output of the operational amplifier, and then the performance of the VCCS will be affected.

$$V_L = R_L \cdot I_L \leq \frac{V_{CC}}{G} \tag{2.8}$$

Lu (1995) used an improved Howland generator, which is shown in Fig. 2.11. It can be observed in this circuit that the non-inverting input voltage of the op-amp is a voltage divider of the load voltage. As a result, the fraction of the load voltage acting at this point as a common-mode voltage is decreased. Therefore, the maximum voltage across the load can be increased and thus the current generator's compliance.

It can be derived from Fig. 2.11 that the output current I_o driven into load R_L is given by

$$I_o = \frac{R_4R_2 - R_1 \cdot (R_3 + R_5)}{R_1R_3 \cdot (R_4 + R_5)} \cdot V_L - \frac{R_2}{R_1R_3} \cdot V_i \tag{2.9}$$

where V_L is the voltage across the load.

Equation (2.9) assumes that the operational amplifier is ideal, which means that the output impedance of the op-amp is zero and both input impedance and open-loop gain are infinite. By definition, the output current I_o of a current source should not depend on the load voltage V_L. This is possible by making the value of R_4 equals to the sum of R_3 and R_5 and the values of R_1 and R_2 identical. If these conditions are satisfied, the output current I_o will be determined by R_3, and the output impedance will be infinite. In practice, it is very difficult to keep the output impedance high over the whole frequency range without causing the circuit to oscillate, especially at higher frequencies (e.g. >1 MHz). Many circuit approaches have been used for improving the performance of the enhanced Howland current source (Bertemes-Filho et al. 2013, 2015; Bertemes-Filho and Vincence 2016).

Problems arise when multifrequency systems are required. Careful adjustment of the feedback networks must be done in order to retain stability and to raise

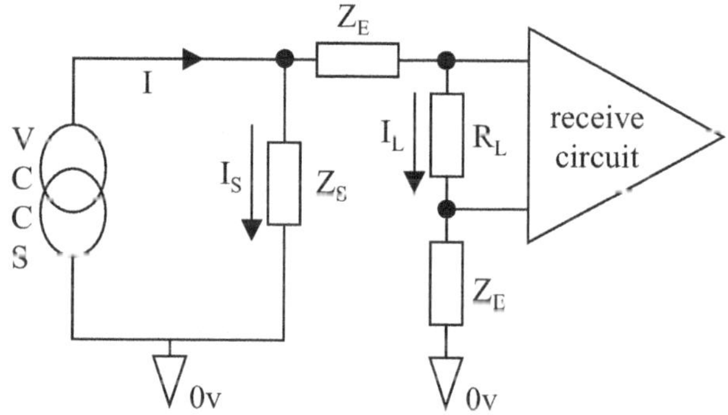

Fig. 2.12 Single-end four-electrode system, where I is the constant driven current, Z_E is the electrode impedance and R_L is the load under study

the output impedance. Some authors have suggested the inclusion of a buffer configured as a voltage follower in the positive feedback loop of the Howland generator (Cusick et al. 1994; Bertemes-Filho et al. 2000). Bertemes-Filho et al. (2000) have concluded that the design of symmetrical current sources should take into account the tolerances of the components used and the output impedance should be calculated under real load condition.

In practice, current generators neither have constant output current nor high output impedance over a wide frequency range. These nonideal characteristics lead to errors in the current injected into the load. The main errors found in a four-electrode system are due to the output impedance Z_S of the current source and the common-mode voltage.

Figure 2.12 shows the effect of the output impedance of a monopolar current source on the load current I_L. It can be seen that the output impedance Z_S of the current source provides a path to some current I_S to flow. Therefore, the accuracy of the load current I_L depends on the output impedance Z_S of the current source, as shown in Eq. (2.10).

$$I_L = \frac{I}{1 + (2Z_E + R_L)/Z_S} \tag{2.10}$$

If the output impedance Z_S of the current source tends to infinity, then the load current I_L will equal that of the current source. Equation (2.10) can be algebraically manipulated in or to show the values of the current source's output impedance Z_S to achieve a load current I_L with an accuracy of at least k%, as shown in Eq. (2.11). Assuming a typical load of 1 kΩ, an electrode impedance of 2 kΩ and an output impedance of 40 kΩ at 1 MHz for the current source (Bertemes-Filho et al. 2000), it can be calculated from Eq. (2.11) that the error in the load current is approximately 11% ($k = 11$) at 1 MHz.

$$Z_S > (2Z_E + R_L) \cdot \frac{100 - k}{k} \tag{2.11}$$

The second source of errors found in drive circuits is related to the common-mode voltage, which appears as an undesired potential across the grounded electrode impedance Z_E and the load R_L (see Fig. 2.12). This potential between the inputs of the receive circuit and ground is usually amplified by a differential amplifier. The common-mode (CM) voltage can have an amplitude much higher than the differential signal across the load to be measured, and hence it must be rejected by the amplifier. Further details about CM voltage are later discussed in this section.

The simplest method of minimising the undesirable CM voltage is by a bipolar current source, which is composed of two-phase opposition controlled current sources (Bertemes-Filho et al. 2009). However, mismatch between the output currents may produce CM voltage at the inputs of the receive circuit. In addition, CM voltage can also be generated by mismatch between drive electrode impedances and by different stray capacitances at each side of the current source.

Also cable capacitances from the receive circuit and the input impedance of the receive circuit produce errors in the driven current and induce common-mode voltages. There are three major approaches to reducing these errors, as described in the following:

1. *Screen driving*: This method consists in driving the cable shield (the second shield for the triaxial cable case) with the signal which is on the cable core by using a buffer configured as a voltage follower, as shown in Fig. 2.13. Then, the cable shield is at the same potential as the signal core itself. In theory, there should then be no current flow through the cable capacitance and hence no attenuation of the signal. Additionally, the first shield of the triaxial cable, which is also known as the outer screen, is earthed to minimise capacitive coupling

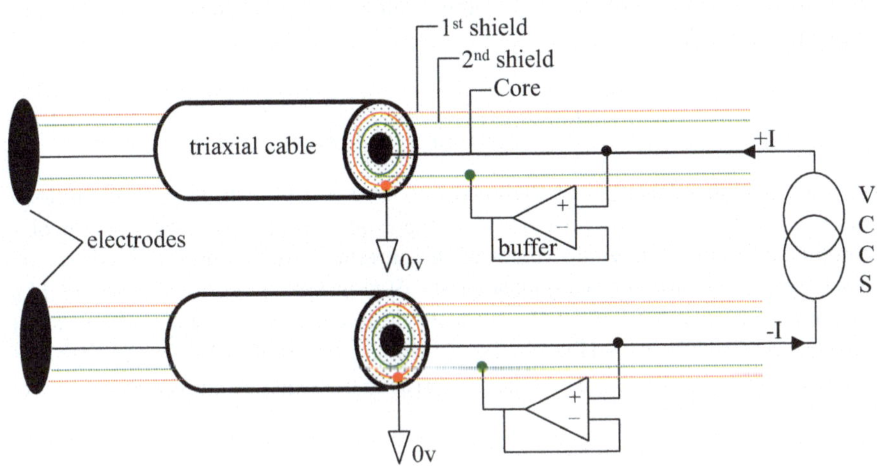

Fig. 2.13 Diagram schematic of the screen driving technique when triaxial cables are used to attach the bipolar current source to the electrodes, where $+I$ is the source current and $-I$ is the sink current

between cables (Lu and Brown 1994). However, the buffer input capacitances to ground cannot be compensated by this technique.

2. *Active electrodes*: The concept is to minimise cable length by placing drive and sense electronics very close to the electrodes. This minimises stray capacitance at the current source output and thus increases the frequency range (Rigaud et al. 1996). Jossinet et al. (1994a, b) connected the output of the current source to the electrode-skin interface. The current sources were encapsulated within the electrode shells. They concluded that the encapsulation removes lead capacitances but the components of the VCCSs should be well matched. Nevertheless, active electrodes need to be small and power has to be supplied to the electronics.
3. *Negative impedance circuit*: It is used to cancel the stray capacitance at the current-source output. This is achieved by producing an equivalent negative impedance in parallel with the stray capacitance. Cook et al. (1994) optimised the output impedance of a current source using a calibration network. Instability due to overcompensation of stray capacitance is the main drawback of this technique. Other attempts using this compensation technique can also be found in Bertemes-Filho et al. (2003, 2004).

In addition to the mentioned approaches, measurement of output current as a method of reducing errors caused by imperfections in the current generator has also been used. However, it is difficult to make the measurements without affecting the circuit performance.

2.4 Basic Voltage Measuring Concepts

The voltage between two electrodes placed on tissue depends mainly on the current which has been injected and the tissue structure. However, electrode construction and position will also affect the measurements. It is desirable that the voltage be measured without interference and noise. The signal needs to be amplified in order to make it compatible with an A/D converter. This chapter will only be concerned with the signal processing prior to A/D converter.

Interference usually means 50 Hz, which is generated by the main power supply, and cross talk such as EMG (ranging between 2 and 500 Hz) and ECG (ranging between 0.05 and 100 Hz). 50 Hz is rejected by the CMRR of the amplifier whereas cross talk by filters. On the other hand, artefacts and noise arise from electrode impedance and the movement of electrodes. The effect of electrode impedance is minimised by tetrapolar measurements. However, noise cannot be totally rejected by a tetrapolar system. The amplified noise may have a higher level than the desired signal. The level of signals to be measured will depend on the distance between the measuring electrodes and on the type of tissue under study, typically ranging between 100 μV and 100 mV.

The majority of the amplifiers used for measuring the potential across tissue have differential inputs and a third input used as the reference potential for the amplifier. The reference is usually the ground of the amplifier and is connected to a third electrode placed on the tissue. The reference to ground causes small differences in the desired signal between the measuring electrodes. The mean of the voltage on the two differential inputs measured with respect to the reference electrode is called the common-mode voltage. It is very important that the differential amplifier rejects common-mode voltage. There are three important characteristics for a good differential amplifier: high input impedance, high CMRR (common-mode rejection ratio) and low noise.

In practice, most biomedical signals are measured by instrumentation amplifiers. The main components of an instrumentation amplifier are two preamplifiers and one amplifier with a differential input and a single-ended output. High input impedance is provided by the two input preamplifiers, and a substantial reduction of the common-mode signal is provided by the output differential amplifier.

Figure 2.14 shows a typical instrumentation amplifier. Resistors R_p set the bias current, which flows into the non-inverting inputs of the operational amplifiers A and B. These resistors define the maximum input impedance of the amplifier, typically greater than 10 MΩ. The preamplifiers A and B offer differential gain but with unity gain to CM signals. The resistors R_f and R_1 determine the differential gain.The inputs V_1 and V_2 are connected to electrodes, which are placed on tissue. Electrodes are characterised by their half-cell potentials, which depend on the electrode properties. For example, gold electrodes present a half-cell potential of 1.68 V. Electrode half-cell potentials limit the gain of the preamplifiers and may lead to saturation. Therefore, the capacitor C_i is introduced to block any DC potential

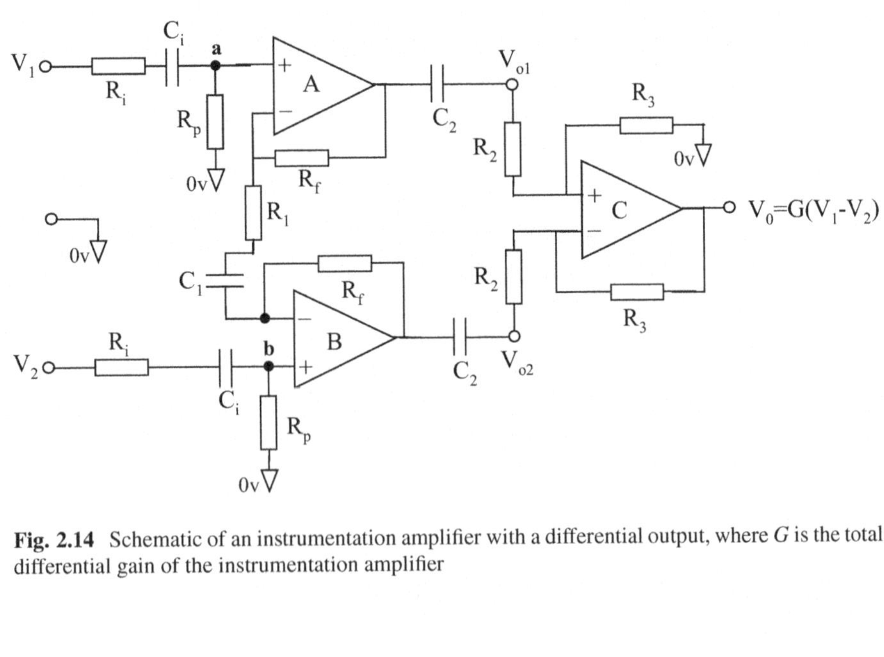

Fig. 2.14 Schematic of an instrumentation amplifier with a differential output, where G is the total differential gain of the instrumentation amplifier

by acting as high-pass filter with the resistor R_p. This capacitor also has a safety function in that it prevents any DC current flow to the electrode in the event of a failure of the preamplifiers A and B. The corner frequency f_0 of the filter is typically 0.05 Hz when $C_i = 33$ nF and $R_p = 100$ MΩ are used, assuming that R_p is much greater than R_i in Eq. (2.12). The AC coupling capacitor C_i does not attenuate the injected sine wave current over the frequency range 1 kHz to 1 MHz covered by this chapter.

$$f_0 = \frac{1}{2\pi \quad C_i \cdot R_p \sqrt{1 - 2R_i/R_p - (R_i/R_p)^2}} \tag{2.12}$$

The output voltage V_0 of the differential amplifier shown in Fig. 2.14 can be evaluated according to Eq. (2.13), assuming ideal op-amps.

$$V_0 = \frac{R_p}{R_p + R_i + 1/j\omega C_i} \cdot \left[1 + 2 \cdot \frac{R_f + 1/j\omega C_2}{R_1 + 1/j\omega C_1}\right] \cdot \frac{R_3}{R_2} \cdot (V_1 - V_2) \tag{2.13}$$

Instrumentation amplifiers are not perfect and undesired signals are also amplified. In practice, the output of the instrumentation amplifier shown in Fig. 2.14 consists of the differential input signal (V_1–V_2), an undesired component due to incomplete rejection of common-mode signals and an undesired component due to electrode impedance imbalance. Figure 2.15 shows a model for the receive circuit taking into account both common-mode signal V_{cm} and electrode impedances Z_{e1} and Z_{e2}.

The total output voltage V_0, which is calculated according to Eq. (2.14), can then be evaluated taking into account the CMRR of the amplifier, the electrode impedances and the equivalent input impedance of the amplifier, which is determined by resistor R_p in Fig. 2.14.

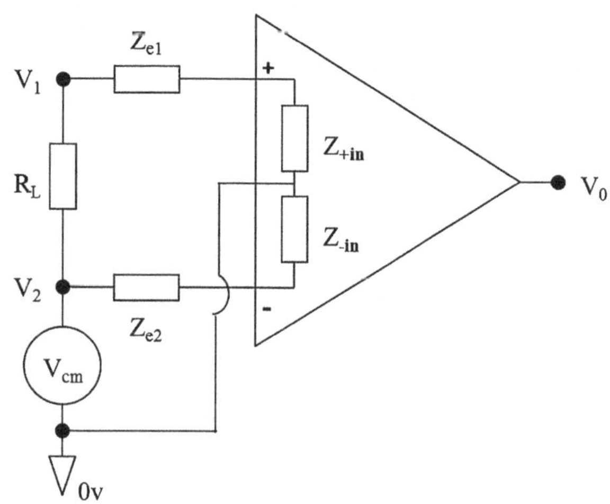

Fig. 2.15 Common-mode signal represented as a voltage generator V_{cm} in the receive circuit, where R_L is the load, Z_{e1} and Z_{e2} are the electrode impedances and $Z_{+in} = Z_{-in} = Z_{in}$ is the equivalent input impedance of the receive circuit

$$V_0 = G \cdot \left[(V_1 - V_2) + V_{cm} \cdot \left(1 + \frac{1}{\text{CMRR}} - \frac{Z_{in}}{Z_{in} + Z_{e1} - Z_{e2}} \right) \right] \tag{2.14}$$

The equivalent input impedance Z_{in} of an instrumentation amplifier used in EIS systems is typically much greater than the electrode impedance imbalance Z_{e1}–Z_{e2}, and hence the amount of undesired signal at the output of the instrumentation amplifier is mostly dominated by the CMRR. However, stray capacitances connected to ground reduce both the CMRR and the input impedance Z_{in} of the amplifier, especially at frequencies of up to 1 MHz. Reduction of the output voltage V_0 by stray capacitances is not taken into account in Eq. (2.14).

In addition to common-mode voltage and electrode impedance imbalance, noise is also an undesired source of error which appears at the output of the amplifier. Noise can be generated by the amplifier itself. This is usually specified by the manufacturer of the op-amp as an internal voltage noise source e_n and internal current noise generator i_n. Noise is also generated by movement of the charge carriers, which is represented by an electrical current flowing along a wire of resistance R_S. J B Johnson in 1928 showed that the noise power generated in a resistor R_S is proportional to the temperature and the value of the resistance; hence it is called *Johnson thermal noise*. The noise components are frequency-dependent. However, the external components of the Howland current source, for example, dominate the total noise leading to an unstable output current and, consequently, resulting in a fully noise voltage across the load under study (Santos and Bertemes-Filho 2017).

Although noise is a big concern in EIS, the use of switches in EIS system gets it even worse, especially at higher frequencies. Usually, multichannel measuring circuits use analogue multiplexers (MUX). MUX has a high effect in the circuitry at higher frequencies due to its input and output capacitances. Furthermore, there are feedthrough stray capacitances due to the use of switches, which link the drive and receive circuits. It is not uncommon to have typical feedthrough capacitances of 10 pF (Lu, 1995). These capacitances degrade the performance of the measurement system, especially at high frequencies. Yet the use of different multiplexers for drive and receive circuits does not eliminate feedthrough and capacitances between cables.

It is common in many receive circuits designed for EIS to use buffers at the receive electrodes as a preamplifier. This increases the input impedance of the receive circuit and reduces the effect of currents through cable and multiplexer stray capacitances. It can also reduce the effect of electrode impedance mismatches. However, the input capacitance of the amplifier degrades the input impedance of the receive circuit, especially at higher frequencies. Nevertheless, using an op-amp with a very low bias current allows R_p (see Fig. 2.14) to be increased. Hence, some researches use op-amps with field-effect transistor (FET) input, whose bias current is about tens of picoamperes. FET op-amps also offer low-noise voltage figures.

Input capacitance can be reduced and hence CMRR improved by using screen driving in the same way as we have described previously in this chapter. Negative impedance converters can also be used for cancelling stray capacitances, which may be encountered at the input of receive circuits. Cook et al. (1994) used a negative capacitance circuit to obtain high accuracy in a single-frequency and adaptative EIS system. They concluded that the circuit must be constantly trimmed in order to achieve accurate measurements.

2.5 Desired Hardware Specifications

For a giving sine wave input voltage, the output current magnitude of the monopolar VCCS should be constant with respect to the frequency range 1 kHz to 1 MHz with a maximum deviation of 1% from the nominal value. The magnitude of the output current should also remain constant with respect to the load ranging between 50 Ω and 5 kΩ. In the case of a bipolar current source, both current source and sink should be balanced with a maximum difference of 0.1% between the nominal values.

It is also important that the output impedance of both monopolar and bipolar current source be much greater than the total load between the injecting electrodes. The total load is approximately the resistance of the material under study summed with the impedance of both electrodes. The output impedance should be at least 100 times greater than the total load at the highest frequency, i.e. 1 MHz.

The gain of the receive circuit must be able to handle input voltage ranging between 1 $mV_{p\text{-}p}$ and 80 $mV_{p\text{-}p}$, when 1 $mA_{p\text{-}p}$ current is applied to the biological material under study. The input impedance is specified to be higher than 100 kΩ within the frequency range 1 kHz to 1 MHz.

The common-mode signals should be kept as low as possible by obtaining a CMRR greater than 80 dB over the frequency range, which means to have CM voltages smaller than 10 $\mu V_{p\text{-}p}$ with respect to the smallest measured signal of 1 $mV_{p\text{-}p}$. The instrumentation amplifier is implemented for obtaining a gain of 100. The noise voltage must be at maximum 1% of the measured biological signal. The total normalised gain must be flat within the specified frequency range, for example, as shown in Fig. 2.16.

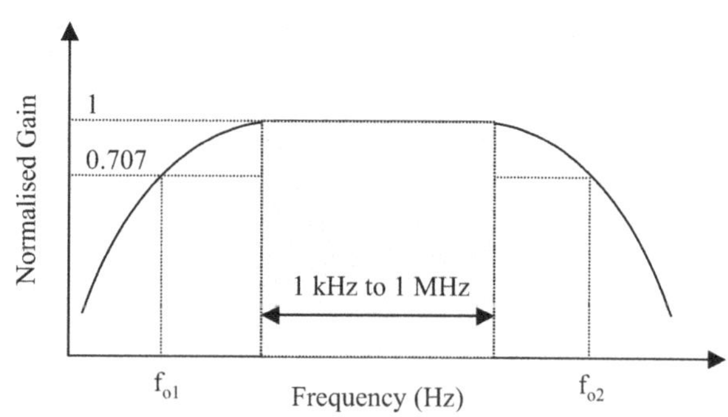

Fig. 2.16 Bode plot of the theoretical gain for the receive circuit, where f_{o1} and f_{o2} are the low and high corner frequencies, respectively

2.6 Discussions

The majority of VCCSs in EIS systems are designed to work over a wide frequency range and to have high output impedance. Both Howland and current mirror-type circuits can achieve an adequate performance. However, the components have to be very well matched in both of circuits. Output impedance greater than 100 kΩ at 1 MHz is very difficult to achieve, unless an additional compensating circuit is used.

A wide bandwidth Howland current generator may suffer from instability problems, unless very accurate components are used and care is taken with circuit layout. Instability seems not to be a problem in current mirror circuits although layout will still be important to minimise stray capacitances. However, the accuracy of the output current depends on the accuracy of the current copied by the transistors.

Nevertheless, stray capacitances degrade the output impedance of all VCCSs and also the voltage measurement circuits. Undesired stray capacitances can be significantly reduced by a negative impedance converter. However, instability can be a problem in multichannel EIS systems.

Although an accurate constant current can be driven into the tissue under study, the measured voltage will always contain undesired components which cannot be completely rejected. Most voltage measuring circuits are based on the three-op-amp configuration as an instrumentation amplifier. High input impedance is required for accurate measurements. However, mismatch of feedback resistors in the differential amplifier stage degrades the CMRR of the amplifier and hence the input impedance.

Common-mode voltage generated by mismatching between electrode impedances is the main source of error in this type of measuring system. Mismatches of this kind are encountered at the electrode-electrolyte interface. Cable stray capacitances also introduce errors in the voltage measurements.

There is a significant reduction of cable stray capacitances when part of the drive and receive circuits are built very close to electrodes. This technique increases significantly the output impedance of the drive system, and hence improvement in the CMRR can be achieved.

Although accurate measurements can be achieved, the performance of the VCCS and the instrumentation amplifier is not well controlled due to the unknown tissue impedance and to stray capacitances which might be encountered in the input circuitry. Therefore, calibration is very important when tissue characterisation is required. The overall gain of the system needs to be measured when the probe is applied to saline solutions of known conductivities. These solutions must be chosen to have similar conductivities to the tissue of interest.

References

Bertemes-Filho, P. (2002). *Characterisation using an impedance spectroscopy probe*. Ph.D. Thesis, Sheffield, England.

Bertemes-Filho, P., Brown, B. H., & Wilson, A. J. (2000). A comparison of a modified Howland circuits as current generators with current mirror type circuits. *Physiological Measurement, 21*(Suppl. 1A), 1–6.

Bertemes-Filho, P., Felipe, A., & Vincence, V. C. (2013). High accurate Howland current source: Output constraints analysis article. *Circuits and Systems, 4*(7), 451–458.

Bertemes-Filho, P., Lima, R. G., Amato, M. B. P., & Tanaka, H. (2004). Capacitive-compensated current source used in electrical impedance tomography. In *International conference on electrical bio-impedance*, Gdansk, Poland, 20–24 June 2004.

Bertemes-Filho, P., Lima, R. G., & Tanaka, H. (2003). An adaptive current source using a negative impedance converter (NIC) for electrical impedance tomography (EIT). In *International Congress of Mechanical Engineering*, June, Sao Paulo, Brazil.

Bertemes-Filho, P., Negri, L. H., & Vincence, V. C. (2015). Designing a mirrored Howland circuit with a particle swarm optimisation algorithm. *International Journal of Electronics, 1*, 1029–1037.

Bertemes-Filho, P., Paterno, A.S., & Pereira, R. M. (2009). Multichannel bipolar current source used in electrical impedance spectroscopy: Preliminary results. In O. Dössel, & W. C. Schlegel (Eds.), *World Congress on Medical Physics and Biomedical engineering*, September 7–12, 2009, Munich, Germany. IFMBE proceedings, Vol 25/7. Berlin, Heidelberg: Springer.

Bertemes-Filho, P., & Vincence, V. C. (2016). Howland current source for wideband bioimpedance application. In *XVI International Conference on Electrical Bioimpedance*, June.

Blad, B., Lindstrom, K., Bertenstam, L., Person, B. R. R., & Holmer, N. G. (1994). A current injecting device for electrical impedance tomography. *Physiological Measurement, 15*(Suppl. 2A), 69–77.

Bragós, R., Rosell, J., & Riu, P. J. (1994). A wide-band AC-coupled current source for electrical impedance tomography. *Physiological Measurement, 15*(Suppl. 2A), 91–99.

Brown, B. H., Smallwood, R. H., Barber, D. C., Lawford, P. V., & Hose, D. R. (1999). *Medical physics and biomedical engineering*. Bristol: Institute of Physics Publishing.

Casas, O., Rosell, J., Bragós, R., Lozano, A., & Riu, P. J. (1996). A parallel broadband real-time system for electrical impedance tomography. *Physiological Measurement, 17*(Suppl. 4A), 1–6.

Cole, K. S., & Cole, R. H. (1941). Dispersion and absorption in dielectrics. *The Journal of Chemical Physics, 9*, 341–351.

Cook, R. D., Saulnier, G. J., Gisser, D. G., Goble, J. C., Newell, J. C., & Isaacson, D. (1994). ACT3: A high-speed, high-precision electrical impedance tomograph. *IEEE Transactions on Biomedical Engineering, 41*(8), 713–721.

Cusick, G., Holder, D. S., Birquett, A., & Boone, K. (1994). A system for impedance imaging epilepsy in ambulatory human subjects. *Innovation et technologie en biologie et médecine, 15*, 33–39.

Denyer, C. W., Lidgey, F. J., McLeod, C. N., & Zhu, Q. S. (1994). Current source calibration simplifies high-accuracy current source measurement. *Innovation et technologie en biologie et médecine, 15*(Spécial 1), 47–55.

Denyer, C. W., Lidgey, F. J., Zhu, Q. S., & McLeod, C. N. (1993). High output impedance voltage controlled current source for bio-impedance instrumentation. *Proceedings of the 15th annual international conference of the IEEE engineering in medicine and biology society, San Diego, 15*(2), 1026–1027.

Denyer, C. W., Lidgey, F. J., Zhu, Q. S., & McLeod, C. N. (1994). A high output impedance current source. *Physiological Measurement, 15*(Suppl. 2A), 79–82.

Foster, R. F., & Schwan, H. P. (1989). Dielectric properties of tissues and biological materials: A critical review. *Critical Reviews in Biomedical Engineering, 17*(1), 25–104.

González-Correa, C. A., Brown, B. H., Smallwood, R. H., Kalia, N., Stoddard, C. J., Stephenson, T. J., et al. (1999). Virtual biopsies in Barrett's esophagus using an impedance probe. *Annals of the New York Academy of Sciences, 873*, 313–321.

Griffiths, H., & Ahmed, A. (1987). A dual-frequency applied potential tomography technique: Computer simulations. *Clinical Physics and Physiological Measurement, 8*(Suppl. A), 103–107.

Hart, B. L., & Barker, R. W. (1976). D.C. matching errors in the wilson current source. *Electronics Letters, 12*(15), 389–390.

Hollas, J. M. (1998). *High resolution spectroscopy*. Chichester: John Willey & Sons.

Jossinet, J., Tourtel, G., & Jarry, R. (1994a). Performance and operation of a set of wideband current generators for EIT. *Innovation et technologie en biologie et médecine, 15*, 40–46.

Jossinet, J., Tourtel, C., & Jarry, R. (1994b). Active current electrode for in vivo electrical impedance tomography. *Physiological Measurement, 15*(Suppl. 2A), 83–90.

Land, R., Cahill, B. P., Parve, T., Annus, P., & Min, M. (2011). Improvements in design of spectra of multisine and binary excitation signals for multi-frequency bioimpedance measurement. *Conference proceedings of the IEEE engineering in medicine and biology society*, Boston, MA, USA.

Li, J., Joppek, C., & Faust, U. (1994). An isolated wideband current source used in multifrequency electrical impedance tomography. *Innovation et technologie en biologie et médecine, 15*, 63–68.

Lozano, A., Rosell, J., & Pallás-Areny, R. (1990). Two-frequency impedance plephysmography real and imaginary parts. *Medical & Biological Engineering & Computing, 28*, 38–42.

Lu, L. (1995). *Aspects of an electrical impedance tomography spectroscopy (EITS) system*. Ph.D. Thesis, Sheffield: University of Sheffield, UK.

Lu, L., & Brown, B. H. (1994). The electronic and electronic interface in an EIT spectroscopy system. *Innovation et technologie en biologie et médecine, 15*(1), 97–103.

McAdams, E. T. (1987). *A study of Electrode-Tissue Impedance Encountered in Cardiac Pacing*. Thesis, lEEDS: University of Leeds, UK.

Min, M, Pliquett, U., Nacke, T., Barthel, A., Annus, P., & Land, R. (2007). Signals in bioimpedance measurement: Different waveforms for different tasks. In H. Scharfetter & R. Merwa (Eds.), *13th International Conference on Electrical Bioimpedance and 8th Conference on Electrical Impedance Tomography*, Graz, Austria.

Nahvi, M., & Hoyle, B. S. (2009). Electrical impedance spectroscopy sensing for industrial processes. *IEEE Sensors Journal, 9*(12), 1808–1816.

Paavle, T., Min, M., & Parve, T. (2012). Aspects of using chirp excitation for estimation of bioimpedance spectrum. In S. Salih (Ed.), *Fourier transform—Signal processing* (pp. 237–256). Rijeka: InTech.

Paterno, A. S., Negri, L. H., & Bertemes-Filho, P. (2012). Efficient computational techniques in bioimpedance spectroscopy. In G. R. Naik (Ed.), *Applied biological engineering—Principles and practice*. Rijeka: InTech. https://doi.org/10.5772/36307

Pethig, R. (1984). Dielectric properties of biological materials: Biophysical and medical applications. *IEEE Transactions on Electrical Insulation, 19*, 453–474.

Pethig, R. (1987). Dielectric properties of body tissues. *Clinical Physics and Physiological Measurement, 8*(Suppl. A), 5–12.

Pliquett, U., Frense, D., Schönfeldt, M., Frätzer, C., Zhang, Y., Cahill, B., et al. (2010). Testing miniaturized electrodes for impedance measurements within the E-dispersion—A practical approach. *Journal of Electrical Bioimpedance, 1*, 41–55.

Raghed, A. O., Geddes, L. A., Bourland, J. D., & Tacker, W. A. (1992). Tetrapolar electrode system for measuring physiological events by impedance. *Medical & Biological Engineering & Computing, 30*, 115–117.

Record, P., Gadd, R., & Vinther, F. (1992). Multifrequency electrical impedance tomography. *Clinical Physics and Physiological Measurement, 13*(Suppl. 2A), 67–72.

Riu, P. J., Rosell, J., Lozano, A., & Pallás-Areny, R. (1992). A broadband system for multifrequency static imaging in electrical impedance tomography. *Clinical Physics and Physiological Measurement, 13*(Suppl A), 61–65.

Rigaud, B., & Morucci, J. P. (1996). Bioelectrical impedance techniques in medicine. Part III: Impedance imaging, first section: General concepts and hardware. *Critical Reviews in Biomedical Engineering, 24*(4–6), 467–597.

Santos, S. F., & Bertemes Filho, P. (2017). Note: Temperature effects in the modified Howland current source for electrical bioimpedance spectroscopy. *Review of Scientific Instruments, 88*(7), 076103.

Smith, R. W. M. (1990). *Design of a real-time impedance imaging system for medical applications*. Ph.D. Thesis, Sheffield: University of Sheffield, UK.

Toumazou, C., & Lidgey, F. J. (1989). Novel current-mode instrumentation amplifier. *Electronics Letters, 25*, 228–230.

Waterworth, A., Brown, B. H., Smallwood, R., & Milnes, P. (2000). Cole equation modelling to measurements made using an impulse driven transfer impedance system. *Physiological Measurement, 21*, 137–144.

Webster, J. G. (1990). *Electrical impedance tomography*. Bristol: Adam Hilger Press.

Wilson, B. (1981). A low-distortion bipolar feedback current amplifier technique. *Proceedings of the IEEE, 69*(11), 1514–1515.

Yang, Y., Kang, M., Lu, Y., Wang, J., Yue, J., & Gao, Z. (2010). Design of a wideband excitation source for fast bioimpedance spectroscopy. *Measurement Science and Technology, 22*, 013001.

Chapter 3
Basic Electrical Impedance Tomography

Richard Bayford

3.1 Introduction

3.1.1 Basic Principles

Electrical impedance tomography (EIT) in its simplest form is based on the use of small alternating electric currents (up to 5 mA rms) flowing between pairs of electrodes placed around or on the region under investigation. The subsequent voltage measurements made at the surface of the imaging domain are processed to produce a map that represents the variation in impedance across the region of interest. The important features to recognize of this imaging technique are that it is based on nonionizing radiation, portable and inexpensive; a typical system can be produced for relatively low cost. The current injected can be a single frequency or varied as in the extension to EIT known as electrical impedance tomographic spectroscopy (EITS). This approach uses a range of frequencies (typically 9.6 kHz to 1.2 MHz) to produce static images of the thorax (Mayer et al. 2005). Frequency ranges greater than this have been reported (10 MHz) for imaging breast and prostate cancers (Murphy et al. 2017; Mahara et al. 2015) where most of the ion flow is within the intercellular region. For thorax imaging, EITS enabled changes in tissue impedance with respect to frequency to be analysed after reconstruction in subjects where the breath was held at maximum inspiration and expiration. The resulting tissue characterization is consistent with an electrical model of the lungs based on the Cole equation (Cole and Cole 1941; Cole 1942), where parameters were related to alveolar structure and composition. One of the first systems developed for use on humans is shown in Fig. 3.1.

R. Bayford (✉)
Biophysics and Engineering, Middlesex University, London, UK
e-mail: r.bayford@mdx.ac.uk

F. Simini, P. Bertemes-Filho (eds.), *Bioimpedance in Biomedical Applications and Research*, https://doi.org/10.1007/978-3-319-74388-2_3

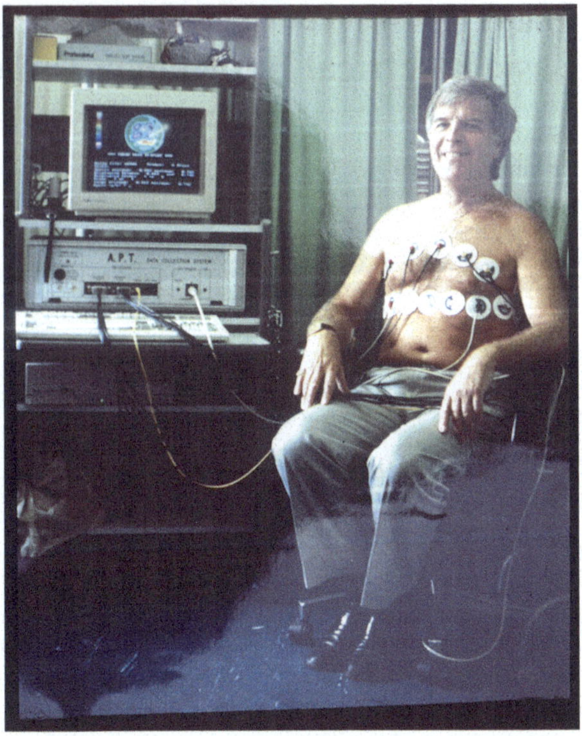

Fig. 3.1 Chest imaging with the Sheffield Mk 1 system (Provided by Prof Brian Brown)

3.2 Hardware

3.2.1 Injection Protocols

The method used to inject current into the domain under investigation can vary. These can be classified as adjacent, opposite and optimum injection pattern assuming the current is injected and voltage is measured, although other variations are possible. The chosen method again depends on the application. For example, when imaging the human brain, the criterion has been to maximize the current density in the brain region. One would assume that an optimum injection pattern would be the choice; however, the complexity of the hardware increases and any errors on each channel that injects current or voltage must be minimized. It is relativity simple to construct an EIT system, but to obtain the dynamic range, accuracy and sample rate that couple with fast change in the physiology is not easily achieved. Another key requirement is the need for these systems to be CE or FDA approved for use on humans, which requires considerable effort.

3.2.2 *Calibration, Validation and Sources of Error*

There are a number of sources of error in EIT in terms of hardware and image reconstruction. Some of these can be dealt with by improving the hardware design, for example, improving the common-mode rejection ratio (CMRR), thereby reducing the effects of stray capacitance. Many EIT systems apply current or voltage and measure in-phase and quadrature components. It is of key importance that the hardware systems are calibrated to obtain the maximum signal and minimum error. For example, a number of systems are based on one of the earliest systems developed for lunging imaging, known as the Sheffield Mark 1 APT system. This makes serial in-phase voltage measurements and is only used for dynamic images. It uses a constant current generator that is connected to electrodes on the body, usually via a multiplexer, and supplies currents to the object to be imaged. The resulting sinusoidally varying voltage on the surface of the body is sampled on additional electrodes and differentially amplified. The signal can be demodulated by rectification using the reference waveform and filtered to obtain a voltage level that is a measure of the impedance. The principal sources of error in the front end arise from common-mode effects such as skin-electrode problems, contact impedance and stray capacitances. The final error for each measurement is dependent on the complex interaction of these effects and will differ for each electrode combination. New approaches are being developed to overcome these problems using active electrodes (Wu et al. 2016).

3.3 Imaging

EIT can produce images that are physiologically useful; however images on their own can have limitations. The image needs to provide full information in regard to the clinical state of the patient in terms of clinical parameters and care need. Two types of image can be produced. The first of these is known as difference imaging, where a change in ratio or percentage represents some physiological parameter like blood volume or volume of air in the lung. The Sheffield Group produced the first lung images using this method, and other systems still provide this type of image, for example, Swisstom AG (Fig. 3.2) and Draeger. The images are obtained by measuring two data sets at different times.

A system that can produce image based on different frequencies at the boundary of the object has also been developed. The difference between two frequencies is divided by a reference data set resulting in frequency difference images. The second type of image that can be obtained is known as absolute imaging, which produces a representation of the absolute conductivity or permittivity used in applications such as breast imaging. This absolute method of imaging is technically more difficult than the difference method, as the contact impedance of the electrodes cannot be accurately characterized when making clinical measurements.

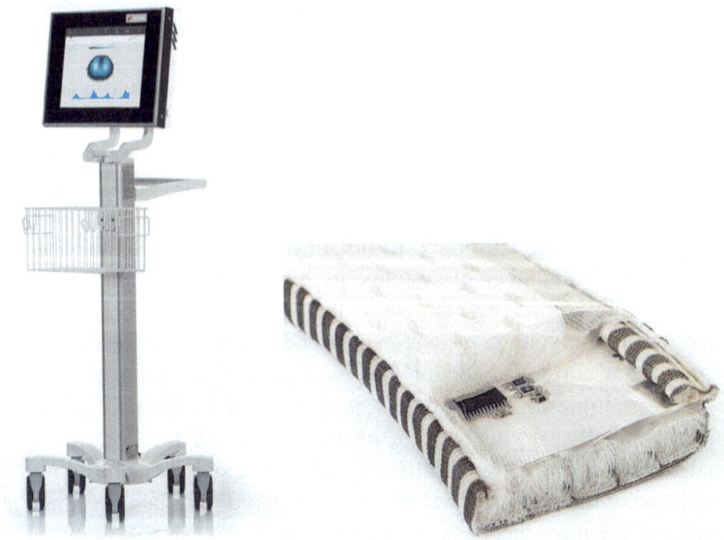

Fig. 3.2 Swisstom AG EIT system (Images kindly provided by Swisstom AG (http://www.swisstom.com/))

It should be noted that changes in conductivity anywhere in the domain can affect all measurements, not just those on a direct path. In the early days of EIT development, assumptions were made to simplify the mathematics of the reconstruction process, though this assumption limits the accuracy of the reconstructed images. A simple so-called back project algorithm was developed based on the Radon transform used in CT imaging. The back-projection algorithm was developed by Barber and Brown (1984), Barber (1989) for the original Sheffield EIT system and has seen numerous variants and improvements. Another assumption sometimes used, which again was an analogue of those used in CT scanning, was the idea that current path was in a single plane and one could assume a 2D solution would be acceptable. However, as ionic current flows in 3D, then it is essential to develop a 3D solution rather than trying to fit a 2D solution into a 3D problem. A key development to overcome this issue, along with the goal of having a so-called gold standard algorithm based on a consensus linear reconstruction algorithm for lung EIT, has been named GREIT (Graz consensus Reconstruction algorithm for EIT) (Adler et al. 2009). A unified approach to linear image reconstruction has been developed for GREIT. The framework for the linear reconstruction algorithm consists of three points, detailed finite element models of a representative adult and neonatal thorax, a consensus on the performance figures of merit for EIT image reconstruction and a systematic approach to optimize a linear reconstruction matrix to the desired performance measures. In achieving the figures of merit, the requirements in order of importance are (a) a uniform amplitude response, (b) a small and uniform position error, (c) small ringing artefacts, (d) a uniform resolution, (e) limited shape deformation and (f) high resolution. Such figures of merit must be attained while maintaining small noise amplification and small sensitivity to electrode and boundary movement. The approach represents the consensus of a large and representative group of experts in EIT algorithm design and clinical applications for pulmonary monitoring. GEIT is designed for lung measurements and the above requirements may vary for other clinical applications.

The EIT community has also developed a resource known as EIDORS: electrical impedance tomography and diffuse optical tomography reconstruction software. It is an open source software suite for image reconstruction in electrical impedance tomography and diffuse optical tomography, designed to facilitate collaboration, testing and research in both electrical and optical tomography (Adler and Lionheart 2006).

EIT is governed by the Maxwell equations model (some time termed the full model) as a forward model, with many assumptions used to simplify this. However, some research has been developed to formulate the problem in terms of the Maxwell equations expressed in scalar and vector potentials. This approach leads to boundary conditions that naturally align with the quantities measured by EIT instrumentation. The effect of frequency on the field distribution is illustrated using the high-frequency model and is compared with Laplace solutions. Numerical simulations and experimental results are also presented to illustrate image reconstruction over a range of frequencies using the new implementation. The results show that scalar/vector potential reconstruction produces images which are essentially indistinguishable from a Laplace algorithm for frequencies below 1 MHz but superior at frequencies reaching 10 MHz (Soni et al. 2006).

Traditionally, image reconstruction has been based on Laplace's equation. However, at high frequencies the coupling between electric and magnetic fields requires solution of the full Maxwell equations (Barber and Brown 1984), which are shown below in the simplified approach. The first of these is to assume the reconstruction is quasistatic. This means that the effect of magnetic induction can be ignored in Maxwell's equations so that the governing equations across a medium can be defined as:

$$\nabla\left(\sigma + i\omega\varepsilon\right)\nabla\phi = 0 \tag{3.1}$$

where σ is the electric impedance of the medium, ϕ is the electric potential, ω is the frequency and ε is the electric permittivity. Equation 3.1 can be reduced to the standard governing equation for EIT when we assume that a low frequency or a direct current is used which would be the case in human applications:

$$\nabla\cdot(\nabla\sigma\phi) = 0 \tag{3.2}$$

This can be considered as determining the conductivity distribution from the measurements made at the surface of the domain. For voltage measurements made at the surface of the head, the reconstruction problem can be defined as

$$V = A\sigma \tag{3.3}$$

where A is known as the sensitivity matrix which relates voltages to image pixels. The sensitivity matrix can be considered as the Jacobian of the forward mapping (matrix of partial derivatives). The inverse problem, to find σ given A and V, becomes $\sigma = A^{-1}\ V$ where A^{-1} is the inverse of A and can be found by a number of methods. However, approximations are sometime used to reduce complexity and computational time of A the forward problem. For example, for analytical brain applications, by assuming the head to be a homogeneous sphere, this does not use accurate geometrical information regarding the shape of the human head and so will introduce errors in the solution.

The aim is to obtain an accurate solution to the forward problem by incorporating the shape and the structure of the boundary form and hence obtain an improved estimation of the values of the elements in the sensitivity matrix A. The reconstruction algorithm involves a solution to the forward and inverse problems. The forward problem can be solved in two ways. The first is analytical, for which a number of solutions are available (Bayford 2006; Kolehmainen et al. 1999); the second is by the use of numerical methods like FEM or boundary element method (BEM) (Pidcock et al. 1995a). The analytical methods are preferable as computation time is reduced, but these have been of limited practical value because solutions only exist for a small number of idealized geometries (Pidcock et al. 1995b; Menin et al. 2013).

Table 3.1 gives examples of approaches used to solve the forward problem.

There are a number of approaches to finding a solution to the inverse problem which is ill-posed and can be linear or nonlinear. Many are based on linearization and regularization as in the GREIT algorithm; however if computation limitation is not an issue, then it is possible to achieve a nonlinear solution. Typically for a linear approach to solving the inverse problem, the truncated singular value decomposition (Penrose-Moore pseudo inversion) or Tikhonov regularization is commonly used. Nonlinear approaches are a much wider choice, such as Gauss–Newton, Global Convergence using first-order Taylor expansion approximation, Levenberg–Marquardt involving a second-order Taylor expansion regularized approximation, variable metric with Brent line search using a Hessian's inverse gradual approximation and Newton types and conjugate gradients Polak-Ribiere, with Brent line

Table 3.1 Examples to solve the forward problem

	Real	Complex
Preconditioner	Incomplete Cholesky or algebraic multigrid (Borsic and Bayford 2010)	Incomplete LU (Jehl et al. 2015)
Solver	Preconditioned conjugate gradients (PCG) (Borsic et al. 2010)	Bi-conjugated gradients (BiCG) (Dehghani et al. 2005) or generalized minimal residual (GMRES) (Horesh 2006)

search and regularized search direction—residual error minimization by conjugated search directions (Krylov space) (Lionheart et al. 2004).

None of these methods are ideal; conjugate gradient has the disadvantage that it requires many more iterations than Newton-based algorithms. A line search has to be implemented and scaling should be carefully treated. Newton–Raphson/Gauss–Newton requires assumptions. The forward solution has continuous first and second derivatives in the data and image domains. The benefits of the approach are the fast convergence and easy implementation, but the disadvantage is repeated computation of Cholesky decomposition of a large dense sensitivity matrix. It requires first and second derivatives, and the system matrix is not necessarily symmetric and positive definite away from the solution, which could cause convergence failure. Levenberg–Marquardt benefits are faster convergence and easy implementation, and the solution can be refined due to proper adjustment of the regularization coefficient. However, it requires repeated computation of Cholesky decomposition of a large dense sensitivity matrix which needs first and second derivatives. The system matrix is not necessarily symmetric and positive definite away from the solution which could cause convergence failure.

New methods are still being developed, for example, D-bar (Hamilton 2017) methods solve the nonlinear problem directly, bypassing the need for detailed and time-intensive forward models, to provide absolute (static) as well as time-difference EIT images. It is possible to use coupling the D-bar methodology with the inclusion of high confidence a priori data results in a noise-robust regularized image reconstruction method (Hamilton 2017). This is the first time a priori D-bar method for complex admittivities for absolute imaging has been used. Additionally, the method can be adjusted for, and tested on, time-difference imaging scenarios. The ability of the method to be used for conductivity, permittivity and absolute as well as time-difference imaging provides the user with great flexibility without a high computational cost. More details on the EIT Reconstruction Algorithms: Pitfalls, Challenges and Recent Developments can be found in Lionheart's paper (Lionheart 2004).

It is worth noting that the electrode model is of key importance in both the image reconstruction and the hardware. At present, we do not have a model that satisfies the requirements of the image reconstruction and the hardware. Mathematical models, like the complete electrode model, are used to develop the reconstruction algorithms, but estimating the load impedance on the current deriver is addressed by using equivalent SPICE models. Ideally a model needs to be developed to address both requirement of the imaging and hardware.

3.4 EIT Systems

Most EIT systems are developed for specific clinical use, for example, the Dartmouth system is designed for breast imaging and is shown in Fig. 3.3.

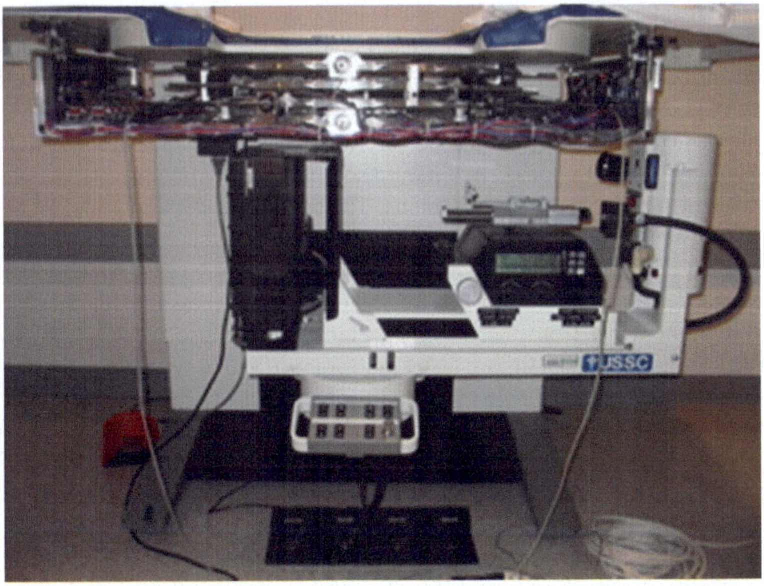

Fig. 3.3 Dartmouth breast imaging system (Image supplied by Prof Alexander Hartov)

Also there are a number of variations which combine EIT with other modalities, for example, magnetic resonance electrical impedance tomography (MREIT) which aims to improve the limitation of EIT (Fig. 3.4).

3.5 Clinical Applications

EIT in its various forms has been applied to a wide range of clinical applications with varying degrees of success. These include thoracic, gastrointestinal function, breast tumour localization and brain function. Those areas that are showing the most promising results are the brain, breast, gastrointestinal tract, lung, rectum and prostate, though only one is getting close to clinical use.

3.5.1 Pulmonary Function

One of the key areas that shows resurgence is monitoring pulmonary function, which was first developed by the Sheffield Group (Barber and Brown 1984). Although there are many other imaging approaches, such as planar X-ray, X-ray CT, MRI and radioisotope scintigraphy, that can all be used to monitor static images of pulmonary function with high spatial resolution so providing information on

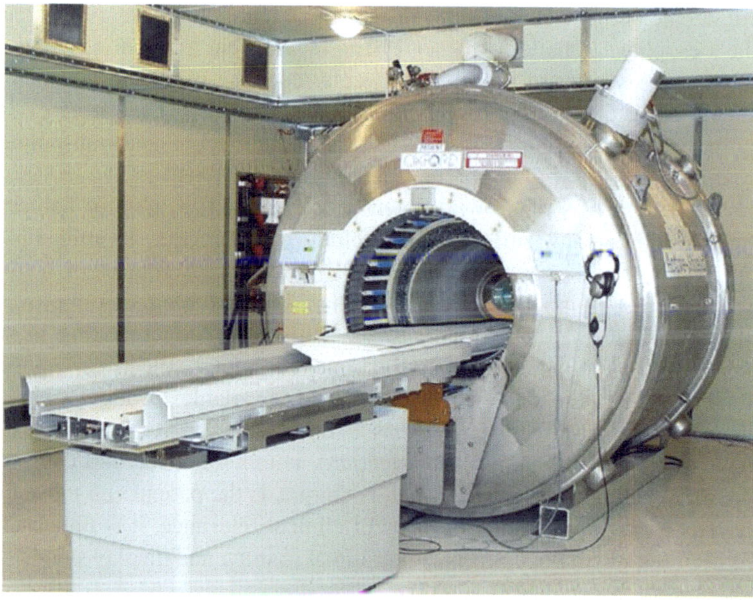

Fig. 3.4 Korean MREIT system (Image supplied by Prof EJ Woo)

ventilation and perfusion. The patient must be brought to the device and repeatedly exposed to radiation with imaging modalities that require the use of X-rays, which is undesirable. MRI offers the advantage that it does not use ionizing radiation; however, as in the case with all the above methods, the subject must be stationary and MRI is considerably more costly. There has been considerable interest in the development of EIT for imaging ventilation and detection of blood clots in the lungs or pulmonary emboli, a common and often serious complication of surgery. Thoracic EIT examinations have been shown to track changes in regional lung volumes at scan rates of up to 40–50 scans/s. It may also be used to monitor the drainage of a pneumothorax caused by pulmonary lesions because the impedance of air is very high, providing high contrast with the surrounding tissue. It has even been used during parabolic flights to determine regional ventilation and fluid shift signals in lateral posture during normo-, hyper-, and microgravity. However, one problem is that as well as being sensitive to lung impedance changes, the technique is also sensitive to the ribs and to tissue movement, which can be large sources of artefact.

EIT also shows promise as a valuable tool for bedside data-driven EIT lung tissue classification for monitoring of mechanically ventilated patients in the intensive care unit (ICU) (Wolf and Arnold 2006). They use a system marketed originally by Viasys (now CareFusion) which is based on the Sheffield Mk1 (Barber and Brown 1984). Lung applications have continued to be developed, for example, Draeger developed clinical systems; these are capable of looking at positive end-expiratory pressure (PEEP). Setting the optimal level of PEEP in the ICU is still considered a matter of debate. Talmor et al. (2008) used the transpulmonary pressure calculated from the oesophageal balloon to set PEEP in a recent randomized controlled study. This strategy aims at preventing alveolar collapse by counterbalancing the gravitational force of the lung by an equal or higher PEEP. They have evaluated the relation between ventilation distribution measured by EIT and transpulmonary pressure during a PEEP trial in porcine acute lung injury (ALI).

A key area, which is of growing interest, is the need to monitor babies born prematurely where many suffer from respiratory failure due to immaturity of the lung (respiratory distress syndrome, RDS) and lack of control of breathing. Although respiratory support, especially mechanical ventilation, can improve their survival, it also causes severe injury to the vulnerable lung resulting in severe and chronic pulmonary morbidity lasting into adulthood. Electrical impedance tomography (EIT) is the only technology that is available to address this need as well as being cost-effective. EIT has a high temporal resolution (real-time continuous imaging of dynamic lung function), is radiation-free and can be used at the bedside. Another advantage, which is unique, is the patient can move during imaging (cradlproject.org). For EIT to be of clinical relevance, it is important that it can be used reliably by operators with minimal training and display standardized, clinically meaningful results.

A number of research groups have developed hardware systems for research into the use of EIT for pulmonary function, for example, Rensselaer, New York. The adaptive current tomography (ACT) system imaged a simulated pulmonary embolus in a dog, where the first one lung at a time was ventilated and then a major branch of the pulmonary artery was occluded. The region of the lungs that was ventilated but not perfused by blood had different electrical properties and, therefore, was observable in a time-varying conductivity map of the thorax. This group also reconstructed conductivity changes during lung ventilation in human volunteers using a ring of electrodes placed around the thorax and a 4×4 rectangular array placed on the subject's chest (Gisser et al. 1988).

Swisstom AG has developed a CE clinical EIT system for adults, using a belt that that can be easily applied. They are also working on a system for neonates as part of an EU CRADL project (cradlproject.org). Disorders ranging from lung growth, maturation and control of breathing are among the most important problems faced by the neonatologist. Premature birth occurs in 5–10% of all pregnancies and is frequently accompanied by complications due to lung immaturity. Many preterm infants exhibit lung dysfunction characterized by arrested lung development and interrupted alveolarization. This immature lung phenotype accounts for 75% of early mortality and long-term disability in infants delivered prematurely. Despite improved survival of extremely premature (EP) infants, i.e. those born <27w gestational age (GA), the prevalence of chronic lung disease in infancy has remained high over the last decade. Chronic lung disease of infancy (CLDI) is associated with long-term, and possibly life-long, respiratory morbidity. Objective, non-invasive measures of lung maturity and development, oxygen requirements and lung function, suitable for use in small, unsedated infants, are urgently required to define the nature and severity of persisting lung disease and to identify risk factors for developing chronic lung problems. Recent progress with EIT has made pulmonary function testing available for this age range and opened up new possibilities for monitoring changes in disease processes affecting the respiratory system. This may improve medical management of infants and children with lung and heart diseases, including those requiring intensive care.

3.5.2 *Breast Imaging*

Early detection and treatment of tumours in the breast still remains the best method to increase chance of survival among patients. The properties of many tumours, in particular those exhibiting malignancy, differ significantly from the surrounding tissue, and separation and characterization of contrasting tissues can deliver early diagnosis. At present, women are screened for breast cancer using X-ray mammography (MG), although some cancers of the breast cannot be seen using this technique and there are restrictions of the groups of patient that can undergo regular scanning. During this procedure, their breast is compressed flat to visualize all the tissue and minimize the required radiation dose, which can be uncomfortable and sometimes painful for the patient. Another disadvantage of the technique is the high (40%) false-positive rate. A positive test is highly alarming and requires the patient to undergo further testing, which could include fine-needle aspiration or biopsy. Meanwhile, the false-negative rate is 26%. Regular screening is recommended for women over 40 years old, but the technique is difficult with younger women whose breast tissue is generally denser. In a study at Dartmouth College, USA, participants lay prone on a customized table where up to four rings consisting of 16 radially adjustable electrodes are put into contact with the pendant breast. In 6 out of 26 subjects who had ACR 4–5 suspicious lesions, inspection of EIS images using visual criteria resulted in 83% detection specificity and numerical criteria 67% (Bayford and Tizzard 2012). An example is shown in Fig. 3.5.

3.5.3 *Brain Imaging*

There is also potential for imaging aspects of brain function; the first application is in focal epilepsy which is a functional abnormality often related to a causal structural abnormality, for example, a lesion, and an individual seizure always originates from the same cerebral focus. Patients with intractable epilepsy can require surgery to remove that part of the brain, and accurate localization of the epileptic focus is crucial before surgical excision. The repetitive activity owing to a focal seizure can cause local ischemia, detectable using EIT (Pidcock et al. 1995a). Continuous monitoring is necessary because the occurrence of seizures is so unpredictable; therefore, EIT may be better suited than MRI for localization of epileptic foci. EEG inverse dipole modelling has been tried for this purpose but was not adopted on

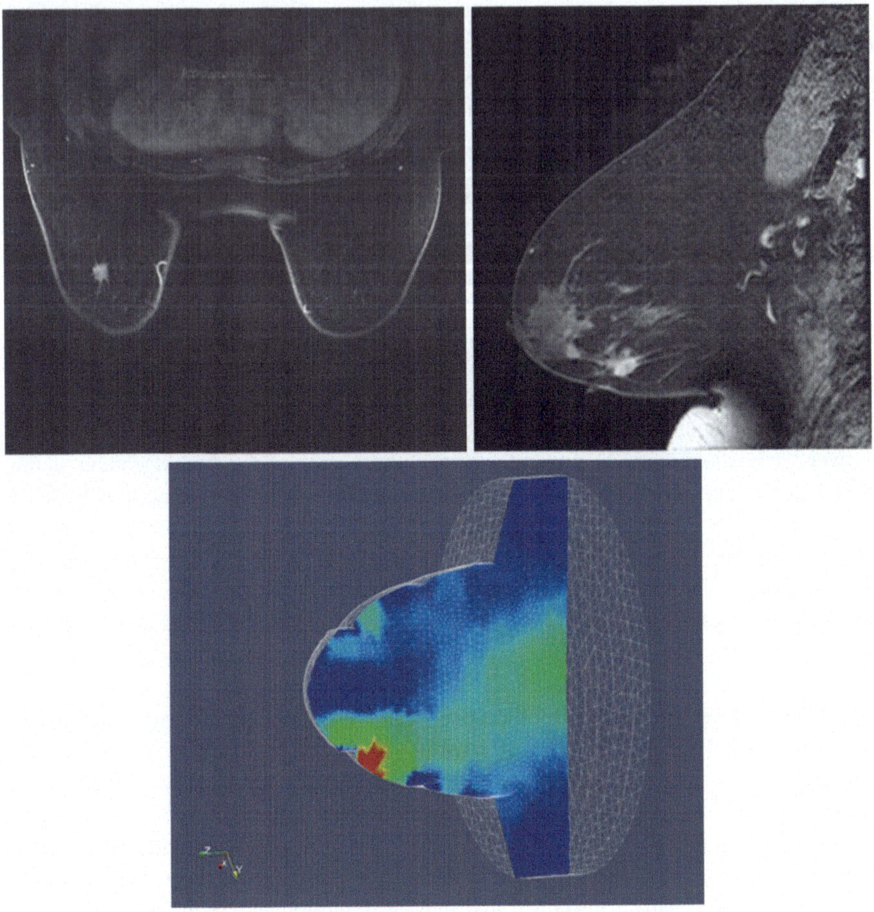

Fig. 3.5 Contrast-enhanced MR images, 1.3 cm cancerous mass, 5 cm from the nipple, axial, sagittal planes and EIT reconstruction

its own, and a combined set of methods is normally undertaken including MEG. Patients have their epilepsy-suppressing medication suspended for a week prior to surgery, during which time they are monitored with EEG to gain as much electrical information as possible during seizure activity. A technique known as telemetry is quite successful for localization of superficial foci, but less so for foci situated deeper in the brain. For these, EEG electrodes are often implanted deep in the brain, a process that can cause irreversible damage. EIT may provide a more powerful, as well as less invasive, alternative for their localization. An example of this is the use of 31 electrodes applied to the scalp in the same way as in EEG, based on the standard 10–20 configuration with some additional electrode positions. Two hundred and fifty-eight individual voltage measurements were taken to produce one impedance tomogram (Fig. 3.6). Studies have also shown changes of the order of

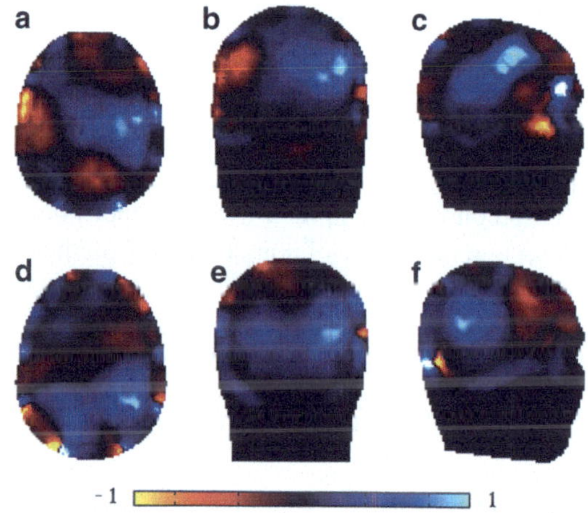

Fig. 3.6 Reconstructions of data collected approximately 6 s prior to two right temporal complex partial seizures in patient 1. In both cases, three orthogonal planes are shown through the maximum conductivity increase. The axial slices (**a**) and (**f**) are 8 and 21 mm, respectively, above the $z = 0$ plane; the coronal slices (**b**) and (**e**) are -44 and -61 mm, respectively, from $y = 0$; and the sagittal slices (**c**) and (**f**) are 57 and 44 mm, respectively, from $x = 0$ (1)

0.5–1.0% were observed in the raw voltage data during visual evoked responses in healthy volunteers (Bayford and Tizzard 2012; Tidswell et al. 2001).

3.5.4 *Hyperthermia*

Hyperthermia, which offers the possibility of reducing tumour size, can be used for certain malignancies by applying temperature treatment. This involves concentrated heating of the tumour tissue to approximately 47 °C without damaging the surrounding tissue. EIT has the potential to be useful in this application. However, it has been found that there are artefacts in the resistance images owing to other physiological factors and that they are significant compared with the temperature-dependent changes. The feasibility of EIT for thermometry has also been questioned in light of changes in tissue fluid content during hyperthermic treatment and of the different impedance spectra of skeletal muscle and tumour tissue in rats during this approach (Paulsen et al. 1996).

3.5.5 *Gastrointestinal*

EIT of gastrointestinal function has also had some success in imaging changes in gastric emptying which can be an indicator of many functional disorders of the gastrointestinal tract, e.g. pyloric stenosis, and of stress. Also, slow gastric emptying can be a problem in the management of newborn babies recovering from intensive care (Bayford and Tizzard 2012).

3.6 Variations of EIT

There is an increased interest in a number of new variations on EIT. Two approaches are currently being developed, and though they are a long way from clinical applications, considerable progress is being made in their development. The first is based on the principle of using another image modality combined with EIT, and the second makes use of non-contacting methods using electrical inductors to induce the current or measure the field.

3.6.1 Magnetic Resonance Electrical Impedance Tomography

The first variation is a combination of MRI and EIT known as MREIT. In MREIT, a reconstruction of the cross-sectional current density and conductivity (or resistivity) images of human body with high spatial and contrast resolution can be obtained. Using LF-MRCDI (low-frequency magnetic resonance current density imaging) technique where MR images are acquired with external current injection through surface electrodes, the first images obtained are of internal magnetic flux density distribution. This internal information transforms the ill-posed inverse problem of EIT into a well-posed one in MREIT (Kim et al. 2009).

3.6.2 Magnetic Induction Tomography

The second of these variations on EIT is known as magnetic induction tomography (MIT). In principle, this can detect very small changes using planar gradiometers (PGRAD), low-noise amplifiers and digital signal processing for achieving high SNR. Systems like Gaze (Scharfetter et al. 2005) currently use one single excitation coil (EXC). The sample is rotated in 16 steps of 22.5°. The main advantages of this approach are that the boundary shape is well defined and, because there are no electrodes, the effect of contact impedance is eliminated.

3.7 Summary

In conclusion, until a sufficient level of investment is devoted to the development of improving both hardware and reconstruction algorithms, EIT will remain a research tool. This is not to say that it does not have the capability to produce useful clinical results; this review has illustrated a number of these such as continuous monitoring of lung functison and breast tumour localization. However, equipment developed on limited resources cannot hope to achieve the results obtained by other imaging modalities, such as MRI. EIT also needs to produce useful clinical parameters and not simply images. However, there are promising areas, which are emerging, which could see EIT been adopted in the clinical enrolment (Bayford and Tizzard 2012).

References

Adler, A., Arnold, J. H., Bayford, R., Borsic, A., Brown, B., Dixon, P., et al. (2009). GREIT: A unified approach to 2D linear EIT reconstruction of lung images. *Physiological Measurement, 30*(6), S35–S55. https://doi.org/10.1088/0967-3334/30/6/303. Epub 2009 Jun 2.

Adler, A., & Lionheart, W. R. B. (2006). Uses and abuses of EIDORS: An extensible software base for EIT. *Physiological Measurement, 27*(5), S25.

Barber, D. C. (1989). A review of image reconstruction techniques for electrical impedance tomography. *Medical Physics, 16*, 162–169.

Barber, D. C., & Brown, B. H. (1984). Applied potential tomography. *Journal of Physics E: Scientific Instruments, 17*, 723–733.

Bayford, R., & Tizzard, A. (2012). Bioimpedance imaging: An overview of potential clinical applications. *Analyst, 137*(20), 4635–4643. https://doi.org/10.1039/C2AN35874C

Bayford, R. H. (2006). Bioimpedance tomography (electrical impedance tomography). *Annual Review of Biomedical Engineering, 8*, 63–91.

Borsic, A., & Bayford, R. (2010). Forward solving in electrical impedance tomography with algebraic multigrid wavelet based preconditioners. *Journal of Physics: Conference Series, 224*(1), 012053.

Borsic, A., Halter, R., Wan, Y., Hartov, A., & Paulsen, K. D. (2010). Electrical impedance tomography reconstruction for three-dimensional imaging of the prostate. *Physiological Measurement, 31*(8), S1–16. https://doi.org/10.1088/0967-3334/31/8/S01

Cole, K. S. (1942). Dispersion and absorption in dielectrics II. Direct current characteristics. *The Journal of Chemical Physics, 10*, 98. https://doi.org/10.1063/1.1723677

Cole, K. S., & Cole, R. H. (1941). Dispersion and absorption in dielectrics I. Alternating current characteristics. *The Journal of Chemical Physics, 9*, 341. https://doi.org/10.1063/1.1750906

Dehghani, H., Soni, N., Halter, R., Hartov, A., & Paulsen, K. D. (2005). Excitation patterns in three-dimensional electrical impedance tomography. *Physiological Measurement, 26*, S185–S197. https://doi.org/10.1088/0967-3334/26/2/018

Draeger. https://www.draeger.com/en_uk/Hospital/Productselector/Ventilation-and-Respiratory-Monitoring/ICU-Ventilation-and-Respiratory-Monitoring.

Gisser, D. G., Isaacson, D., & Newell, J. C. (1988). Current topics in impedance imaging. *Clinical Physics and Physiological Measurement, 9*, 35.

Hamilton, S. J. (2017). EIT Imaging of admittivities with a D-bar method and spatial prior: Experimental results for absolute and difference imaging. Published 22 May 2017. Institute of Physics and Engineering in Medicine. Physiological Measurement, Volume 38, Number 6. In: *16th International Conference on Electrical Bioimpedance and 17th International Conference on Biomedical Applications of Electrical Impedance Tomography* (pp. 19–23). Stockholm.

Horesh, L. (2006). *Some novel approaches in modelling and image reconstruction for multi-frequency electrical impedance tomography of the human brain*. PhD thesis. London: University College London.

Jehl, M., Dedner, A., Betcke, T., Aristovich, K., Klofkorn, R., & Holder, D. (2015). A fast-parallel solver for the forward problem in electrical impedance tomography. *IEEE Transactions on Biomedical Engineering, 62*(1), 126–137.

Kim, H. J., Kim, Y. T., Minhas, A. S., Jeong, W. C., Woo, E. J., Seo, J. K., et al. (2009). In vivo high-resolution conductivity imaging of the human leg using MREIT:

The first human experiment. *IEEE Transactions on Medical Imaging, 28*, 1681–1687. https://doi.org/10.1109/TMI.2009.2018112

Kolehmainen, V., Arridge, S. R., Lionheart, W. R. B., Vauhkonen, M., & Kaipio, J. P. (1999). Recovery of region boundaries of piecewise constant coefficients of an elliptic PDE from boundary data. *Inverse Problems, 15*(5), 1375.

Lionheart, W., Polydorides, N., & Borsic, A. (2004). Part 1 of electrical impedance tomography: Methods, history and applications. In D. Holder (Ed.), (pp. 3–64). Bristol: Institute of Physics Publishing. ISBN: 0750309520.

Lionheart, W. R. B. (2004). EIT reconstruction algorithms: Pitfalls, challenges and recent developments. *Physiological Measurement, 25*(1), 125.

Mahara, A., Khan, S., Murphy, E., Schned, A., Hyams, E., & Halter, R. J. (2015). 3D microendoscopic electrical impedance tomography for margin assessment during robot-assisted laparoscopic prostatectomy. *IEEE Transactions on Medical Imaging, 34*(7), 1590–1601. https://doi.org/10.1109/TMI.2015.2407833

Mayer, M., Brunner, P., Merwa, R., & Scharfetter, H. (2005). Monitoring of lung edema using focused impedance spectroscopy: A feasibility study. *Physiological Measurement, 26*, 185–192. https://doi.org/10.1088/0967-3334/26/3/004

Menin, O. H., Rolnik, V., & Martinez, A. S. (2013). Boundary element method and simulated annealing algorithm applied to electrical impedance tomography image reconstruction. *Revista Brasileira de Ensino de Física, 35*(2), 1–7.

Murphy, E. K., Mahara, A., Wu, X., & Halter, R. J. (2017). Phantom experiments using soft-prior regularization EIT for breast cancer imaging. Published 22 May 2017, Institute of Physics and Engineering in Medicine. Physiological Measurement, Volume 38, Number 6. In: *16th International Conference on Electrical Bioimpedance and 17th International Conference on Biomedical Applications of Electrical Impedance Tomography* (pp. 19–23). Stockholm.

Paulsen, K. D., Moskowitz, M. J., Ryan, T. P., Mitchell, S. E., & Hoopes, P. J. (1996). Initial in vivo experience with EIT as a thermal estimator during hyperthermia. *International Journal of Hyperthermia, 12*, 573–591. https://doi.org/10.3109/02656739609027666

Pidcock, M. K., Kuzuoglu, M., & Leblebicioglu, K. (1995a). Analytic and semi-analytic solutions in electrical impedance tomography: I. Two-dimensional problems. *Physiological Measurements, 16*, 77–90.

Pidcock, M. K., Kuzuoglu, M., & Leblebicioglu, K. (1995b). Analytic and semi-analytic solutions in electrical impedance tomography: II. Three-dimensional problems. *Physiological Measurements, 16*, 91–110.

Scharfetter, H., Merwa, R., & Pilz, K. (2005). A new type of gradiometer for the receiving circuit of magnetic induction tomography (MIT). *Physiological Measurement, 26*, S307–S318.

Soni, N. K., Paulsen, K. D., Dehghani, H., & Hartov, A. (2006). Finite element implementation of Maxwell's equations for image reconstruction in electrical impedance tomography. *IEEE Transactions on Medical Imaging, 25*(1), 55–61.

Swisstom AG. http://www.swisstom.com/.

Talmor, D., Sarge, T., Malhotra, A., O'Donnell, C. R., Ritz, R., Lisbon, A. et al. (2008) Mechanical ventilation guided by esophageal pressure in acute lung injury. The New England Journal of Medicine 359:2095–2104.https://doi.org/10.1056/NEJMoa0708638.

Tidswell, A. T., Gibson, A., Bayford, R. H., & Holder, D. S. (2001). Three-dimensional electrical impedance tomography of human brain activity. *NeuroImage, 13*, 283–294.

Wolf, G. K., & Arnold, J. H. (2006). Electrical impedance tomography: Ready for prime time? *Intensive Care Medicine, 32*, 1290–1292. https://doi.org/10.1007/s00134-006-0253-z

Wu, Y., Langlois, P. J., Bayford, R., & Demosthenous, A. (2016). Design of a CMOS active electrode IC for wearable electrical impedance tomography systems. ISBN Information: Electronic ISSN: 2379-447X. INSPEC Accession Number: 16226602. doi: https://doi.org/10.1109/ISCAS.2016.7527373.

Chapter 4
Electrical Impedance Tomography to Detect Trends in Pulmonary Oedema

Franco Simini, Eduardo Santos, and Martín Arregui

4.1 Introduction

Electrical impedance tomography (EIT) has evolved over the last four decades since the Sheffield pioneering works (Barber and Brown 1984) until present-day clinical equipment. The ability to obtain "slices" of parts of the body non-invasively is what makes EIT so attractive despite its relatively slow development and limited adoption in clinical settings. As described in the chapter by Bayford, EIT employs harmless currents with no physiological nor functional consequences on the patient. If one compares the high-definition anatomical details available in present-day computer tomogram (CT) scans, magnetic resonance imaging (MRI) or even positron emission tomography (PET) images—which all involve hundreds of processed radial attenuation signals—with the result of a series of 16 electrodes applied on the patient's skin for high-frequency current injection and voltage measurement, elementary concepts of signal analysis theory imply that EIT is not competitive in terms of accuracy. This is because paired combinations of 16 measurement points are outnumbered by the thousands of pixels available in the other imaging techniques.

Although complex inverse problems are mathematically tackled in EIT making a number of theoretical assumptions, the tomographic images with colour-coded conductivity are secondary to a limited data set and therefore have a relatively modest information content. Since air and water have very different resistivities, water can be spatially distinguished from air, with the potential to give clinicians a powerful tool to monitor and record the presence of both elements in body segments of critical patients where there should be only air, such as the airways and alveoli.

F. Simini (✉) · E. Santos · M. Arregui
Núcleo de Ingeniería Biomédica de las Facultades de Medicina e Ingeniería, Universidad de la República, Montevideo, Uruguay
e-mail: simini@fing.edu.uy

F. Simini, P. Bertemes-Filho (eds.), *Bioimpedance in Biomedical Applications and Research*, https://doi.org/10.1007/978-3-319-74388-2_4

In this chapter we discuss the strength of EIT over other imaging techniques, based upon its fundamental "weakness", i.e. giving only "rounded-up figures" of air/fluid occupation, without high anatomical accuracy, at a much lower price in terms of costs, untoward effect of ionizing radiation (CT or PET) or size of the equipment (MRI). EIT is a low-cost, non-invasive, continuous measurement method used to obtain images of the distribution of pleuropulmonary fluids and air.

Estimation of alveolar fluid content and distribution is essential in the management of conditions such as cardiogenic pulmonary oedema, pleural effusions, pneumonia and acute respiratory distress syndrome (ARDS). Electrical impedance of tissues can be estimated by measuring voltages on the skin while applying high-frequency currents whose amplitudes are far below perception thresholds. Processing electrical impedance matrices yields tomographic images (Barber and Brown 1984).

This chapter (1) reviews the basic concepts of EIT with emphasis on precision and consistency which are related to the potential of becoming a clinical monitoring technique; (2) shows design decisions made by several authors, comparing them to existing commercial devices with similar characteristics; and (3) describes the main features of an EIT prototype as a starting point for interested readers and developers.

4.2 Physics and Clinical Wish List

The behaviour of water and air differs greatly with respect to ionic current travelling through them. Different tissue types or different tissue physiopathological conditions also react differently to the passage of electrical current. The electrical current in man-made conductors turns into ionic displacements in biological tissue; the electrodes are the key factors of our measurement capacity. This physical property suggests that an instrument capable of displaying the relative content of air and tissue in the thorax could be clinically useful. If in addition the geometrical distribution of air-filled alveoli and fluid-filled alveoli was drawn, then a practical evaluation tool could be devised. In case the processing time of such inner distribution was quick enough, we would have a suitable monitoring equipment. The injection of high-frequency current (>20 kHz) at amplitudes not perceived by the human body (of the order of a few mA) and the resulting measurements of voltages on the chest's skin allow to estimate the electrical impedance. This estimation can be displayed as a tomographic image, by applying one of a set of possible reconstruction techniques.

Treatment of pulmonary oedema is based, among other clinical parameters, upon the estimation of alveolar volume occupied by liquid. This estimation of the fluid occupation of the chest would be greatly facilitated by the availability of a graphical representation of its size and density to guide therapeutic interventions. Currently the clinician has only indirect estimations because X-ray imaging or computed tomography cannot be performed often enough, due to the difficulties of repeatedly carrying patients to radiological facilities and to the limitations imposed by the side effects of ionizing radiations. EIT is a promising alternative to moving critically ill patients and at the same time gives continuous information. Following the publications of several research groups in the 1980s, the Nucleo de Ingenieria Biomedica (nib) has developed since 1995 a number of circuits (Ferreira et al. 2002), tomographic reconstruction software products (Hartman et al. 2002) and prototypes (Gonzalez et al. 2005, Quinteros and Simini 2007; Santos and Simini 2012) under the name of IMPETOM (IMPEdance TOmography Montevideo) with test results in phantoms and healthy volunteers.

Circuit designs for EIT are reviewed, including wave generators, current sources, differential amplifiers, synchronous voltmeters, multiplexers and control modules. Nine designs are studied; the characteristics of five of them are presented (England 2005, Uruguay 2002, Iran 2006, China 2007 and Switzerland 2012). Three different solutions are compared, and an optimal design is proposed which includes a direct digital synthesizer (DDS) for signal generation, a modified Howland configuration for current source, 16 bits for the analog-to-digital conversion and a digital signal processor (DSP) for the synchronous demodulation as well as to process the measurements for the reconstruction algorithm. This allows us to design low-cost, gross graphical representations for oedema extent monitoring, with little anatomical accuracy. It is important for physicians to have a real-time representation, over time (hours and days), of the trend in fluid accumulation in the lungs. Figure 4.1 shows a colour-coded representation of normal lungs, detected by an EIT system.

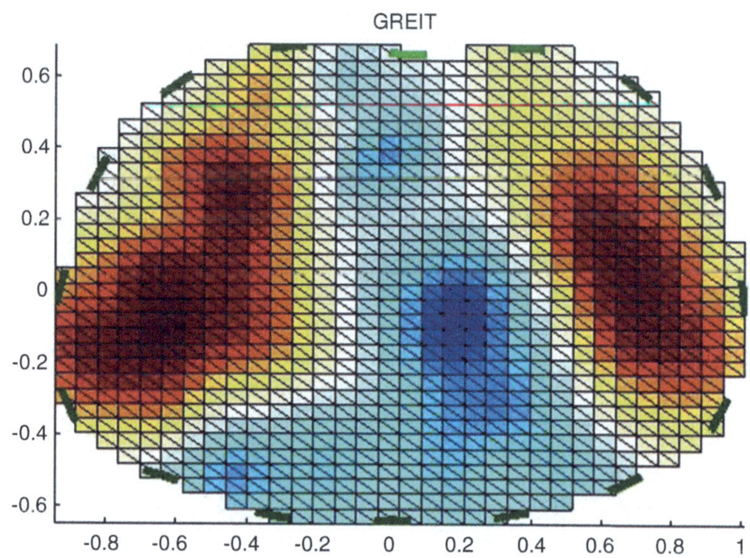

Fig. 4.1 EIT representation of normal lungs. Red is high bioimpedance, yellow is intermediate impedance, and blue is low impedance, i.e. organs and tissues. Taken from Santos (2014) with permission

4.3 Hardware Architecture

The main blocks of an EIT system, (1) waveform synthesis, (2) current source, (3) differential amplifier, (4) synchronous voltmeter, (5) multiplexers and (6) control circuitry, are shown in Fig. 4.2.

Waveform Synthesis The waveform generator creates the reference signal for the sinusoidal current source, besides generating the synchronizing signal to be used by the demodulator. There are analog and digital solutions; the latter is the most commonly used like in the circuit described by Saulnier (2004). This is followed by a digital-analog converter with stored samples. The performance of waveform synthesis is measured in terms of spectral purity and its signal-to-noise ratio (SNR).

Current Source The current source is driven by the waveform generator signal to convert it into a current to be injected into the body through the electrodes. Either single-ended or floating input can be used; the former is appropriate for multiple-source systems, while the latter configuration is best adapted to single-source systems (Saulnier 2004). All current source designs published (discrete op amps or transistors) face the same problem: i.e. output impedance is difficult to be kept high (Saulnier 2004). This impedance consists of an output resistance and stray capacitance: the output impedance thus affects the value of the injected current, depending on the load seen by the source, in this case the human body (including the electrode/skin interface), which is constantly changing. That is why the design of the source must take into account load variability in order to obtain an output impedance that allows the current to vary at most within the A/D least significant bit (Saulnier 2004).

Differential Amplifier Either differential or single-ended voltages may be measured in principle, but the vast majority of systems include differential configurations, with as a consequence a reduced voltage dynamic range (Saulnier 2004). Designers face here a second problem: the common-mode voltage of the amplifiers. To reduce it, the components and the operating point must be chosen very carefully. In addition to this, one may use specific techniques, such as an additional electrode

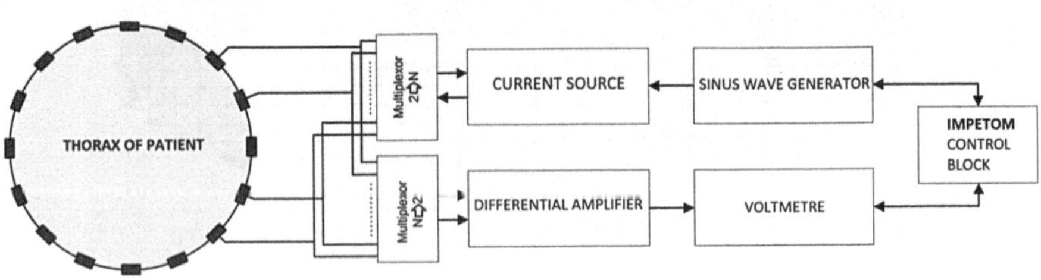

Fig. 4.2 EIT system structure with typically 16 electrodes affixed onto the skin of the patient's body. After Santos (Santos and Simini 2013) with permission

located away from the area where the measurement is taken to feedback a "driven-right-leg" current much in the same way as in ECG circuits (Webster 2010).

Synchronous Voltmeter To reconstruct a tomographic image, it is necessary to take phase-sensitive voltage measurements, either to measure the resistivity and permittivity or just the resistivity. This is why it is necessary to have a coherent signal from the wave generator. This demodulation can be either in the analog or digital domain; the latter is the most used recently (Saulnier 2004). One way to perform the digital demodulation is multiplying the A/D-converted input by a cosine function of the same frequency, and the samples are then integrated over a number of cycles of the original wave.

Multiplexers Multiplexers are needed in systems with a single current source or with fewer voltmeters than electrodes. Multiplexers are used to select channels through which current is injected and/or where the voltage is measured. Multiplexers have nonideal characteristics and exhibit, for instance, an "on-resistance" which distorts the desired signal and, above all, introduces input and output capacitances.

Control and Data Storage Block Control is necessary to synchronize all other blocks, to store the measured data and to transfer it to a computer that will perform the image reconstruction. Possible designs span from basic microcontroller (Xu et al. 2005) to high-performance digital signal processors (DSPs) (Xu et al. 2005; Wang et al. 2005a, b; Hamidi et al. 2010) to include more complex data preprocessing before computer involvement.

4.4 Review of EIT Systems

Several authors have published original work on EIT hardware systems, some of which are described here. A great variety of technical solutions are presented, giving the reader hints for personal developments.

Ferreira et al. (2002) proposed an EIT circuit with sinusoidal signal generation using the function generator MAX038 (Maxim). The current source includes the operational amplifier AD844 (Analog Devices). Two multiplexers MAX336 (Maxim) select the electrodes that carry the current. In the measurement stage, 16 identical circuits avoid the use of a multiplexer. The INA114 (Burr-Brown) operates as an instrumentation amplifier, and its output is fed to a bandpass filter made with the MAX274 chip (Maxim). A sample and hold circuit captures the filter output with the LTC1043 (Linear Technology) IC.

González et al. (2005) used the PC-LPM-16PnP (National Instruments) board as control stage, with 16 A/D 12-bit conversion input channels. The results measured show a SNR (signal-to-noise ratio) of 40 dB, a CMRR (common-mode rejection ratio) of the order of 50 dB and a current source output impedance of 560 k at 50 kHz.

Wang et al. (2005a, b) used a TMS320F206 (Texas Instruments) DSP for control and data preprocessing. A voltage-controlled current source (VCCS) generates the signal that is injected by the electrodes. To measure voltages, a multiplexer is implemented with the AD734 (Analog Devices) IC, a low-pass filter with MAX275 chip (Maxim) and AD1674 (Analog Devices) converter, reaching a rate of 30 frames per second.

Soleimani (2006) designed a low-cost equipment where the waveform is digitally generated using a 23 kHz stored sinusoid in an EPROM (27C258), fed to a D/A converter (DAC-0808 National Semiconductor). 1.3% of harmonic distortion was measured. The signal is fed to a buffer which in turn is connected to a VCCS, (implemented using AD644 by Analog Devices). To measure voltages, a synchronous demodulator is used due to its ability to remove noise. An AD625 instrumentation amplifier (Analog Devices) is included as input stage to be fed to the PCL-812PG I/O (Advantech) board.

Xu et al. (2005) created a system using 128 electrodes for three-dimensional (3D) images of the human thorax. A multiple frequency current (1 kHz–2 MHz) is injected using the IC AD9852 (Analog Devices). The control module is implemented with TMS320F2812 (Texas Instruments). The multiplexer is a MAX306 (Maxim). Data acquisition is done by AD624 (Analog Devices) used as a preamplifier, followed by a fourth-order Butterworth filter (MAX275 by Maxim). The synchronizing signal (AD9852) is used in the DSP to implement the synchronous demodulator. Reconstructed images are shown, but no SNR nor any other performance figures are given.

Bera and Nagaraju developed in 2009 (Bera and Nagaraju 2009) a system to study calibration issues. For current injection they use a voltage-controlled oscillator (VCO) built with the MAX038 (Maxim), which feeds a modified Howland-type current source (Saulnier 2004). This module is constructed with two operational amplifiers AD811 (Analog Devices) and generates a current of 1 mA at 50 kHz. A differential amplifier stage and filter is used for the voltage measurement with a multimeter and a digital oscilloscope. A central electrode is used to reduce the common-mode signal. Command mode is reduced at best to 67 mV.

Hamidi et al. implemented in 2010 (Hamidi et al. 2010) a synchronous demodulator with a DSP (Texas Instruments MS320C6713) which is mounted on the development board TLV320AIC23 (Texas Instruments). This board has two 16-bit ADCs (AIC23 Texas Instruments) 96 kbps sampling rate. With simulated data, the system had a phase error of 0.12° and a signal-to-noise ratio (SNR) of 130 dB.

Gaggero et al. (2012) developed a system to address two important problems affecting EIT reliability and usability: (a) electrode contact impedances, large and varying due to movement, and (b) placing of the electrodes for each individual patient. To solve these problems, Gaggero implemented a method of active electrodes, which uses a voltage buffer in the proximity of the electrodes to stabilize the contact impedance. A multiplexing method reduces the length of wires from the electrode belt to the central block. The signal generator, the power source, the electrode handling and the communication with the PC are in the central block of the equipment. Two standard boards are used for this: an Altera Stratix II Development

Kit (Altera) which includes both AD9433 ADC (TI) and DAC904 DAC (TI) and an Ethernet interface. A custom developed board is responsible for the analog stage, with differential amplifier and high-pass filter implemented with an AD8221 (Analog Devices), signal conversion into a differential signal and low pass filtering (THS4502 by IT). The sinusoidal signal is generated in the FPGA by a numerically controlled oscillator (NCO) and a modified Howland current source.

Dixtal Biomedical developed a commercial system DX 1800, seen in operation in Montevideo in 2010, which provides real-time EIT chest images. It has two modules: the DX1800 itself and a monitoring software that runs on a computer. The DX 1800 has 32 electrodes to be placed around the patient's chest. The software handles the user interface and displays the reconstructed chest images.

PulmoVista 500 was first introduced by Dräger Medical (2011) to display lung ventilation in real time. A very practical belt with 1 row of 16 electrodes is placed around the chest and injects currents between 80 and 130 kHz, generating images at a rate of 10–30 per second with 1440 × 900 pixel resolution.

Santos in 2014 (Swisstom) suggested to use standard evaluation boards for an efficient EIT system design, as described in the following paragraph.

Alfaro et al. (2015) and Arregui et al. in (2016) and 2016 include an "RC phantom" in the design of an EIT system, as a calibration feature for both development time and normal use.

4.5 Suggested Design Options

After our own and projects by others (which we just reviewed), there are three possible options for EIT hardware: one is a completely analog signal processing design, and the other two solutions use a DSP, and decisions depend upon cost factors, building difficulty, performance and speed.

4.5.1 Discrete Components with No Digital Processing

The first option is to design using discrete components without a DSP, but with a basic microcontroller (e.g. ATmega168, Atmel, costs $3, the one used in Arduino boards) which would be responsible for control and synchronization of the remaining blocks. Signal demodulation would be carried out using discrete components, as well as filtering and amplification. An A/D converter would be used to send signals to the computer for image reconstruction. A precaution for improving system performance would be to use at least 16-bit ADC resolution. The data acquisition cards (DAQs) have multiple input channels, making it possible to have a separate channel for each pair of electrodes, enabling measurements in parallel. This also means to have to build 16 input channels (same as IMPETOM C (Ferreira et al. 2002)), increasing the cost and difficulty of design. A DAQ with several input channels is expensive, such as the NI USB-6211 by National Instruments (approx. $690). The main problem with this solution is electrical noise, which is difficult to reduce, as well as the imperfection of discrete electronic components that dominate the design giving in the end poor reconstructed images. The advantage of this option is its low cost, except for the expensive DAQ included.

4.5.2 *Standard Evaluation Board*

To minimize signal processing in analog form, and thus noise interference, A/D conversion must be included in the design as near as possible to the measurement point. As much processing as possible is performed in digital format, including basic first filtering and gain stages, using a DSP standard evaluation board. Demodulation is also done in the digital domain. This option gives great processing flexibility, in terms of giving designers the possibility to obtain both resistivity and permittivity measurements, in addition to reducing discrete component-originated errors. By using a DSP board, the measurement vector ready to be used in the reconstruction is transmitted to the computer.

Examples of boards are the Starter Kit TMS320C6713 (IT, $ 395) or the Evaluation System ADSP-BF533 EZ-Kit (Analog, $ 490). Despite that these boards are limited, the advantage of using them is that of shortening prototype development time because they are complete systems with DSP, memory, demodulators, A/D with programmable gain amplifiers (PGAs), communication with the computer, ports for general use as well as a circuit to program the DSP. Figure 4.3 shows the block diagram for a system implemented with an evaluation board.

The disadvantage is that standard boards have a fixed set of components, some of them not suited to solve the problem faced. Most of these cards, at least those found in a price range between $300 and $600, are designed to work with audio, and therefore the A/D and PGAs are available within a codec (Alfaro et al. 2016) specifically designed for audio processing. The most notorious limitation is that codecs operate at frequencies used in standard audio, the higher sampling frequency being 96kSPS (samples per second). According to the Nyquist theorem, a maximum limit for our EIT frequency left is 48 kHz, which is not enough to get good SNR values, because the operating frequency available in practical design turns out to be below 20 kHz, due to the several simultaneous tasks. A pity considering the DSP within the TMS320C6713 board has the potential to operate at 225 MHz and to perform 1350 million floating-point operations per second (MFLOPs) (Alfaro et al. 2016). Another disadvantage is the relatively high cost.

4.5.3 *Discrete Components with Digital Processing*

To overcome the limitations of using a Starter Kit, we have the option to work with DSPs and to choose the components that are most suited to our problem. The use of DACs with a higher sampling frequency, such as the AD7892 (Analog, $ 23)

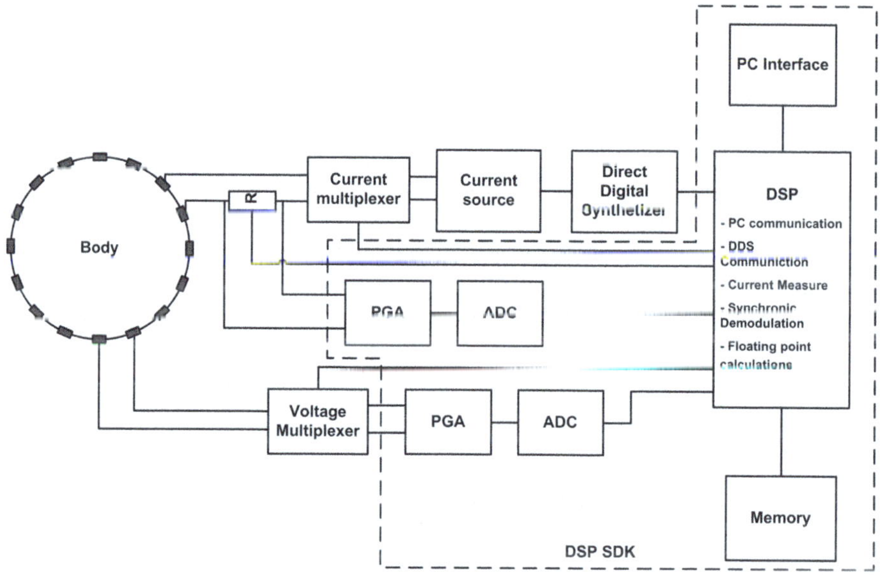

Fig. 4.3 EIT design using an evaluation board. After Santos and Simini (2012), with permission

which has 18 bits of resolution and the ability to get up to 1MSPS, which would theoretically allow us to work up to a frequency of 500 kHz. The advantage of a lower cost is greatly offset by the difficulty to achieve a truly compatible and harmonious design (including clock speed, type of communication, supply voltages) as well as to assemble in the laboratory successive working prototypes (component layout, soldering techniques, handling).

4.5.4 Suggested Design

The proposed design for IMPETOM improves over previous implementations, overcoming the most outstanding problems by digitizing the signals in early stages. By doing this we can take advantage of the precision of digital filtering and demodulation. Our suggested design is based on an evaluation board, such as OMAP-L137/TMS320C6747 Floating-Point Starter Kit ($ 415, Spectrum Digital). The most important characteristics are the inclusion of an OMAP-L137 processor, with a C6747 VLIW DSP floating-point processor and an ARM926EJ-S processor operating up to 300 MHz, 64 MB SDRAM, USB2 2.0 interface, TLV320AIC3106 stereo codec and expansion connectors for daughter cards.

The codec has four input channels, and considering that a channel will be used to measure the injected current, there would be three input channels available to measure voltages in parallel. This system enables to perform two significant improvements over previous systems (Ferreira et al. 2002; Gonzalez et al. 2005; Quinteros and Simini 2007):

1. To measure the injected current with a digital metre controlled by the control block which ensures that the signal obtained makes the most of the range of the ADC.
2. To have a single voltage measurement channel with a multiplexer. Despite the increased acquisition time (due to parallel channels read in sequence) and the multiplexer sources of noise and errors (Saulnier 2004), this change reduces the complexity and cost of 16 measuring channels of the IMPETOM project (Ferreira et al. 2002).

4.6 Inverse Problem

4.6.1 Mathematical Background

Since EIT is a technique used to derive conductivity within a region from measurements taken on its boundary, the region of interest is the space delimited by the patient's skin. The same electrodes used to inject currents are also used to measure voltages as "boundary conditions", i.e. the contour of what the technique is interested in. There are several possible configurations for an EIT system, including a variety of frequencies and waveforms of injected currents, the number of electrodes and their positioning, e.g. adjacent or opposite electrodes described elsewhere in this book and within this chapter. The problem to solve when extracting information from the unknown "inside" from data measured on the border is essentially a mapping problem. In very much the same way as cardiologists reconstruct the electrical activity of the heart from "boundary" conditions, i.e. skin-detected electrical activity in the form of an ECG, EIT reconstruction methods give a conductivity pattern inside the region that maps into boundary values. When the calculated skin values—calculated from the supposed internal pattern—map into values very similar to the one actually measured in real life, then the supposed pattern is given as the tomographic reconstruction.

Let V be the boundary values, resulting of an unknown mapping U, which is a function of the injected currents I and the conductivity distribution s, then:

$$\mathrm{V} = \mathrm{U}\ (\mathrm{s}, \mathrm{I})$$

The Jacobian matrix J is somehow a derivative dV/ds which if inverted could help determine the distribution s. But it is not possible because the procedure is an "ill-conditioned problem", i.e. little noise in the measured data lead to unacceptable errors in the reconstructed conductivity. To overcome the ill-conditioning, a method is needed and one is the Tikhonov regularization.

4.6.2 *First Part of the Solution: Forward Problem-Solving*

To solve the inverse problem, it is necessary to know the forward problem, which is the mapping of the conductivity inside the body to the potentials along the boundary. Since it is a non-linear mapping, we assume there must be a linear equivalent. The determination of the forward operator by mathematical analysis is only possible if the geometry and the conductivity of the EIT problem are simple (e.g. uniform conductivity in a cylinder) (Saulnier 2004). In all other cases, it is necessary to numerically discretise both the region and the conductivity with finite elements. It is also necessary to feed the method with initial experimental data to obtain a linear formula. For the saline-filled tank, a circular geometry is acceptable, while for chest reconstruction it must be replaced by closer anatomical approximations, sometimes adapted to the size and age of the patient.

4.6.3 *Second Part of the Solution: Regularization*

The inverse problem consists in determining the parameters m corresponding to the measurements d knowing the mapping H. The measurement errors and uncertainties involved are such that in practical terms, it does not hold. This is called an ill-conditioned problem in Hadamard's sense (Borsic et al. 2007) which implies that at least one of the following is true:

1. The solution does not exist.
2. The solution is not unique.
3. The solution is a noncontinuous function of the data, such that small data perturbations cause arbitrarily large errors in the reconstructed parameters.

The Tikhonov method confines the solution to a space of expected solutions: this is done by adding a term introducing information about where the solution might be, starting from an initial estimation. Different modifiers can be used, such as Tikhonov's standard regularization or NOSER's regularization which takes into account an initial conductivity estimate (Santos 2014).

4.7 Comparison of Reconstruction Implementations

To obtain a tomographic reconstruction, we consider three different implementations: the original IMPETOM (Hartman et al. 2002) and GREIT and NOSER regularizations, both obtained using the EIDORS package (Borsic et al. 2007) which is an open-source software suitable for EIT image reconstruction with several algorithms to solve both forward and inverse problems. IMPETOM original reconstruction software uses the Newton-Raphson method with finite elements (Hartman et al. 2002).

4.7.1 Phantom and Volunteer Experimental Data

The original data was obtained using IMPETOM circuitry (Gonzalez et al. 2005) and was tested with both the IMPETOM system and EIDORS (EIDORS: Electrical Impedance Tomography and Diffuse Optical Tomography Reconstruction Software (Version 3.5)). This comparison is important for prototype development and deciding the best algorithm to adopt.

For development purposes an original cylindrical tank phantom was build. It consists of a cylindrical tank 21 cm in diameter, filled with saline solution. Inside the tank an empty 8 cm diameter plastic bottle was initially moved along a diameter of the tank and subsequently around its boundary (Ferreira 2002).

A healthy volunteer was harnessed with the 16 electrodes of IMPETOM to image his chest (Gonzalez 2005).

A/D converter of IMPETOM allowed to acquire images at a rate of 15 frames per second. The experiment with the phantom included 300 frames; 400 frames of the volunteer were acquired.

4.7.2 GREIT Reconstruction Algorithm

We used the GREIT "Graz consensus linear Reconstruction algorithm for lung EIT" (Adler et al. 2009). The specifications of GREIT are:

1. Single-ring electrode configurations with adjacent current injection and measurement.
2. Linear reconstruction of a 2D conductivity difference image, based on a forward model, which allows real-time processing.
3. Reconstruction into a 32 by 32 pixel array for a single ring of 16, 12 or 8 electrodes.
4. Predefined shapes of (a) neonatal chest, (b) male and female adult chest and (c) cylindrical tank.

GREIT is useful because it includes performance figures of merit for EIT image reconstructions (Adler et al. 2009). Consensus figures of merit, in order of importance, are:

1. Uniform amplitude response.
2. Small and uniform position error.
3. Small ringing artefacts.
4. Uniform resolution.
5. Limited shape deformation.
6. High resolution.

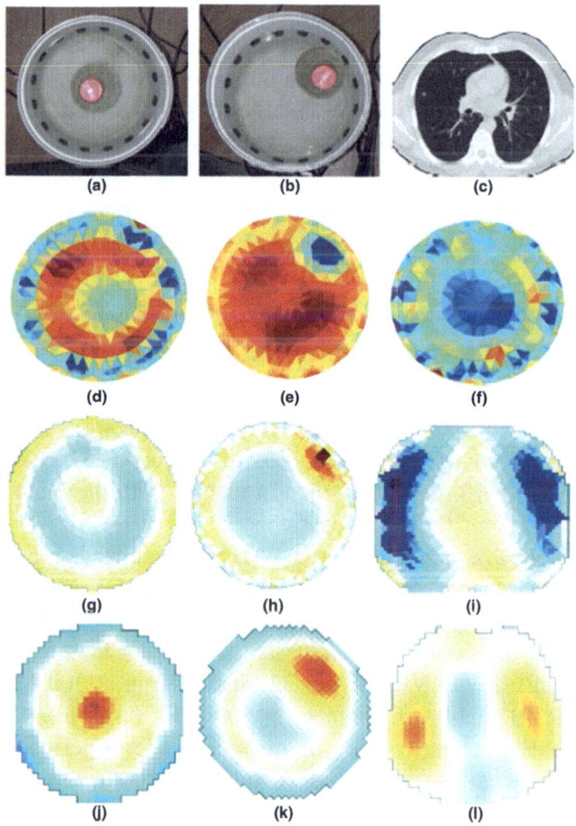

Fig. 4.4 Comparison of reconstruction methods: the first row shows the phantom or patient to be imaged. The lines below show the output of IMPETOM (492 pixels), NOSER (equivalent to 2048 pixels) and GREIT (1024 pixels) algorithms. The first column is an empty plastic bottle in saline. The second column shows the bottle placed near the edge of the tank. The third column is a sample CT scan and tomographic sections of a normal volunteer

4.7.3 Comparison of Reconstructions

Standard IMPETOM parameters were used: 492 elements, 2D circular mesh. Regularization NOSER parameter was set by trial and error to 0.07 for phantom imaging (Fig. 4.4g, h) and to 1.0 for the healthy volunteer (Fig. 4.4i). For the cylindrical phantom, 256 element 2D circular mesh was used. A thorax-shaped mesh was selected for the volunteer. The GREIT algorithm was trained with a uniform distribution of points for the chest and a heavy-centred distribution of points (Hamidi et al. 2010) for the phantom. The initial image for the cylindrical tank is a uniform circle which gives a frame for simulated measurements with the forward solution. The frame difference was then used to feed the inverse problem in order to obtain tomographic sections (Fig. 4.4). The initial frame is the mean of 400 real skin measurements. The difference between the mean frame and the real frame is fed to the inverse problem algorithm producing the tomographic image.

The results of reconstructions are shown in Fig. 4.4. Considering that our goal was to develop a prototype for quantitative air/fluid chest distribution (Santos and Simini 2012) and not an anatomically accurate image, the results are promising.

IMPETOM Fig. 4.4d, e, f were not as good as GREIT Fig. 4.4j, k, l but were better than NOSER Fig. 4.4g, h, except for the lung reconstruction in Fig. 4.4i. With standard desktop computing power, all three methods create images in less than a second. The GREIT algorithm included a 3D mesh to account for current leakages to adjacent planes, which is a factor of quality. There are two details that will enable to obtain better results for chest reconstruction (Fig. 4.4i, l) with respect to the original IMPETOM prototype (Fig. 4.4f): a chest-shaped model and starting with the mean of 400 frames to feed the inverse problem-solver.

4.8 EIT Using a Standard Board

Evaluation module OMAP-L137 from Spectrum Digital can be used as the core part of an EIT system. Its audio codec AIC3106 is used as a signal generator and to perform voltage measurements, with the included DSP C6747 from Texas Instruments for data processing.

The evaluation module (EVM) is an independent platform that allows developing and studying several applications for the OMAP-L137 integrated circuit. It has incorporated a multicore system on chip (SoC) that permits to work simultaneously the two processors nested within: the floating-point digital signal processor (DSP) C6747 VLIW with 32 bits, running at 300 MHz, and the RISC processor ARM926EJ-S also running at 300 MHz (both from Texas Instruments). Besides, the EVM has a series of microcontrollers, peripherals, inputs and outputs for diverse purposes. The audio codec AIC3106 has two separate input channels with a resolution of 24 bits and 48 kHz sampling rate, a multichannel serial port McASP, stereo inputs and outputs Line IN and Line OUT and 32 megabytes of SDRAM.

The system we implemented has a 16 kHz digital sine wave generated using a table within the DSP, which is sent to the audio codec through the serial channel McASP. The internal digital-to-analog converter (DAC) in the AIC3106 transmits the converted signal through the Line OUT port. The voltage signal Vin feeds a Howland current source, whose output impedance depends upon resistor matching and non-linearities.

Input signals to the audio codec VREF and VOUT are amplified by a programmable gain amplifier (PGA). Then signals pass through an antialiasing filter. Resultant signals are converted to digital samples at a sample rate of 48 kHz by means of an ADC. These samples are sent from AIC3106 to DSP C6747 through the McASP channel. A synchronous demodulator is implemented in the DSP to obtain the real part of the voltage VOUT. After digitizing, both input signals are multiplied sample by sample and inverted (since the current source has an inverter configuration). The real part of VOUT is proportional to the DC component of the resulting signal. An infinite impulse response (IIR) filter is used to filter the signal and the result is the DC component.

In order to test the audio codec behaviour as ADC input channel, we used a Tektronix CFG253 signal generator directly connected to one of the Line IN

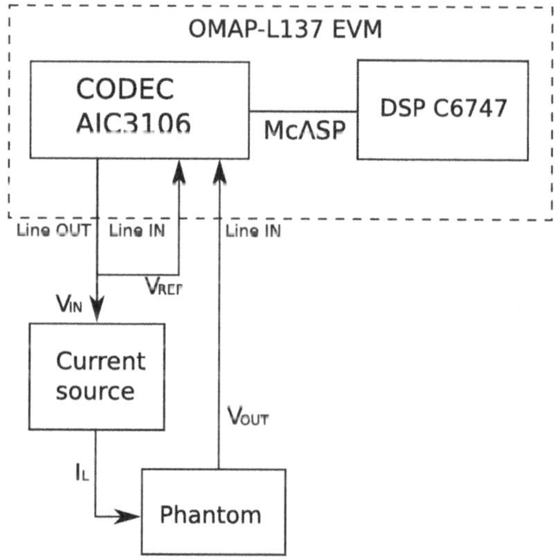

Fig. 4.5 Block diagram of an EIT system implemented with the evaluation module (EVM) OMAP-L137 and an external current source. Taken with permission from Alfaro et al. (2016)

channels in the EVM. The sinusoidal input signals were simultaneously measured with oscilloscope Tektronix TDS210. The results show a linear response from the codec with differences under 5% for values in the range from 42 mV to 1.5 V (Alfaro et al. 2016).

The system in Fig. 4.5 was then implemented with a Howland current source. A RC parallel circuit worked as our load impedance: a 100 nF capacitor placed in parallel with a variable resistor. An algorithm to obtain the real part of the impedance was implemented in the DSP, given the linearity between Vref and the current generated. By comparing the estimated real part calculated by the DSP with the oscilloscope, it was noticed that the EVM responds correctly to changes in the load impedance and the mean difference between values was 8.5% (every measurement was repeated five times to assess repeatability, giving 0.3% mean deviation).

The EVM (OMAP-L137) module with audio codec AIC3106 has proven to be suitable as an EIT system core. Since measurements are reliable above 42 mV, improving the current source is essential and should be such as to have a 42 mV effect for the smallest bioimpedance to be detected (Fig. 4.6).

4.9 Electrical Phantom to Test EIT Systems

Before using circuits on volunteers, we designed a simple saline-filled tank such as the one in Fig. 4.7, as a test phantom. This allows to perform qualitative tests. But we also need detailed calibration tests. Since most of the information is derived from the real part of the impedance values measured on a patient, but there is also a capacitive component, we decided to simulate a simple measurement by connecting a phantom made of a resistance $R = 350$ to $R = 600$ ohm in parallel with a capacitor of $C = 47$ nF. This circuit is fed by the modified Howland current

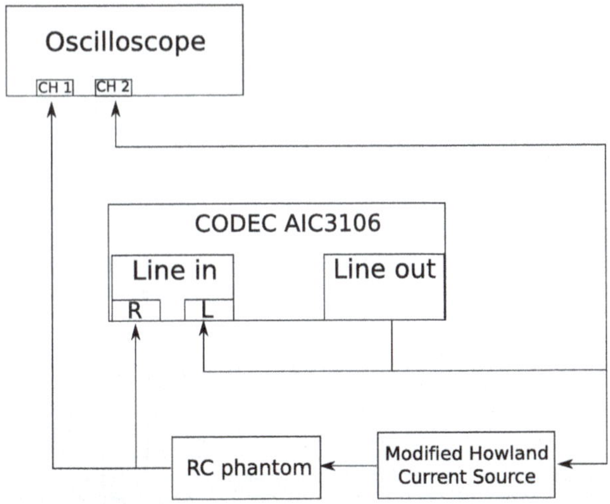

Fig. 4.6 Connection diagram of the test bench. The modified Howland current source feeds a phantom (parallel RC: $C = 47$ nF and a variable resistor). The oscilloscope allows to compare the original codec voltage signal to the phantom-shaped voltage. Taken with permission from Arregui et al. (2016)

Fig. 4.7 Test tank with three electrode rows for saline solution. After Quinteros and Simini (2007) with permission

source (Saulnier 2004) controlled by one of the codec output signals, proportional to a DSP-originated 16 KHz sinusoidal wave. We suggest to include the test RC "phantom" within the actual EIT instrument, to be available as "self-test" or "self-calibration".

The codec was configured to work with a 48 kHz sampling frequency in slave mode, and the data transfer protocol was "I2S to DSP" and 16-bit word length. No offset is added to the data. The PGA gain of each channel was set to 0 db. Communication between DSP and codec is through McASP serial port communication settings synchronized with two-slot TDM format.

The current source of 16 KHz (in addition to give rise to currents to be injected) is measured by the oscilloscope, as a phase reference for all current measurements. The same oscilloscope is also connected to the output of the RC phantom and to the input of the current source, where the voltage is in phase with the current injected

Table 4.1 Real part of the RC phantom and relative error

R Phantom (Ω)	Impedance R//C real part @ 16 KHz (Ω)	Oscil. measurement (Ω)	IMPETOM measurement (Ω)	IMPETOM adjusted R//C (Ω)	Relative error (%)
350	93.7	107.6	108.3	92.6	1.23
400	87.5	111.3	104.5	88.2	0.82
450	81.5	101.3	98.8	81.8	0.38
500	76.0	94.7	94.6	77.1	1.43
550	70.9	85.5	88.7	70.4	0.74
600	66.4	78.9	84.7	66.0	0.66
Average error					0.876

into the phantom. This allows to compare measurements made by the EIT system with a reliable instrument such as a laboratory oscilloscope.

The 2016 IMPETOM software uses the phase of the reference voltage signal to compare it to the phase of the resulting phantom voltage. The output voltage signal of the phantom circuit allows to calculate the real part of the phantom impedance. We do not use as a time reference the original DSP generated signal which is fed to the codec to control the current source, because it would introduce a lag from the time the DSP sends the signal to the time the codec effectively outputs the signal at its output channel. To avoid this delay, we use a second codec input channel to measure the signal of its own output port. In this way we reduce the time overhead derived from causes other than the phase change to be measured. Sampled signals and data are then sent from the codec to the DSP. We have tested our circuit with several values of R and a fixed C value, all reported in Table 4.1.

In our previous IMPETOM designs, time control was performed by a single analog signal multiplexed into 16 current sources. But when a decision was made to use digital circuits and audio kits to cut development costs, real part impedance measurement errors became high. Debugging gave as a result an unexpected phase lag between signals. An internal digital signal, given by the manufacturer as a vector in codec memory, was initially used as a reference to measure real-world hardware signal phases, but this also produced uncontrolled errors. A way to overcome this difficulty in future designs is to use the same codec-generated D/A signal as a reference for measurement as well as to create the signal to submit to the "patient simulator" or phantom, with reliable measurement values to be expected. This was actually the case, and thus the inclusion of the "phantom" with variable R and fixed C can be of help during development as well as a "self-test" feature of prototypes and future instruments alike. An error of 1% was measured, which is a reliable enough test bench to let prototypes inherit the precision of laboratory instruments. Without the "RC phantom", the impedance real part would be estimated with large errors due to lack of common time reference. This special measurement requirement is necessary because modern designs include software signal generation, as opposed to traditional layouts with a central analog oscillator to command all current sources. Future additional enhancements could be the inclusion of measuring bridges to improve sensitivity and 32-bit digital encoding to reduce quantification errors.

4.10 Current Leakage to Adjacent Chest Sections

EIT consists of current injection along a line on the skin of the patient and voltage measurements taken at the same electrodes. The assumption that all injected current is confined to the section under study, i.e. the section defined by the electrodes, is plausible but should be confirmed.

A first set-up with three rows of electrodes on the sides of a saline-filled tank is shown in Fig. 4.7. Similar to the "one row phantom tank" in all aspects, this phantom allows both to inject current in 16 electrodes and to measure 48 voltages (actually 46, since the two injecting electrodes are not used simultaneously). The distance from the central to the adjacent sections is 5 cm, which can eventually allow to measure voltage induction up to 10 cm away by "stimulating" either the top or bottom row. The EIT images of cylindrical empty bottles obtained were very similar irrespective of the row at which the voltages were measured, using the IMPETOM system (Quinteros and Simini 2007).

Manufacturers of real-time EIT system (Dräger 2011) publish estimated distributions of current away from the injection section which fade away to 10% at 10 cm of the central section.

4.11 Discussion and Conclusion

At a time of growing investment in health and of increasing concern on ionizing radiation, "soft" instrumentation such as EIT has the potential of a great development and adoption in clinical settings. Surprisingly only a small fraction of the effort devoted to classical imaging technology such as MRI, CT and PET is published in bioimpedance under all its forms, spectroscopy, EIT, organ composition analysis or new applications. This chapter contains clear evidence that sound electronics and mathematical methods can be successfully applied to EIT development and testing to match the clinical need of simple reliable fluid occupancy monitoring of intensive care patients. Thirty years have passed since the initial work of the Sheffield group to commercial dissemination in an incipiently meaningful proportion of medical units. All is set for EIT to fully occupy its market share for critically ill patient management in the coming years.

Acknowledgements Interest in electrical bioimpedance and motivation to develop EIT systems was originated by a clinical request by Professor Javier Hurtado and associates Dr. Walter Olivera and Dr. Cristina Santos in 1986, at that time active in the foundation of the Pulmonary

Functional Testing Laboratory of the University Hospital of Uruguay. Pioneering work by Dr. Fernando Nieto—prematurely deceased 2001—with one of the authors gave as a result our first bioimpedance spectroscopic tetra-electrode instrument in 1994 which prompted us to tackle the development of an EIT device with successive prototype versions. Editorial assistance with intensive care insight and counselling by Dr. Bruno Simini of Ospedale di Lucca, Italy, was very much appreciated and is greatly thanked for. The work of the following students since 1994 is acknowledged here: Cecilia Frugoni, Ramiro Escuder, Lauro Artía, Raúl Hartman, Jorge Lobo, Mateo Ruétalo, Adriana Ferreira, Alfredo Rodriguez, Santiago González, Andrés Liguori, Walter Quinteros, Nicolás Alfaro and Fernanda Martinucci. Special thanks are due to Professor Roberto Markarian and to Dr. Pablo Musé for their interest and suggestions during their 2013 graduate course on Real and Functional Analysis. Several colleagues have given substantial help to this work including, but not limited to, Pablo Mazzara, Rafael Canetti, Linder Reyes and Daniel Geido.

References

Adler, A., et al. (2009). GREIT: A unified approach to 2D linear EIT reconstruction of lung images. *Physiological Measurement, 30*(6), S35–S55.

Alfaro, N., Arregui, M., Martinucci, F., Santos, E., & Simini, F. (2016). Audio codec and digital signal processor for an electrical impedance tomography system. In *II Lat. Am. Conf. Bioimpedance, CLABIO 2015* (pp. 16–19).

Arregui, M., Santos, E., & Simini, F. (2016). Test bench to validate audio codec kit as EIT complex voltage measurement circuit. In *VII Latin American Congress on Biomedical Engineering CLAIB 2016, Bucaramanga.*

Barber, D. C., & Brown, B. H. (1984). Applied potential tomography. *Journal of Physics E: Scientific Instruments, 17*, 723–733.

Bera, T. K., & Nagaraju, J. (2009). A Simple instrumentation calibration technique for Electrical Impedance Tomography (EIT) using a 16-electrode phantom. In *Automation science and engineering, 2009. CASE 2009. IEEE International Conference on* (pp. 347–352).

Borsic, A., Graham, B., Adler, A., & Lionheart, W. (2007). *Total variation regularization in electrical impedance tomography* (techreport). Manchester, UK.

Dräger. (2011). PulmoVista 500 Data Sheet.

EIDORS: Electrical Impedance Tomography and Diffuse Optical Tomography Reconstruction Software (Version 3.5).

Ferreira, A., Rodríguez, A., & Simini, F. (2002). *IMPETOM-C Tomógrafo de Impedancia Eléctrica.* Academic thesis, Universidad de la Republica, Montevideo.

Gaggero, P. O., Adler, A., Brunner, J., & Seitz, P. (2012). Electrical impedance tomography system based on active electrodes. *Physiological Measurement, 33*(5), 831.

Gonzalez, S., Liguori, A., & Simini, F. (2005). IMPETOM Academic thesis.

Hamidi, S. A., Jafari, R., Nia, A. M., & Soleimani, M. (2010). Design and implementation of a DSP-based digital phase sensitive demodulation for an EIT system. *Journal of Physics Conference Series, 224*(1).

Hartman, R., Lobo, J., Ruétalo, M., & Simini, F. (2002). *IMPETOM-I Tomógrafo de Impedancia Eléctrica.* Academic thesis, Universidad de la Republica, Montevideo.

Quinteros, W., & Simini F. (2007) *IMPETOM-48 Tomografo de Impedancia Electrica con tres hileras de electrodos.* Academic thesis, Universidad de la Republica, Montevideo.

Santos, E. (2014). *Alternativas de proyecto tendientes a un tomógrafo por impedancia eléctrica para la presentación del estado edemático en tiempo real.* Tesis Maest, Univ. la República, Uruguay, p. 199.

Santos, E., & Simini, F. (2012). Electrical impedance tomography for pulmonary oedema monitoring: Review and updated design. *Journal of Physics Conference Series, 407*, 12024.

Santos, E., & Simini, F. (2013). *Comparison of EIT reconstruction techniques applied to IMPETOM*. 13th International Conf. Bioinformatics and Bioengineering, Chania, Greece

Saulnier, G. (2004). EIT instrumentation. In D. Holder (Ed.), *Electric impedance tomography: Methods, history and applications* (p. 65). London: IOP Publishing.

Soleimani, M. (2006). Electrical impedance tomography system: An open access circuit design. *Biomedical Engineering Online, 5*(28).

Swisstom, A. G. Swisstom EIT System.

Wang, C., Liu, J., & Wang, H. (2005a). Pipeline data acquisition method in the EIT system. In *Instrumentation and measurement technology conference, 2005. IMTC 2005. Proceedings of the IEEE* (Vol. 1, pp. 437–440).

Wang, M., Ma, Y., Holliday, N., Dai, Y., Williams, R. A., & Lucas, G. (2005b). A high-performance EIT system. *Sensors Journal, IEEE, 5*(2), 289–299.

Webster, J. (2010). *Medical instrumentation. Application and design*. Hoboken, NJ: Wiley.

Xu, G., Zhang, S., Wu, H., Yang, S., Geng, D., Yan, W., et al. (2005). The acquisition hardware system with direct digital synthesis and filtered back-projection imaging in electrical impedance tomography. In *Engineering in medicine and biology society, 2005. IEEE-EMBS 2005. 27th Annual International Conference of the* (pp. 7758–7761).

Chapter 5
Electrical Impedance Signal Analysis for Medical Diagnosis

Francisco Miguel Vargas Luna, Marco Balleza-Ordaz, María Raquel Huerta Franco, and Pere Riu

5.1 Introduction

The application of electrical impedance measurements and modeling in living tissues is a systematic practice, which started in the 1920s and 1930s, with the works of Cole (1928), Cole and Cole (1936), Fricke (1925), Fricke and Curtis (1935) and more fundamentally in the 1950s, with the works of Schwan (1957), Schwan and Kay (1957a, b). Nowadays, in medical applications, the most common (or known) topic related to this technique is the estimation of body composition, starting roughly during the 1960s (Hoffer et al. 1969).

Electrical bioimpedance (EBI) measurements can be performed at a single frequency, using frequency sweeps or through time-domain option. The EBI of a material, in this case, tissue, organs, or human body parts, is presented in a raw form as magnitude and phase angle, real and imaginary components, resistance and reactance, or the dielectric (physical) parameters like complex permittivity or conductivity. The EBI can also be represented by the electrical parameters from a circuit partner in a related electrical model analysis (Grimnes and Martinsen 2014). Each option has the purpose of observing special features, changes compared with controls, different stages, change in time, or any other issue to establish

F. M. Vargas Luna (✉) · M. Balleza-Ordaz
Physical Engineering Department, Universidad de Guanajuato, Guanajuato, Mexico
e-mail: mvargas@dci.ugto.mx; marco.balleza@fisica.ugto.mx

M. R. Huerta Franco
Applied Science to Work, Universidad de Guanajuato, Guanajuato, Mexico

P. Riu
Departament d'Enginyeria Electrònica, Universitat Politecnica de Catalunya, Barcelona, Spain
e-mail: pere.riu@upc.edu

F. Simini, P. Bertemes-Filho (eds.), *Bioimpedance in Biomedical Applications and Research*, https://doi.org/10.1007/978-3-319-74388-2_5

the identification of the sample/subject. In medicine, this identification should be associated with disease stage, physiological, physical, chemical, or biological differentiation or evolution over time.

In this chapter, some examples of the big amount of work published that have been performed to address these relationships are included and considered as representative samples of possibilities as in body composition, skin cancer, breast tumor/cancer, kidney problems, swelling, edema, muscle/fat/water quality and quantity, lung function/condition, gastrointestinal problems, knee injury, device possibilities, methodologies, etc. The question is why have not EBI widely achieved apparently to be part of the standard clinical techniques and methodologies? Why do these results seem to remain in most cases in the research lab without maybe convincing the physicians (or regulators) to be safe, sensitive, and specific? Maybe the question is exactly that do bioimpedance devices and techniques applied in medicine have all these characteristics, according to the research published? EIB has application for non-imaging characterization and as a technique to produce images. These two options are addressed below.

The outline of this chapter is focused mainly on the application of EBI technique and its relation or usefulness for medical diagnosis. The signal analysis performed for this purpose is mentioned in each case, but the progress of signal analysis methodologies in time is not addressed. Final remarks about the signal analysis are revisited at the end of the chapter.

The Use of EBI in Medicine for Non-imaging Characterization.

EBI single-channel measurements, independently if it is 2, 3, or 4 electrodes configuration, are very common for the assessment of physical parameters of samples, either detached from the human body as in vitro/ex vivo measurements of tissue samples (biopsies, surgery samples etc.), human body fluids, etc. or attached them to the human body as in the search for skin anomalies or inner localized features. They are often multifrequency measurements (impedance spectroscopy), in order to determine physical parameters, directly or associated with specific electronic circuit models. This is performed to characterize the samples and look at local changes to infer conclusions about, for example, cell cultures, swelling or edema, vascularity, subcutaneous fat and body composition, skin cancer, micro-fluids, or in the case of knee injuries, among others. Some of these attempts are seeking for clinical recognition and validation.

As a first attempt to apply EBI in medical problems, it can be mentioned in vitro essays or in vitro sample characterization. A good example of in vitro EBI applications, using one channel multifrequency assessment, is the cell characterization to detect cancer cells. As far as we know, most of the works have been performed in cell cultures. For example, Hong et al. (2012) characterized four cell types (HeLa, A549, MCF-7, and MDA-MB-231) in a trap device, using equivalent circuit model, reporting the voltage tolerance for each kind of cells. An intermediate situation can be the detection of abnormal cells in fluids or processed smear samples, for example, in the case of cervix cancer test as was performed by Das et al. (2015) that after processing the sample, to get a pellet of cells in suspension, and using a commercial device (Applied Biophysics, USA), made a frequency scanning to get an average impedance, finding impedances three order of magnitude lower in abnormal cells. The detection directly in fluids like blood is already ongoing (e.g., the work of Han and Han 2015) that used impedance cytometry with lateral magnetophoresis to get a 94% detection of colon cancer cells in a lab trial. This work uses the enhanced phase change produced by a highly conducting material bound mainly to the tumor cell surface. Despite these encouraging results, the researchers did not speak about sensitivity or final conclusions for diagnosis, waiting for more quantitative and detailed research. Nevertheless, these attempts have resulted in manufactured devices for abnormal cell detection using bioimpedance (e.g., ICsense Belgium).

One of the most promising achievements in single-channel EBI, local measurements, and directly in the body is the skin cancer detection. The researchers have been involved in evaluating their clinical usefulness for more than 10 years. For example, Har-Shai et al. (2005) already studied in 2005 dermal melanoma evaluation of trunk and extremities by EBI presuming specificity of 64% and sensitivity of 91%, increasing up to 100% when combined with imaging techniques. In this case, also a device was tested together with the methodology.

Nowadays, there are commercial devices, for example, Nevisense device (DermNet, New Zealand), that uses multi-frequency, and in fact multichannel recording, in a four depths assessment by means of five bar electrode configuration. The feasibility of the device is supported in studies that promise more than 98% of sensitivity in melanoma detection and specificity, discriminating pathologies that resemble melanoma, 20% higher than that obtained from clinicians (Mohr et al. 2013), and finally assuring around 20% in positive predictive value and 98% in negative predictive values in a subsequent work by the same research group (Malyehy et al. 2014). These works used a nearby healthy skin assessment as the individual control sample. Different classification analysis was performed using the real and imaginary part of each measurement for each frequency and each pair of electrodes. That is, no physical parameters or partner circuit parameters were obtained, and no absolute electric description was determined for affected tissue.

Other related attempts toward cancer detection by dermal evaluation was performed in 2007 by Weitzen et al. (2007) who, by local dermal resistance measurements, used an already commercial device (Medex Screen Ltd) aimed to diagnose internal organ pathologies (regarding the so-called biological active points); the work presumes to have a sensitivity higher than 75% and specificity higher than 90% in lung, breast, and prostate cancer diagnosis. However, the mechanism of cancer detection is not clear as in other similar alternative techniques, so the work concluded the need for further research to support the reliability of this technique.

Additional efforts toward cancer diagnosis are, for example, the prostate cancer studies. Ex vivo sample EBI evaluation using a sensing biopsy needle in multifrequency tests and contrasted resistivity and reactance parameters with those from benign tissue has proven to have around 76 and 75% of sensitivity and specificity, respectively, when using directly the resistivity recorded and 79 and 71% of the same indicators when using reactance (Mishra et al. 2013). These investigators determined the best frequency for the highest level of classification in each case.

Another example applying multivariate classification analysis, using multifrequency results, has been tested for prostate malignant tissue differentiation from benign tissue (Khan et al. 2016). Indirect assessments in prostate cancer patients have been performed related to body composition, as the work done by Tyagi et al. (2016); these authors search for a relation between the phase angle in a whole-body measurement and the prostate cancer severity. In this trend, a previous work by Davis et al. (2009) studied, in general, the relation between EBI phase angles, hydration, and survival time in cancer patients. This general use of bodily EBI parameters related to a specific pathology is also discussed below.

Similar works were performed related to colorectal cancer. Gupta et al. (2004), knowing that EBI phase angle has been tested as a prognostic of several pathological conditions such as HIV, liver cirrhosis, chronic obstructive pulmonary disease, lung cancer, and in patients receiving dialysis, searched for the same hypothesis for colorectal cancer patients related the EBI parameter with the survival of the subjects. Four years later, the same research group (Gupta et al. 2008) made another research again with colorectal cancer patients but realizing the importance of the nutrition factor in the EBI phase angle results.

The other cancer more addressed by researchers is the uterine cervix cancer already mentioned in the cell analysis diagnosis. Das et al. (2015), trying to improve the well-known false negative rates of conventional Pap smear test (30–40%), reported the three order of magnitude lower resistance in a wide range of frequencies for the abnormal cells compared with healthy samples but did not report a quantitative analysis of these findings to enhance the sensitivity in a combined test. In a previous work, Brown et al. (2005) used a narrow probe and measured impedance at eight frequencies to fit a Cole model. The lower low-frequency resistance of the sample was of course reported, but a discussion of other electrical parameters and its influence in the strata classification was also analyzed getting better results if they combine parameters for classification.

In breast cancer, there are two main routes that researchers have followed with EBI technique. The first one is about one of the main secondary effects of the breast cancer treatment, the breast cancer-related lymphedema (BCRL). In this case, the EBI usage resembles the body composition evaluation when bodily fluids change in an edema condition. The qualification of EBI for this purpose seems to be quite reasonable. Recent publications realized that the usefulness of the technique in BCRL lies in pre-existing ill condition, but it needs validation in the case of early detection (Seward et al. 2016). Nevertheless, in another publication in the same year, a guideline for the clinical use of EBI in BCRL is provided (Shah et al. 2016) although the topic is still considered under investigation. Of course, this application is not directly related to breast cancer diagnosis or its monitoring.

Single measurements in biopsy breast samples were also reported mainly focusing on the measurement system capabilities (Pandya et al. 2014). The electrical circuit model parameters are considered for the discrimination, and although helpful results were obtained, the authors are aware that the heterogeneity, setting conditions and environment must be studied with detail before any conclusive availability of the device.

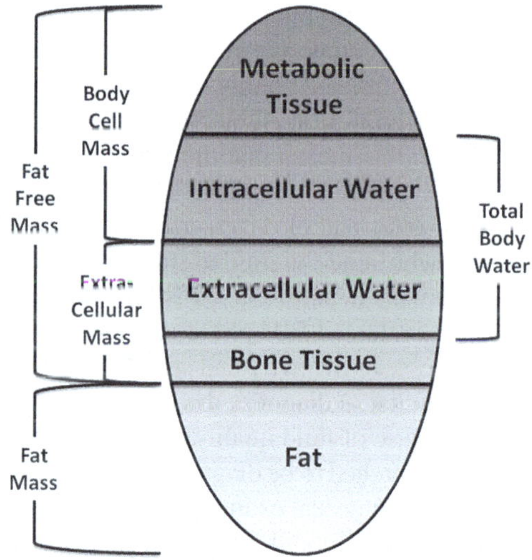

Fig. 5.1 Parameters considered for EBI body composition evaluation

Maybe one of the most known single-channel impedance application in clinics is the assessment of body composition. Body composition is a generic name that includes the estimation of different parameters: total body water (TBW), extracellular/intracellular water (ECW, ICW), fat-free mass (FFM), etc. (Fig. 5.1).

Since 1985, total body impedance parameters have been used as predictors of body composition. Lukaski et al. (2016) used the so-called impedance index, stature^2/resistance (Ht^2/R), as the best linear predictor for FFM. Confirmed later by Kushner et al. (1992), this index has been studied as the main factor to predict FFM, adding other anthropometrical parameters for improvement and to get equations in particular groups (Eq. 5.1), stratifying by gender, age, childhood, physical activity, race, occupation, etc. (vg Matias et al. 2016; Sun et al. 2003; Wickramasinghe et al. 2008; Masuda and Komiya 2004; Langer et al. 2016; Hastuti et al. 2016; Aleman-Mateo et al. 2010):

$$\mathrm{FFM} = \mathrm{A} + \mathrm{B}\ \left(\mathrm{Ht}^2/\mathrm{R}\right) + \mathrm{C}\ (\mathrm{weight}) + (\mathrm{gender}) + \ldots \tag{5.1}$$

There is a wide range of commercial devices available, either for medical or domestic use, while the research, up to now, is mainly concerned about specific applications. Body composition assessments sometimes are used as a predictor of a disease or to monitor its evolution over time, by assessing changes in body fluids, or ECW/ICW balance, and lean body mass (FFM). Applications include assessment of fluid removal or dry weight in hemodialysis patients, fluid accumulation after surgery, nutrition or eating disorder, or monitoring the effects of exercise training, among others. It must be considered that anthropometric parameters are important factors for these assessments.

Khalil et al. (2014) made a revision on this aspect concluding the need for more research in the correlation of EBI with health deviations. The same authors in 2016

(Khalil et al. 2016) published a research about bioimpedance vector analysis (BIVA) used to differentiate between healthy and severe and non-severe dengue patients. The very significant results in this case, when a specific pathology is considered to compare with healthy subjects, seem to be explained by the hydration differences. Of course, this means that the sensitivity could be very high, but the specificity is not, to be useful in diagnosis. Other examples of these high statistical differences for a specific pathology in assessing BIVA are the work by Dorna Mde et al. (2016), who made a similar study with advanced fibrosis in patients chronically infected with hepatitis C virus and the work published by Barrea et al. (2016) that also evaluated EBI phase angle in patients diagnosed with psoriasis. This last work assessed sensitivity and specificity establishing cutoffs to discriminate severity stages, but not to diagnose the disease.

In the case of fluid quality, specifically concerning sodium levels, hyponatremia has been searched to be diagnosed with EBI spectroscopy as an alternative to clinical diagnosis or physical examination (Kim et al. 2014; Kose et al. 2015). However, the authors, having good statistical results, just suggested the better performance of EBI. In particular, the last work mentioned showed gender differences in their results, as it is obtained also in other applications (ventilation) as we discuss below.

A very important case of body fluid evaluation is the case of patients with kidney diseases and mainly those already undertaking dialysis treatment. Here some authors correlate it with clinical results, as in the case of oxidative stress (Zouridakis et al. 2016), concluding the need for further research. Others just compared different methodologies, either inside bioimpedance or with usual techniques, without any diagnosis purposes (Broers et al. 2015).

The results of these works suggest that EBI body composition is a very good indicator of severity in patients already diagnosed with affections that have, as one of the important clinical data, hydration stage, inflammation, swelling, edema, or some alteration in body fluids or in general, despite the redundancy, in body composition. These include of course cardiovascular pathologies. In this vein, some authors remark the importance of body composition application, for example, to measure the body water content in acute heart failure patients, not only in admission, to report disease severity, but also to monitor evolution and therapy response (Di Somma et al. 2014). Even though EBI is not considered the gold standard for fluid determination (it is isotope dilution), in an emergency situation, the easiness and time saving of bioimpedance alternative compensate the possible loss of precision compared with a gold standard. It is in another very similar study (Génot et al. 2015) that sensitivity and specificity are evaluated (67% and 76%, respectively), even considering that it does not improve the results of an important biomarker BNP (brain natriuretic peptide) when EBI is combined with it. All these examples, of course, are used in a self-referenced way, for monitoring after diagnosis and not claiming to replace standard techniques.

Some other works explore the intrinsic limitations of the body composition in routine clinics, because of its dependence on the multiple regression models and the limited physical models proposed (Lukaski 2013). Nevertheless, the research in body composition is concerned still with the validation of the methodologies in diverse-specific situations either pathological or not, as for the challenge of its use in children in growth, (Kyle et al. 2015) where aside the necessity of care stratification and cross-validation, the importance of EBI is still recognized for monitoring, based on self-referenced use. In adolescents, Verney et al. (2016) found the technique not suitable, contrasted with DXA, in severely obese subjects. Similarly, in adults, regarding the body fat loss, Macfarlane et al. (2016) found the poor performance of EBI when contrasted also with DXA in a fat loss evolution experiment. About the use in patients, for example, in surgical and oncological patients, Haverkort et al. (2015), in a systematic review, found large variability among calibration equations, limiting the use of the technique for these patients, unless longitudinal or self-referenced evolution assessments were considered.

There are others works where indirect relationship of body composition (or body part evaluation by EBI) with disease is used, as in lymphedema evaluated by arm volume assessment. Bundred et al. (2015) addressed this issue concluding that the gold standard would be still the perometer because the moderate correlation of it with EBI. In the same trend, a swelling evaluation in knee arthroplasty using EBI was searched; Pichonnaz et al. (2015) made EBI usage validity against direct volume and circumference, which are altered by muscular atrophy and surgical dressing. The authors found that EBI assessment surpasses the effectiveness to evaluate swelling compared with the traditional way to measure volume or circumference, having a sensitivity higher than 80% depending on stage. On the other hand, it is also the swelling, edema, or hydration status that can distort the evaluation of, for example, skeletal muscle quantification for sarcopenia assessment (Rubbieri et al. 2014). In this case, sarcopenia not only includes the lean mass loss, but it also comprises the muscle functionality that is not evaluated by EBI technique (Lardiés-Sánchez et al. 2016).

Plethysmography means volume changes measurement, and it is used to get, for example, blood changes or air lung changes, which can be performed by EBI. The technique has been useful in obstruction or disturbance, in veins and arteries, and in fact changes in fluids volume that is what it is measured in impedance cardiography (addressed below) (Corciovă et al. 2012).

Attempts to record vascular damage in regions as brain have also some insights. For example, the work of Atefi et al. (2013) where brain stroke patient was completely classified using multichannel and multifrequency measurements to get physical parameters (from electronic circuit model) and using statistical classification analysis. Most of the works in ischemia and vascularity damage are reported in animal model, but relatively few works have been performed in human patients.

Several efforts also have been made to establish EBI technique as a practical tool to detect different types of knee injuries. It was demonstrated that some knee pathologies, e.g., in osteoarthritis (OA), may cause the changes of synovial fluid characteristics, and this, in turn, may produce changes in the EBI dielectric parameters (Frank and Grodzinsky 1987; Foster and Schwan 1989), which can be detected and used for diagnostic purpose. Also, it is hypothesized that knee swelling can cause changes in the electrical features of knee tissues, which can also produce variations in EBI determinations. For example, Gajre et al. (2007)

analyzed impedance magnitude changes in OA knees contrasted with healthy knees (control group) by performing two separate tests: active flexion-extension of leg in sitting position and normal walking, finding statistically significant differences between groups in both exercises. Neves et al. (2009) used the impedance spectroscopy to find out if there is a correlation between the impedance parameters and different levels of OA (determined by Dejour scale) in a group of males belonging to parachute military brigade. The results suggest that there is statistically significant correlation between the levels of OA and impedance vector. Another EBI application in this field is the assessment of swelling after the total knee arthroplasty (TKA). Pichonnaz et al. (2015) used the EBI technique to evaluate the level of swelling after TKA in patients 1 day before and 2 and 8 days after TKA. They compared EBI resistance with anthropometric parameters (knee circumference and limb volume). The obtained results evidenced that the impedance determinations obtained 1 day before and those obtained 2 and 8 days after TKA shown statistically significant differences. These data were highly correlated with anthropometry corresponding to each stage of measurements. Based on their results, the authors concluded that EBI is a good option for detecting swelling after TKA. The main problem of these studies and others is that they addressed different knee pathologies using different bioimpedance techniques (bioimpedance spectrometry, plethysmography) that make comparison difficult. Also, since the gold standard technique to evaluate knee damage—nuclear magnetic resonance—is too expensive, statistical comparisons were made with clinical observations. So, more controlled studies should be performed to establish EBI as a useful clinical tool in this area.

The application of EBI in the gastrointestinal system has been performed in many ways. The one that could seem most obvious is the application related to pH measurement. In the gastroesophageal reflux episodes (GERD), the change in acidity in the esophageal tract is monitored.

Moawad et al. (2013) assessed reflux episodes, using a catheter, evaluating stomach pH changes in the location. Zhou et al. (2014) made similar research comparing sensitivity and specificity with other standard techniques. They pointed out that EBI technique helps to increase sensitivity up to 20% more than using conventional methodologies alone, but EBI by itself has very poor specificity. More recently, Ravi and Katzka (2016) made a critical review of the use of intraluminal EBI measurements to get information about stomach acid, weak acid or non-acid reflux episodes, bolus transit, gas or liquid movement, esophageal motility, and mucosal impedance changes as consequences of inflammatory conditions. These wide possibilities demand an accurate signal processing to discriminate each fact and avoid the over-interpretation or misinterpretation of the findings. Although gastric intraluminal assessments have these confusing drawbacks, external measurements have also been tested. Giouvanoudi and Spyrou (2008) already investigated the possibility of external or cutaneous evaluation for gastric pH by monitoring the impedance changes after the ingestion of meals with different conductivities.

External cutaneous EBI evaluation of the gastric region has also been considered. In this case, the raw signal obtained (impedance magnitude) has information not only about skin resistance, and local tissue composition, but also about the movements due to respiration, cardiac movement, gastric, intestine, and colon movement that ought to be discriminated before any analysis. Huerta-Franco et al. (2012) analyzed this kind of data using different experimental paradigms, e.g., in fasting state, during stress, and under the effect of prokinetic drugs. These authors proposed an alternative methodology to analyze the parameters to be considered, together with the use of FFT analysis of the signal. About the particular case of gastric motility, comparisons with the more standard technique (although also still under search) electrogastrography, have been performed in many works; nevertheless, no final conclusions about the compatibility between both techniques, limitations, and boundaries of applicability have been established. Gastric motility is, after all, just a sign to be considered in several pathologies.

EBI Imaging for Medical Diagnosis. Due to the high contrast between air and tissue, maybe one of the most promising applications of the imaging alternative of EBI measurements is the monitoring and quantification of ventilation. In pneumology, there is nothing good enough to monitor the respiration in a noninvasive way and, especially, during long periods of time. Among the most known methods, some are not accurate or have calibration problems as the respiratory inductive plethysmography (Stick et al. 1992), or they are invasive and alter ventilation determinations as the pneumotachometer (used as gold standard) (Askanazi et al. 1980; Perez and Tobin 1985; Paek and McCool 1992). The advantage of EBI in this clinical application lies in the fact that the resistivity of lung tissue is five times higher than other thoracic tissues (Brown 2003). Different authors compared statistically the impedance changes versus tidal volume determinations obtained by impedance devices and pneumotachometer, respectively. From the obtained results, a high significant correlation ($r > 0.9$; $p < 0.05$) between determinations was evidenced (Harris et al. 1987, 1988; Holder and Temple 1993; Eyüboglu et al. 1995). This means that the impedance changes (ΔZ) and the volume changes obtained by a pneumotachometer (ΔV_T) are proportional (Eq. 5.2):

$$\Delta Z = A \times \Delta V_T \qquad (5.2)$$

where A is the proportionality coefficient that relates both determinations.

Impedance changes due to respiration have been obtained in different device alternatives, from non-imaging electrode configurations as, for example, the research of Voscopoulos et al. (2013, 2014) that proposed the used of two sets of electrodes, each one comprising three electrodes: one set placed along the sternum and the other across the right midaxillary line at the level of xiphoid process. The authors got minute ventilation, tidal volume, and respiratory rate for obstructive sleep apnea episodes in obese patients showing a high correlation with gold standard results. Other example is the study conducted by Houtveen et al. (2006), which assessed respiratory patterns by inductance plethysmography contrasted to impedance changes obtained by a four-electrode arrangement.

However, imaging seems to be more reliable for ventilation studies by EBI. Electrical impedance tomography (EIT) is the imaging technique aimed to obtain

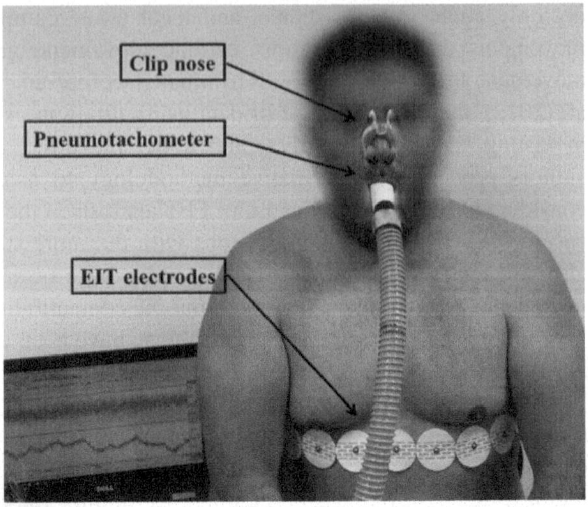

Fig. 5.2 Setup of EIT for tidal volume determination

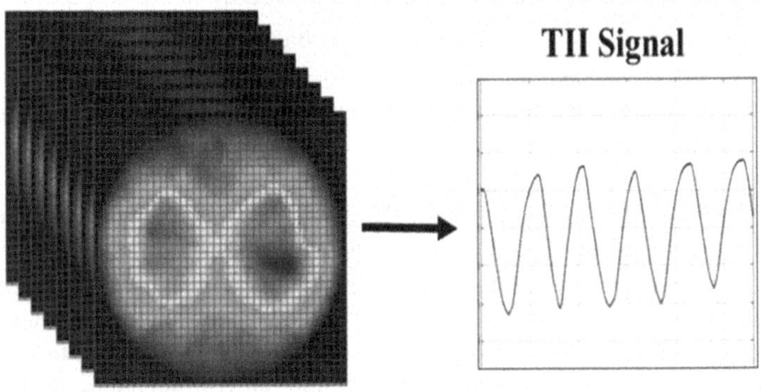

Fig. 5.3 EIT impedance signal obtained by addition of pixels from the ROI. Each point of the signal (right) correspond to one image at corresponding time (left)

transversal (or even volumetric) conductivity images, applied at the thorax level for ventilation studies (Brown et al. 1985). In this case, EIT provides a large impedance information of lung ventilation due to a multielectrode arrangement (usually 16) placed around thorax in the sixth intercostal space (Fig. 5.2). For tidal volume evaluation, the time impedance signal is obtained by adding the pixel values corresponding to that time impedance frame. The addition of pixels (in each image) is called total impedance index (TII). So, the impedance signal is composed by the values of TII at each time for a particular region of interest (ROI) (Fig. 5.3).The adjustment of impedance changes due to lung ventilation obtained by EIT for transforming them into a measurable volume signal is still a task in development. For this case, Balleza et al. (2007, 2009), Balleza-Ordaz et al. (2015) have conducted a research where have adjusted the impedance signal by using all pixels of EIT image and using ROIs. This group, unlike others, decided to

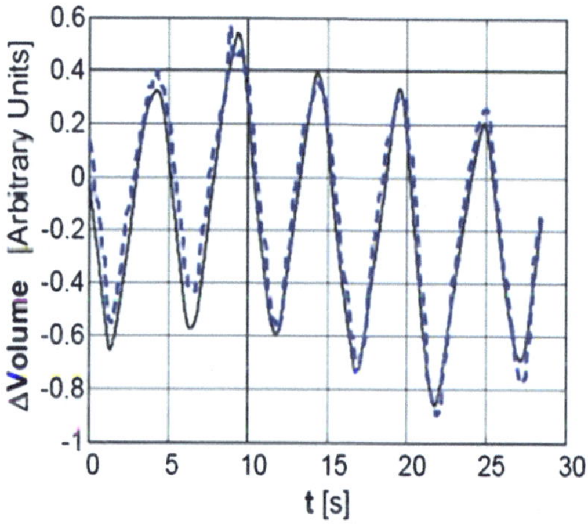

Fig. 5.4 Concordance between pneumotachometer signal (solid) and EBI signal (dashed) with the particular proportionality coefficient "A"

determine the proportionality coefficient "A" from the Eq. 5.2 in terms of different anthropometrical parameters (Eq. 5.3 and Fig. 5.4).

$$A = A\ (\text{age, height, weight, body mass index, etc.}) \tag{5.3}$$

These studies were conducted in groups of males and females and in a group of chronic obstructive pulmonary disease (COPD) patients. From the obtained results, the mathematical adjustment of "A" was better for males than for females because of the obvious anthropometrical differences. In the case of COPD, the adjustment of "A" needs to take into account other parameters because of the anatomical deformation in this kind of patients. The adjustment of impedance changes due to respiration needs a complete statistical study of anthropometry of volunteers and proportional coefficients that relate impedance and volume determinations.

Another multichannel and imaging application of EBI technique is the breast cancer detection and monitoring. There exist devices already developed but still under research as the Sussex MK4 EIM system (Zhang et al. 2014) and its improved version RPEIM (Zhang et al. 2015), in both cases a 3-D breast image is developed from a multichannel configuration. The expertise of the clinician of course is essential in this type of methodologies, because although the image is a conductivity based map of the breast, usually qualitative analysis ought to be performed to get conclusions. Another simpler device is the Electrical Breast Densitometer built by a research company (ScnoSENSE Medical System Inc.) (Maskarinec et al. 2016) that gives images using resistance as the parameter under consideration. An inverse correlation between densitometry and the electrical alternative is reported; however, no conclusions about electrical results and breast cancer risk or classification possibilities between cancer and non-cancer patients were demonstrated. The conclusion about the need of further research for device development in this case was clearly stated.

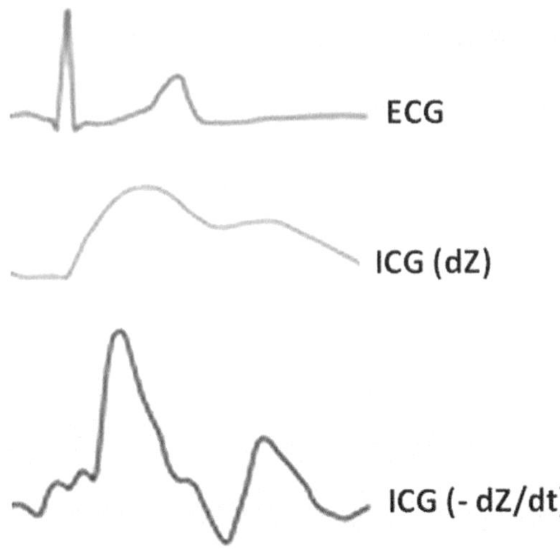

Fig. 5.5 ECG and ICG signal comparison. The delaying features between electric signal and volume assessment by ICG are used as an index of cardiac functional characteristics

Going beyond the habitual radio frequency range used in most EBI devices, some researchers like the group of Porter et al. (2016) and Bahramiabarghouei et al. (2015) have developed a promising device using microwave antenna array to develop in the near future a portable 3-D time-domain cancer screening. The device is still under investigation and has not been proposed for commercial purposes as far as we know.

The impedance cardiography (ICG) has already a long story since it was proposed in 1966 (Kubicek et al. 1966) assuming that the change in the thoracic impedance comes mainly from the change in volume of blood with constant conductance and dominating by far by the thick principal vessels (Fig. 5.5).

Equations, based on a physical model, have been used to estimate left ventricular stroke volume (SV) or cardiac output (CO) using impedance magnitude (Z) and its first derivative dZ/dt, as alternative to the use of gold standard, the thermodilution method (Woltjer et al. 1997; Bernstein and Lemmens 2005). Although noninvasive methods to assess SV does exists as the Doppler echocardiography (DE) and CO_2 breath analysis, ICG is still in the scene, being searched in particular applications and recognized as useful for continuous monitoring of cardiac activity (vg Lorne et al. 2014). A new similar technique, the endotracheal impedance, has been developed but is still in discussion (Maass et al. 2014). The monitoring of SV changes by ICG seems to be quite acceptable, but the SV assessment is still controversial, and in general its clinical validation is still a pending task for some authors (Fellahi and Fischer 2014). Nevertheless, there exist commercial devices widely spread and used in clinics and of course still under development (Yazdanian et al. 2016).

In this topic, EBI measurements using catheters inside the body are in an early development stage. Assessing myocardial state is of major importance in several clinical conditions. Myocardial electrical impedance accounts for the spatial myocardial fiber orientation and the structural integrity of the myocardium. A few

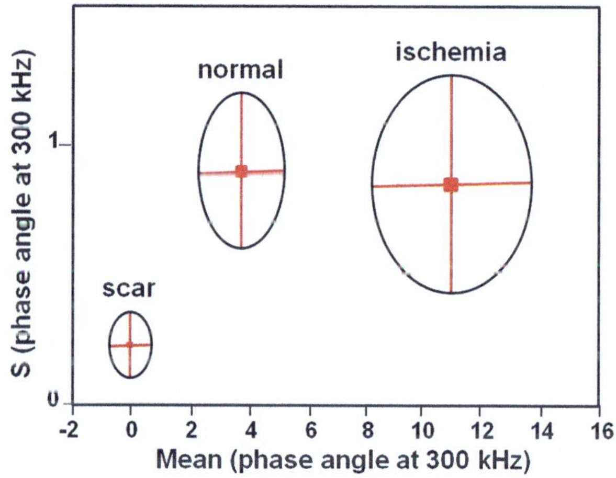

Fig. 5.6 Different states of myocardium tissue can be differentiated by using the phase angle of the impedance measured at 300 kHz. X-axis shows mean values showing intra subject variability; y-axis shows intersubject standard deviation of the mean

studies revealed that electrical impedance of infarct scar is lower than normal myocardium, and this difference allowed recognition of the necrotic region with intracavitary electrocatheters in anesthetized pigs (Salazar et al. 2004). In these studies, the two components of myocardial impedance (i.e., module and phase angle) were calculated according to the behavior of alternating currents of different frequencies (from 1 kHz up to 300 kHz) injected to the myocardium, and the optimal combination of magnitude and phase angle was found to differentiate scar from health myocardium (Fig. 5.6). In a similar study, the feasibility to assess different degrees of rejection after heart transplantation was demonstrated (Cinca et al. 2008). Currently, an attempt to perform left ventricular maps of myocardium state using impedance is being carried out, with the help of a CARTO system to define location.

Let's mention that, in a similar way, a catheter placed inside the lungs, through a bronchoscopy, is being used to assess the different states of pulmonary parenchyma and is able to distinguish fibrosis, emphysema, and health tissue. A fast time-domain measurement system is required to be able to isolate contribution in the measured impedance from ventilation and perfusion processes (Sánchez et al. 2013).

5.2 Final Remarks

We must be clear that a medical diagnosis is a multifactorial task based in clinical data and laboratory results, in which the use of EBI methodologies can be included. Nevertheless, aside from the impedance cardiography and the body composition applications, these tools have not spread out within the physicians for other purposes. The main work to do in the near future is the validation of the EBI technique, either to improve or even to measure the sensibility, specificity, and the predicting values of the whole amount of proposals published in this field. In most cases, the laboratory tests are performed to quantitatively support the clinical-based pre-diagnosis. In the same way as one single parameter, either physical, chemical, biological, or physiological, many times it is not enough to have a solid conclusion about a health disorder; also one single device result is hardly enough to conclude diagnosis. Also, it has been considered that all techniques ought to be validated, that is, contrasted against a gold standard or the accepted methodology in each case to assess some specific, anatomical condition, system function parameter, or symptom quantification. This validation comes with acknowledge of their limits, in sensibility, specificity, and predictive values. If these parameters are not at the level of the gold standard, the shortage should be compensated with other benefits such as saving time, higher comfort of the patient, feasibility to use in children, elderly and patients in bed, the economy, the lower risks to health, and no collateral effects, among many others. Thus, more research should be done in facing the challenges of establish strictly the validation, sensitivity, specificity, and the predictive values of the new medical possibilities suggested or proposed in research published works. And more than that, when these quality aspects are fulfilled, it is still the need of encouraging the commercial aspects of the devices proposed. Of course, promising tasks have already performed in this direction (look at successful cases mentioned in this chapter). The understanding of the need of this work does not diminish the importance of the promising results in many of the medical subjects mentioned above, where EBI methodologies are shown to have evident capabilities to be part of the medical pool of technologies for clinical diagnosis.

If the analysis procedure is reviewed in all the references mentioned, most of them still consider the basic impedance parameters (either magnitude and phase or resistance and reactance), because of the assessment of large portions of the body related to the health problem in question. The changes in time or health condition are the main concern of this case, considering a self-referenced process. Fewer works deal with electric parameters of equivalent circuits or the dielectric characterization of the sample, and this kind of analysis is performed mainly when small sample tissues or localized affections are studied. In these cases, the equivalent circuit is known to be simple for better interpretation and accuracy of the results. In any case, the search for sensitivity and specificity not always is addressed; moreover, sometimes multivariate analyses are performed including several EBI, and non-EBI, parameters to look for concordance with the gold standard. In this vein, the examples are body composition as well as ventilation, which are still looking for the best calibration equation, which includes anthropometric parameters, to predict the corresponding physiological parameter. This means that detailed stratification or case considerations must be taken into account. Biological material characterization has the challenge to consider its complexity in structure and composition. When the whole body or a large portion of the body is evaluated, the number of factors to be included in the equation is large enough to focus the concerns in the best correlation with the clinical conclusions. The image alternative (EIT) is highly dependent on the clinical interpretation; therefore, image devices as skin cancer, lung, or breast tomography are already in the market as a complementary tool for clinical decision.

The EBI technique characteristic more mentioned is the fact that it is a safe methodology, compared mainly with invasive and radiative alternatives, and the more criticized is the poor resolution mainly in imaging and localized assessments.

5.3 Conclusions

Medical diagnosis means to define the cause or nature of a disease or condition and this finally ends in the determination of the precise disease affected the subject. When a methodology or an instrument is used for this purpose, the diagnostic usefulness is more linked to the specificity than the sensitivity, more with the positive predictive value than the negative predicting value. In this vein, if EBI were meaningful for a specific affection or condition, the measurement should be very tightly associated with a very particular alteration (physiological or anatomical) specific for the disease considered. Localized EBI measurements could be associated to specific anatomical or physiological alterations as in vascularization, localized tumor, and melanoma getting in this way, a kind of diagnosis alternative for these pathologies (as one of the most successful examples with a commercial device already in the market). In this case of success, specificity is still the problem as it exists a benign resembling of it, dysplastic nevus, although sensitivity is quite high.

Other set of localized EBI evaluations are those used by catheter for internally body measurements, these of course are very specific in its purpose, giving values for pH, local vascularity, relative local changes compared with neighborhood, etc. In this case, the interpretation is more direct as in the previous cases.

Body composition is the most spread EBI technique among clinicians, in spite of the controversy about its reliability, its easiness makes it ideal for first insight or monitoring purposes. Nevertheless, the attempts to be used directly as specific diagnosis criteria seem to fail. The body composition results seem to join the other clinical parameters to reinforce medical decisions and diagnosis. There exist lots of device alternatives to be used, but it seems that the methodology still has problems to be associated specifically to a clinical parameter. Many formulas and calibration equations have been published to cover as many cases as possible, but there is still work to do. However, the authors in general agree about its usefulness at least in monitoring and as a clinical factor to be considered together with other evaluations for diagnosis.

In imaging, TIE still depends, for the most important applications, on the calibration equations for quantitative purposes, having the need for stratification among other challenges.

As we already mentioned, medical diagnosis has been, is, and maybe will be a multifactorial task; the complexity of the human body anatomy and physiology, the huge number of possible strata, confusing factors, and other personal features to be considered make this work not a single step process. Despite the sophistication of the data post-processing and estimation, raw EBI parameters and measurement's quality assessment must be provided as a part of the diagnosis feature, without viewing this as a handicap, weakness or drawback of the methodologies associated, neither an obstacle to continue the research in this field.

References

Aleman-Mateo, H., Rush, E., Esparza-Romero, J., Ferriolli, E., Ramirez-Zea, M., Bour, A., et al. (2010). Prediction of fat-free mass by bioelectrical impedance analysis in older adults from developing countries: A cross-validation study using the deuterium dilution method. *The Journal of Nutrition, Health & Aging, 14*(6), 418–426. https://doi.org/10.1007/s12603-010-0031-z.

Askanazi, J., Silverberg, P. A., Foster, R. J., Hyman, A. I., Milic-Emili, J., & Kinney, J. M. (1980). Effects of respiratory apparatous on breathing pattern. *Journal of Applied Physiology, 48*, 577–580. PMID:6769880.

Atefi, S. R., Seoane, F., Thorlin, T., & Lindecrantz, K. (2013). Stroke damage detection using classification trees on electrical bioimpedance cerebral spectroscopy measurements. *Sensors, 13*(8), 10074–10086. https://doi.org/10.3390/s130810074.

Bahramiabarghouei, H., Porter, E., Santorelli, A., Gosselin, B., Popović, M., & Rusch, L. A. (2015). Flexible 16 antenna Array for microwave breast cancer detection. *IEEE Transactions on Biomedical Engineering, 62*(10), 2516–2525. https://doi.org/10.1109/TBME.2015.2434956

Balleza, M., Calaf, N., Feixas, T., Gonzalez, M., Antón, D., Riu, P. J., et al. (2009). Measuring breathing pattern in patients with chronic obstructive pulmonary disease by electrical impedance tomography. *Archivos de Bronconeumología, 45*, 320–324. https://doi.org/10.1016/j.arbres.2009.01.013.

Balleza, M., Fornos, J., Calaf, N., Feixas, T., Gonzalez, M., Antón, D., et al. (2007). Monitoring of breathing pattern at rest by electrical impedance tomography. *Archivos de Bronconeumología, 43*, 300–303. PMID:17583638.

Balleza-Ordaz, M., Perez-Alday, E., Vargas-Luna, M., & Riu, J. P. (2015). Tidal volume monitoring by electrical impedance tomography (EIT) using different regions of interest (ROI): Calibration equations. *Biomedical Signal Processing and Control, 18*, 102–109. https://doi.org/10.1016/j.bspc.2014.12.004.

Barrea, L., Macchia, P. E., Somma, C., Napolitano, M., Balato, A., Falco, A., et al. (2016). Bioelectrical phase angle and psoriasis: A novel association with psoriasis severity, quality of life and metabolic syndrome. *Journal of Translational Medicine, 14*(1), 130. https://doi.org/10.1186/s12967-016-0889-6

Bernstein, D. P., & Lemmens, H. J. (2005). Stroke volume equation for impedance cardiography. *Medical & Biological Engineering & Computing, 43*(4), 443–450. https://doi.org/10.1007/BF02344724.

Broers, N. J., Martens, R. J., Cornelis, T., Diederen, N. M., Wabel, P., van der Sande, F. M., et al. (2015). Body composition in dialysis patients: A functional assessment of bioimpedance using different prediction models. *Journal of Renal Nutrition, 25*(2), 121–128. https://doi.org/10.1053/j.jrn.2014.08.007.

Brown, B. H. (2003). Electrical impedance tomography (EIT): A review. *Journal of Medical Engineering & Technology., 27*(3), 97–108. https://doi.org/10.1080/0309190021000059687.

Brown, B. H., Barber, D. C., & Seagar, A. D. (1985). Applied potential tomography: Possible clinical applications. *Clinical Physics and Physiological Measurement, 6*, 109–121. https://doi.org/10.1088/0143-0815/6/2/002.

Brown, B. H., Milnes, P., Abdul, S., & Tidy, J. A. (2005). Detection of cervical intraepithelial neoplasia using impedance spectroscopy: A prospective study. *BJOG, 112*(6), 802–806. https://doi.org/10.1111/j.1471-0528.2004.00530.x.

Bundred, N. J., Stockton, C., Keeley, V., Riches, K., Ashcroft, L., Evans, A., et al. (2015). Comparison of multi frequency bioimpedance with perometry for the early detection and intervention oflymphoedema after axillary node clearance for breast cancer. *Breast Cancer Research and Treatment, 151*(1), 121–129. https://doi.org/10.1007/s10549-015-3357-8.

Cinca, J., Ramos, J., Garcia, M. A., Bragos, R., Bayés-Genis, A., Salazar, Y., et al. (2008). Changes in myocardial electrical impedance in human heart graft rejection. *European Journal of Heart Failure, 10*(6), 594–600. https://doi.org/10.1016/j.ejheart.2008.04.013

Cole, K. S. (1928). Electric impedance of suspensions of arbacia eggs. *The Journal of General Physiology, 12*(1), 37–54. PMID:19872447.

Cole, K. S., & Cole, R. H. (1936). Electric impedance of asterias eggs. *The Journal of General Physiology, 19*(4), 609–623. PMID:19872951.

Corciovă, C., Turnea, M., Matei, D., & Andritoi, D. (2012). Evaluation of cardiac parameters using electrical impedance plethysmography. *Revista Medico-Chirurgicală a Societăţii de Medici şi Naturalişti din Iaşi, 116*(3), 927–932. PMID:23272554.

Das, L., Das, S., & Chatterjee, J. (2015). Electrical bioimpedance analysis: A new method in cervical cancer screening. *Journal of Medical Engineering, 2015*, 636075. https://doi.org/10.1155/2015/636075.

Davis, M. P., Yavuzsen, T., Khoshknabi, D., Kirkova, J., Walsh, D., Laseen, W., et al. (2009). Bioelectric impedance phase angle changes during hydration and prognosis in advanced cancer. *The American Journal of Hospice & Palliative Care, 26*, 180–187. https://doi.org/10.1177/1049909108330028.

Di Somma, S., Vetrone, F., & Maisel, A. S. (2014). Bioimpedance vector analysis (BIVA) for diagnosis and Management of Acute Heart Failure. *Current Emergency and Hospital Medicine Reports, 2*, 104–111. https://doi.org/10.1007/s40138-014-0043-9.

Dorna Mde, S., Santos, L. A., Gondo, F. F., Augusti, L., de Campos Franzoni, L., Sassaki, L. Y., et al. (2016). Phase angle is associated with advanced fibrosis in patients chronically infected with hepatitis C virus. *Life Sciences, 154*, 30–33. https://doi.org/10.1016/j.lfs.2016.02.061.

Eyüboglu, B. M., Öner, A. F., Baysal, U., Biber, C., Keyf, A. I., Yilmaz, U., et al. (1995). Aplication of electrical impedance tomography in diagnosis of emphysema – A clinical study. *Physiological Measurement, 16*, A191–A211. https://doi.org/10.1088/0967-3334/16/3A/018.

Fellahi, J. L., & Fischer, M. O. (2014). Electrical bioimpedance cardiography: An old technology with new hopes for the future. *Journal of Cardiothoracic and Vascular Anesthesia, 28*(3), 755–760. https://doi.org/10.1053/j.jvca.2013.12.026.

Foster, K. R., & Schwan, H. P. (1989). Dielectric properties of tissues and biological materials: A critical review. *Critical Reviews in Biomedical Engineering, 17*, 25–104. PMID: 2651001.

Frank, E. H., & Grodzinsky, A. J. (1987). Cartilage electromechanics—I. Electrokinetic transduction and the effects of electrolyte pH and ionic strength. *Journal of Biomechanics, 20*, 615–627. https://doi.org/10.1016/0021-9290(87)90282-X

Fricke, H. (1925). The electric capacity of suspensions with special reference to blood. *The Journal of General Physiology, 9*(2), 137–152. PMID:19872238.

Fricke, H., & Curtis, H. J. (1935). The electric impedance of hemolyzed suspensions of mammalian erythrocytes. *The Journal of General Physiology, 18*(6), 821–836. PMID:19872891.

Gajre, S. S., Singh, U., Saxena, R. K., & Anand, S. (2007). Electrical impedance signal analysis in assessing the possibility of non-invasive diagnosis of knee osteoarthritis. *Journal of Medical Engineering & Technology, 31*(4), 288–299. https://doi.org/10.1080/03091900600863745.

Génot, N., Mewton, N., Bresson, D., Zouaghi, O., Francois, L., Delwarde, B., et al. (2015). Bioelectrical impedance analysis for heart failure diagnosis in the ED. *The American Journal of Emergency Medicine, 33*(8), 1025–1029. https://doi.org/10.1016/j.ajem.2015.04.021.

Giouvanoudi, A. C., & Spyrou, N. M. (2008). Epigastric electrical impedance for the quantitative determination of the gastric acidity. *Physiological Measurement, 29*(11), 1305–1317. https://doi.org/10.1088/0967-3334/29/11/006.

Grimnes, S., & Martinsen, O. (2014). *Bioimpedance and bioelectricity basics* (3rd ed.). Academic Press. ISBN:9780124114708.

Gupta, D., Lammersfeld, C. A., Burrows, J. L., Dahlk, S. L., Vashi, P. G., Grutsch, J. F., et al. (2004). Bioelectrical impedance phase angle in clinical practice: Implications for prognosis in advanced colorectal cancer. *The American Journal of Clinical Nutrition, 80*(6), 1634–1638.

Gupta, D., Lis, C. G., Dahlk, S. L., King, J., Vashi, P. G., Grutsch, J. F., et al. (2008). The relationship between bioelectrical impedance phase angle and subjective global assessment in advanced colorectal cancer. *Nutrition Journal, 7*, 19. https://doi.org/10.1186/1475-2891-7-19

Han, S. I., & Han, K. H. (2015). Electrical detection method for circulating tumor cells using Graphene Nanoplates. *Analytical Chemistry, 87*(20), 10585–10592. https://doi.org/10.1021/acs.analchem.5b03147.

Harris, N. D., Sugget, A. J., & Barber, D. C. (1987). Applications of applied potential tomography (APT) in respiratory medicine. *Clinical Physics and Physiological Measurement, 8*, 155–165. https://doi.org/10.1088/0143-0815/8/4A/020.

Harris, N. D., Suggett, A. J., Barber, D. C., & Brown, B. (1988). Applied potential tomography: A new technique for monitoring pulmonary function. *Clinical Physics and Physiological Measurement, 9*(A), 79–85. https://doi.org/10.1088/0143-0815/9/4A/014.

Har-Shai, Y., Glickman, Y. A., Siller, G., McLeod, R., Topaz, M., Howe, C., et al. (2005). Electrical impedance scanning for melanoma diagnosis: A validation study. *Plastic and Reconstructive Surgery, 116*(3), 782–790. https://doi.org/10.1097/01.prs.0000176258.52201.22

Hastuti, J., Kaagawa, M., Byrne, N. M., & Hills, A. P. (2016). Proposal of new body composition prediction equations from bioelectrical impedance for Indonesian men. *European Journal of Clinical Nutrition, 70*(11), 1271–1277. https://doi.org/10.1038/ejcn.2016.113.

Haverkort, E. B., Reijven, P. L., Binnekade, J. M., de van der Schueren, M. A., Earthman, C. P., Gouma, D. J., et al. (2015). Bioelectrical impedance analysis to estimate body composition in surgical and oncological patients: Asystematic review. *European Journal of Clinical Nutrition, 69*(1), 3–13. https://doi.org/10.1038/ejcn.2014.203.

Hoffer, E. C., Meador, C. K., & Simpson, D. C. (1969). Correlation of whole-body impedance with total body water volume. *Journal of Applied Physiology, 27*(4), 531–534. PMID:4898406.

Holder, D. S., & Temple, A. J. (1993). Effectiveness of the Sheffield EIT system in distinguishing patients with pulmonary pathology from a series of normal subjects. In D. S. Holder (Ed.), *Clinical and physiological applications of electrical impedance tomography* (1st ed., pp. 277–298). London: CRC Press. ISBN-13: 978–1857281644.

Hong, J.-L., Lan, K.-C., & Jang, L.-S. (2012). Electrical characteristics analysis of various cancer cells using a microfluidic device based on single-cell impedance measurement. *Sensors and Actuators B: Chemical, 173*, 927–934. https://doi.org/10.1016/j.snb.2012.06.046.

Houtveen, J. H., Groot, P. F., & de Geus, E. J. (2006). Validation of the thoracic impedance derived respiratory signal using multilevel analysis. *International Journal of Psychophysiology, 59*(2), 97–106. https://doi.org/10.1016/j.ijpsycho.2005.02.003.

Huerta-Franco, M. R., Vargas-Luna, M., Montes-Frausto, J. B., Flores-Hernández, C., & Morales-Mata, I. (2012). Electrical bioimpedance and other techniques for gastric emptying and motility evaluation. *World Journal of Gastrointestinal Pathophysiology, 3*(1), 10–18. https://doi.org/10.4291/wjgp.v3.i1.10.

Khalil, S. F., Mohktar, M. S., & Ibrahim, F. (2014). The theory and fundamentals of bioimpedance analysis in clinical status monitoring and diagnosis of diseases. *Sensors, 14*(6), 10895–10928. https://doi.org/10.3390/s140610895.

Khalil, S. F., Mohktar, M. S., & Ibrahim, F. (2016). Bioimpedance vector analysis in diagnosing severe and non-severe dengue patients. *Sensors, 16*(6), 911. https://doi.org/10.3390/s16060911.

Khan, S., Mahara, A., Hyams, E. S., Schned, A. R., & Halter, R. J. (2016). Prostate cancer detection using composite impedance metric. *IEEE Transactions on Medical Imaging, 35*(12), 2513–2523. https://doi.org/10.1109/TMI.2016.2578939.

Kim, J. S., Lee, J. Y., Park, H., Han, B. G., Choi, S. O., & Yang, J. W. (2014). Estimation of body fluid volume by bioimpedance spectroscopy in patients with hyponatremia. *Yonsei Medical Journal, 55*(2), 482–486. https://doi.org/10.3349/ymj.2014.55.2.482.

Kose, S. B., Hur, E., Magden, K., Yildiz, G., Colak, D., Kucuk, E., et al. (2015). Bioimpedance spectroscopy for the differential diagnosis of hyponatremia. *Renal Failure, 37*(6), 947–950. https://doi.org/10.3109/0886022X.2015.1040418.

Kubicek, W. G., Karnegis, J. N., Patterson, R. P., Witsoe, D. A., & Mattson, R. H. (1966). Development and evaluation of an impedance cardiac output system. *Aerospace Medicine, 37*(12), 1208–1212. PMID: 5339656.

Kushner, R. F., Schoeller, D. A., Fjeld, C. R., & Danford, L. (1992). Is the impedance index (ht2/R) significant in predicting total body water? *The American Journal of Clinical Nutrition, 56*(5), 835–839. PMID:1415001.

Kyle, U. G., Earthman, C. P., Pichard, C., & Coss-Bu, J. A. (2015). Body composition during growth in children: Limitations and perspectives of bioelectrical impedance analysis. *European Journal of Clinical Nutrition, 69*(12), 1298–1305. https://doi.org/10.1038/ejcn.2015.86.

Langer, R. D., Borges, J. H., Pascoa, M. A., Cirolini, V. X., Guerra-Júnior, G., & Gonçalves, E. M. (2016). Validity of bioelectrical impedance analysis to estimation fat-free mass in the Army cadets. *Nutrients, 8*(3), 121. https://doi.org/10.3390/nu8030121.

Lardiés-Sánchez, B., Sanz-Paris, A., Boj-Carceller, D., & Cruz-Jentoft, A. J. (2016). Systematic review: Prevalence of sarcopenia in ageing people using bioelectrical impedance analysis to assess muscle mass. *European Geriatric Medicine, 7*(3), 256–261. https://doi.org/10.1016/j.eurger.2016.01.014.

Lorne, E., Mahjoub, Y., Diouf, M., Sleghem, J., Buchalet, C., Guinot, P. G., et al. (2014). Accuracy of impedance cardiography for evaluating trends in cardiac output: A comparison with oesophageal Doppler. *British Journal of Anaesthesia, 113*(4), 596–602. https://doi.org/10.1093/bja/aeu136.

Lukaski, H. C. (2013). Evolution of bioimpedance: A circuitous journey from estimation of physiological function to assessment of body composition and a return to clinical research. *European Journal of Clinical Nutrition, 67*(Suppl 1), S2–S9. https://doi.org/10.1038/ejcn.2012.149.

Lukaski, H. C., Johnson, P. E., Bolonchuk, W. W., & Lykken, G. I. (2016). Assessment of fat-free mass using bioelectrical impedance measurements of the human body. *The American Journal of Clinical Nutrition, 41*(4), 810–817. PMID:3984933.

Maass, S. W., Roekaerts, P. M., & Lancé, M. D. (2014). Cardiac output measurement by bioimpedance and noninvasive pulse contour analysis compared with the continuous pulmonary artery thermodilution technique. *Journal of Cardiothoracic and Vascular Anesthesia, 28*(3), 534–539. https://doi.org/10.1053/j.jvca.2014.01.007.

Macfarlane, D. J., Chan, N. T., Tse, M. A., & Joe, G. M. (2016). Agreement between bioelectrical impedance and dual energy X-ray absorptiometry in assessing fat, lean and bone mass changes in adults after a lifestyle intervention. *Journal of Sports Sciences, 34*(12), 1176–1181. https://doi.org/10.1080/02640414.2015.1096416.

Malvehy, J., Hauschild, A., Curiel-Lewandrowski, C., Mohr, P., Hofmann-Wellenhof, R., Motle, R., et al. (2014). Clinical performance of the Nevisense system in cutaneous melanoma detection: An international, multi-centre, prospective and blinded clinical trial on efficacy and safety. *The British Journal of Dermatology, 171*(5), 1099–1107. https://doi.org/10.1111/bjd.13121.

Maskarinec, G., Morimoto, Y., Laguana, M. B., Novotny, R., & Leon-Guerrero, R. T. (2016). Bioimpedence to assess breast density as a risk factor for breast cancer in adult women and adolescent girls. *Asian Pacific Journal of Cancer Prevention, 17*(1), 65–71. https://doi.org/10.7314/APJCP.2016.17.1.65.

Masuda, T., & Komiya, S. (2004). A prediction equation for total body water from bioelectrical impedance in Japanese children. *Journal of Physiological Anthropology and Applied Human Science, 23*(2), 35–39. https://doi.org/10.2114/jpa.23.35.

Matias, C. N., Santos, D. A., Júdice, P. B., Magalhães, J. P., Minderico, C. S., Fields, D. A., et al. (2016). Estimation of total body water and extracellular water with bioimpedance in athletes: A need for athlete-specific prediction models. *Clinical Nutrition, 35*(2), 468–474. https://doi.org/10.1016/j.clnu.2015.03.013.

Mishra, V., Schned, A. R., Hartov, A., Heaney, J. A., Seigne, J., & Halter, R. J. (2013). Electrical property sensing biopsy needle for prostate cancer detection. *Prostate, 73*(15), 1603–1613. https://doi.org/10.1002/pros.22695.

Moawad, F. J., Betteridge, J. D., Boger, J. A., Cheng, F. K., Belle, L. S., Chen, Y. J., et al. (2013). Reflux episodes detected by impedance in patients on and off esomeprazole: A randomised double-blinded placebo-controlled crossover study. *Alimentary Pharmacology & Therapeutics, 37*(10), 1011–1018. https://doi.org/10.1111/apt.12301.

Mohr, P., Birgersson, U., Berking, C., Henderson, C., Trefzer, U., Kemeny, L., et al. (2013). Electrical impedance spectroscopy as a potential adjunct diagnostic tool for cutaneous melanoma. *Skin Research and Technology, 19*, 75–83. https://doi.org/10.1111/srt.12008.

Neves, E. B., Pino, A. V., de Almeida, R. M., & de Souza, M. N. (2009). Knee bioelectric impedance assessment in healthy/with osteoarthritis subjects. *Physiological Measurement, 31*(2), 207–219. https://doi.org/10.1088/0967-3334/31/2/007.

Paek, D., & McCool, D. (1992). Breathing patterns during varied activities. *Journal of Applied Physiology, 73*, 887–893. PMID:1400052.

Pandya, H. J., Kim, H. T., Roy, R., Chen, W., Cong, L., Zhong, H., et al. (2014). Towards an automated MEMS-based characterization of benign and cancerous breast tissue using bioimpedance measurements. *Sensors & Actuators B: Chemical, 199*, 259–268. https://doi.org/10.1016/j.snb.2014.03.065.

Perez, W., & Tobin, M. J. (1985). Separation of factor responsible for change in breathing pattern induced by instrumentation. *Journal of Applied Physiology, 59*, 1515–1520. PMID:4066581.

Pichonnaz, C., Bassin, J. P., Lécureux, E., Currat, D., & Jolles, B. M. (2015). Bioimpedance spectroscopy for swelling evaluation following total knee arthroplasty: A validation study. *BMC Musculoskeletal Disorders, 16*, 100. https://doi.org/10.1186/s12891-015-0559-5.

Porter, E., Bahrami, H., Santorelli, A., Gosselin, B., Rusch, L. A., & Popovic, M. (2016). A wearable microwave antenna Array for time-domain breast tumor screening. *IEEE Transactions on Medical Imaging, 35*(6), 1501–1509. https://doi.org/10.1109/TMI.2016.2518489.

Ravi, K., & Katzka, D. A. (2016). Esophageal impedance monitoring: Clinical pearls and pitfalls. *The American Journal of Gastroenterology, 111*(9), 1245–1256. https://doi.org/10.1038/ajg.2016.256.

Rubbieri, G., Mossello, E., & Di Bari, M. (2014). Techniques for the diagnosis of sarcopenia. *Clinical Cases in Mineral and Bone Metabolism, 11*(3), 181–184. PMCID: PMC4269140.

Salazar, Y., Bragos, R., Casas, O., Cinca, J., & Rosell, J. (2004). Transmural versus nontransmural in situ electrical impedance spectrum for healthy, ischemic, and healed myocardium. *IEEE Transactions on Biomedical Engineering, 51*(8), 1421–1427. https://doi.org/10.1109/TBME.2004.828030.

Sánchez, B., Vandersteen, G., Martin, I., Castillo, D., Torrego, A., Riu, P. J., et al. (2013). In vivo electrical bioimpedance characterization of human lung tissue during the bronchoscopy procedure. A feasibility study. *Medical Engineering & Physics, 35*(7), 949–957. https://doi.org/10.1016/j.medengphy.2012.09.004.

Schwan, H. P. (1957). Electrical properties of tissue and cell suspensions. *Advances in Biological and Medical Physics, 5*, 147–209. PMID:13520431.

Schwan, H. P., & Kay, C. F. (1957a). Capacitive properties of body tissues. *Circulation Research, 5*(4), 439–443. PMID:13447191.

Schwan, H. P., & Kay, C. F. (1957b). The conductivity of living tissues. *Annals of the New York Academy of Sciences, 65*(6), 1007–1013. PMID:13459187.

Seward, C., Skolny, M., Brunelle, C., Asdourian, M., Salama, L., & Taghian, A. G. (2016). A comprehensive review of bioimpedance spectroscopy as a diagnostic tool for the detection and measurement of breast cancer-related lymphedema. *Journal of Surgical Oncology, 114*(5), 537–542. https://doi.org/10.1002/jso.24365.

Shah, C., Vicini, F. A., & Arthur, D. (2016). Bioimpedance spectroscopy for breast cancer related lymphedema assessment: Clinical practice guidelines. *The Breast Journal, 22*(6), 645–650. https://doi.org/10.1111/tbj.12647.

Stick, S. M., Ellis, E., LeSouëf, P. N., & Sly, P. D. (1992). Validation of respiratory inductance plethysmography ("Respitrace") for the measurement of tidal breathing parameters in newborns. *Pediatric Pulmonology, 14*(3), 187–191. https://doi.org/10.1002/ppul.1950140308.

Sun, S. S., Chumlea, W. C., Heymsfield, S. B., Lukaski, H. C., Schoeller, D., Friedl, K., et al. (2003). Development of bioelectrical impedance analysis prediction equations for body composition with the use of a multicomponent model for use in epidemiologic surveys. *The American Journal of Clinical Nutrition, 77*, 331–440. PMID:12540391.

Tyagi, R., Mishra, S., Gaur, N., Panwar, A., Saini, D., Singh, K., et al. (2016). Bioelectric impedance phase angle in carcinoma prostate - a hospital-based study. *International Journal of Medical Science and Public Health, 5*(9), 1826–1830. https://doi.org/10.5455/ijmsph.2016.30122015335.

Verney, J., Metz, L., Chaplais, E., Cardenoux, C., Pereira, B., & Thivel, D. (2016). Bioelectrical impedance is an accurate method to assess body composition in obese but not severely obese adolescents. *Nutrition Research, 36*(7), 663–670. https://doi.org/10.1016/j.nutres.2016.04.003.

Voscopoulos, C., Brayanov, J., Ladd, D., Lalli, M., Panasyuk, A., & Freeman, J. (2013). Evaluation of a novel noninvasive respiration monitor providing continuous measurement of minute ventilation in ambulatory subjects in a variety of clinical scenarios. *Anesthesia & Analgesia, 117*(1), 91–100. https://doi.org/10.1213/ANE.0b013e3182918098.

Voscopoulos, C., Ladd, D., Campana, L., & George, E. (2014). Non-invasive respiratory volume monitoring to detect apnea in post-operative patients: Case series. *Journal of Clinical Medicine Research, 6*(3), 209–214. https://doi.org/10.14740/jocmr1718w.

Weitzen, R., Epstein, N., Shoenfeld, Y., & Zimlichman, E. (2007). Diagnosing diseases by measurement of electrical skin impedance: A novel technique. *Annals of the New York Academy of Sciences, 1109*, 185–192. https://doi.org/10.1196/annals.1398.022.

Wickramasinghe, V. P., Lamabadusuriya, S. P., Cleghorn, G. J., & Davies, P. S. W. (2008). Assessment of body composition in Sri Lankan children: Validation of a bioelectrical impedance prediction equation. *European Journal of Clinical Nutrition, 62*(10), 1170–1177. https://doi.org/10.1038/sj.ejcn.1602835.

Woltjer, H. H., Bogaard, H. J., & de Vries, P. M. (1997). The technique of impedance cardiography. *European Heart Journal, 18*(9), 1396–1403. PMID:9458444.

Yazdanian, H., Mahnam, A., Edrisi, M., & Esfahani, M. A. (2016). Design and implementation of a portable impedance cardiography system for noninvasive stroke volume monitoring. *Journal of Medical Signals and Sensors, 6*(1), 47–56. PMID:27014612.

Zhang, X., Chatwin, C., & Barber, D. C. (2015). A feasibility study of a rotary planar electrode array for electrical impedance mammography using a digital breast phantom. *Physiological Measurement, 6*(6), 1311–1335. https://doi.org/10.1088/0967-3334/36/6/131.

Zhang, X., Wang, W., Sze, G., Barber, D., & Chatwin, C. (2014). An image reconstruction algorithm for 3-D electrical impedance mammography. *IEEE Transactions on Medical Imaging, 33*(12), 2223–2241. https://doi.org/10.1109/TMI.2014.2334475.

Zhou, L. Y., Wang, Y., Lu, J. J., Lin, L., Cui, R. L., Zhang, H. J., et al. (2014). Accuracy of diagnosing gastroesophageal reflux disease by Gerd Q, esophageal impedance monitoring and histology. *Journal of Digestive Diseases, 15*(5), 230–238. https://doi.org/10.1111/1751-2980.12135.

Zouridakis, A., Simos, Y. V., Verginadis, I. I., Charalabopoulos, K., Ragos, V., Dounousi, E., et al. (2016). Correlation of bioelectrical impedance analysis phase angle with changes in oxidative stress on end-stage renal disease patients, before, during, and after dialysis. *Renal Failure, 38*(5), 738–743. https://doi.org/10.3109/0886022X.2016.1158042.

Chapter 6
Tissue Engineering Instrumentation Based on Electrical Impedance Measurements

Marcio Nogueira de Souza

6.1 Overview About Tissue Engineering

One important problem that the next generations must face is how to increase the availability of tissues and organs for transplantation to reduce the number of deaths and injuries. Despite the increasing global awareness of the importance of donation, the availability of compatible donors points out the necessity for a technological solution.

Although the term "tissue engineering" was initially used in the literature to describe cases of surgical manipulation of tissues and organs, it is almost a fact that the currently accepted definition for it was introduced around 1987: "Tissue Engineering is the application of the principles and methods of engineering and life sciences toward the fundamental understanding of structure-function relationships in normal and pathologic mammalian tissue and the development of biological substitutes to restore, maintain, or improve function." Thus, tissue engineering can be thought of as the technological response to the transplantation problems and is one of the most promising and new biomedical disciplines. It is at the interface of medical implant industry and the biological revolution observed in the beginning of the twenty-first century, and it is expected to create an entirely new generation of tissue and organ implants as well as strategies for repair and regeneration.

Despite its promising future, many challenges must be overcome before tissue engineering becomes the solution to reduce the number of deaths and injuries. Some of the concerns are the selection of the type of cell to be used in each engineered tissue; usage of a suitable three-dimensional (3D) growing environment, i.e., the

M. N. de Souza (✉)
Biomedical Engineering Program – COPPE/Federal University of Rio de Janeiro, Rio de Janeiro, Brazil
e-mail: souza@peb.ufrj.br

F. Simini, P. Bertemes-Filho (eds.), *Bioimpedance in Biomedical Applications and Research*, https://doi.org/10.1007/978-3-319-74388-2_6

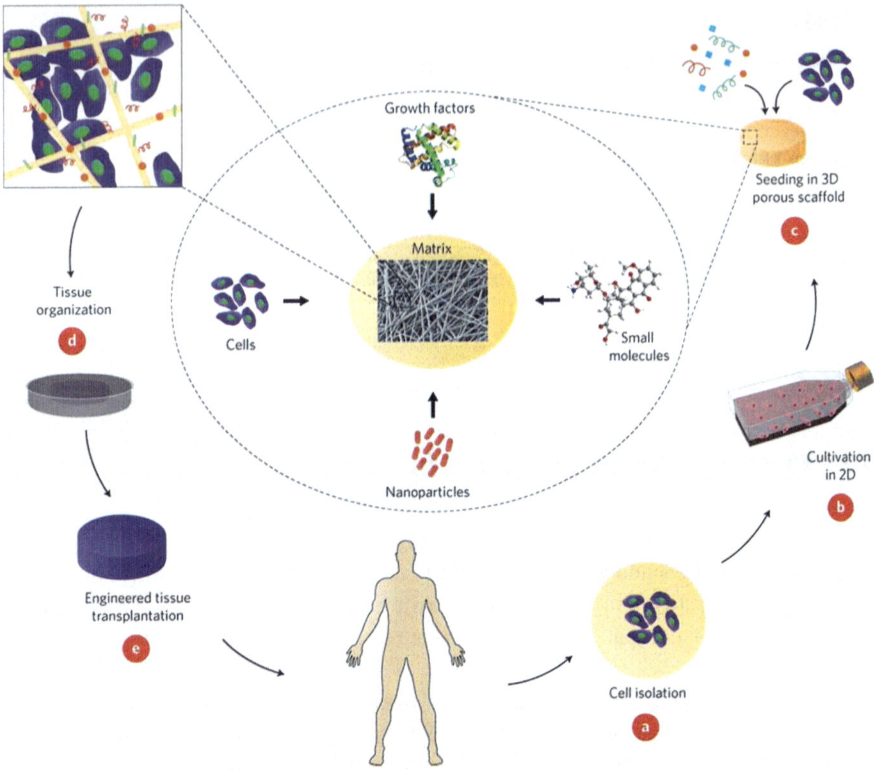

Fig. 6.1 Concept model associated to tissue engineering. (**a**) Cells are isolated from the patient and may be cultivated (**b**) in vitro on two-dimensional surfaces for efficient expansion. (**c**) Next, the cells are seeded in porous scaffolds together with growth factors, small molecules, and micro- and/or nanoparticles. The scaffolds serve as a mechanical support and a shape-determining material, and their porous nature provides high mass transfer and waste removal. (**d**) The cell constructs are further cultivated in bioreactors to provide optimal conditions for organization into a functioning tissue. (**e**) Once a functioning tissue has been successfully engineered, the construct is transplanted on the defect to restore function (Dvir et al. 2011)

scaffold, to supply a mechanical and biochemical environment that mimics the natural environment; vascularization for the 3D-engineered tissue to meet its nutritional and metabolic requirements; good manufacturing practices and preservation; and shipping, among others. Figure 6.1 illustrates a concept model that can be associated to tissue engineering. Considering such a scenario, novel experimental techniques, especially the noninvasive ones, must be studied and developed in order to solve some of the critical challenges associated to the success of tissue engineering within the near future.

6.2 Dielectric Spectroscopy

Basic concepts of bioimpedance and the bioelectrical impedance spectroscopy (BIS) technique can be found in Chap. 1. Dielectric spectroscopy is another noninvasive experimental technique related to the bioelectrical impedance technique, and it is intended to measure the characteristics of the permittivity and conductivity of a biological or nonbiological sample as a function of frequency (Heileman et al. 2013). Thus, it is also helpful to obtain information about the structures and electrical properties at a macroscopic level. As mentioned before, permittivity is a measure of the polarization of the medium, and it is normally frequency dependent. In heterogeneous media composed of complex electrolytes and biological cells, there are interfaces between constituents of different electrical properties, and the characteristic polarization is an interfacial polarization due to the buildup of charge on the interfaces (Asami 2002).

The quantity assessed by dielectric spectroscopy is the complex relative permittivity $\boldsymbol{\varepsilon}^*$ defined as:

$$\boldsymbol{\varepsilon}^* = \varepsilon' - j\varepsilon'' = \varepsilon + \frac{\sigma}{j\varepsilon_0\omega}$$

where ε' and ε'' are the real and imaginary parts, respectively, of ε^*; ε is known in this field as relative permittivity (which is equal to ε'), σ is the conductivity, ε_0 is the permittivity of vacuum, and ω is the angular frequency. Instead of the conductivity σ, some authors use the loss factor $\varepsilon'' = \frac{\sigma}{j\varepsilon_0\omega}$ to characterize the sample.

Figure 6.2 shows the behavior of ε' and ε'' as a function of frequency for a sample that exhibits a single dispersion, by Bode plots (on the left) and Cole-Cole plots (on the right).

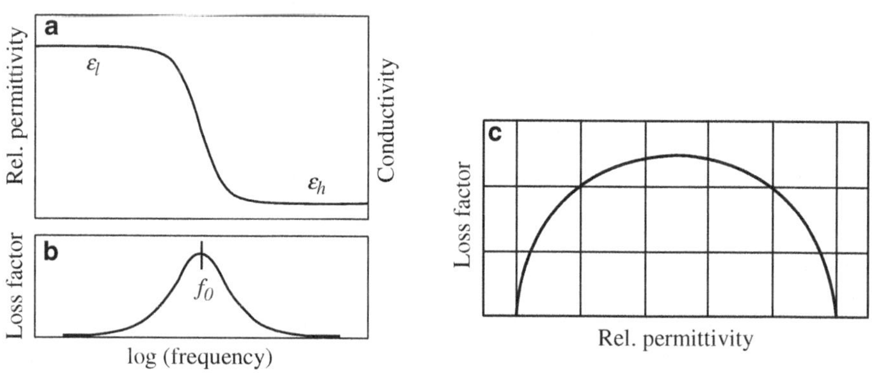

Fig. 6.2 Illustration of the behavior of ε' and ε'' for a single Debye dispersion. ε_l and ε_h stand for the low- and high-frequency limits of relative permittivity $\varepsilon = \varepsilon'$ (Asami 2002)

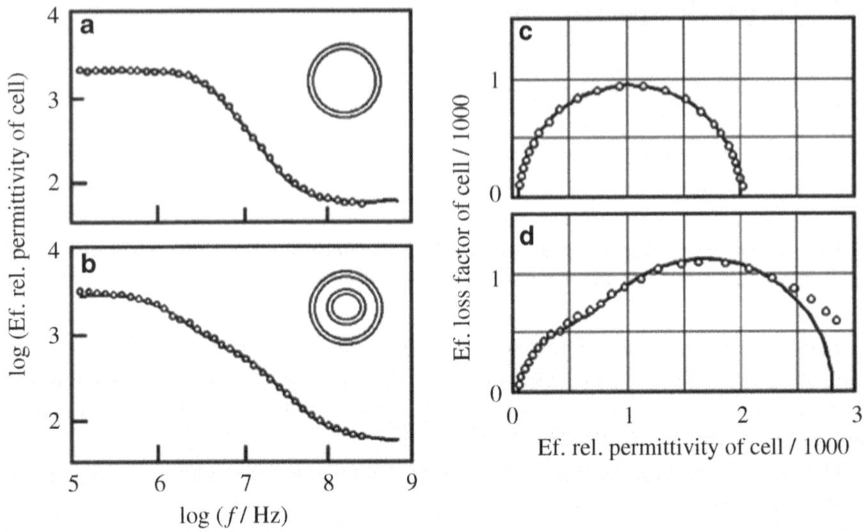

Fig. 6.3 Illustration of the effective relative permittivity of (**a**) spherical erythrocyte and (**b**) lymphocyte calculated from the dielectric relaxation of their cell suspensions using Hanai's mixture equation. (**c**) and (**d**) are the Cole-Cole plots for (**a**) and (**b**), respectively (Asami et al., 1989, cited by Asami 2002)

It is worthwhile to mention that real biological tissue normally exhibits a Cole-Cole plot that deviates from an ideal semicircle, indicating the existence of distribution of relaxation times instead of a single relaxation time or frequency dispersion. This is expected because different types of cells exhibit different behaviors of the relative permittivity and loss factor as a function of frequency, and such characteristics lead to a more complex behavior of ε' and ε''. Figure 6.3 illustrates the simulation of the relative permittivity and loss factor for a spherical erythrocyte and for a lymphocyte performed by Asami et al. (1989).

6.3 Overview of Electrical Impedance Technique Applied to Tissue Engineering

Cells in a natural biological organism normally sense their immediate microenvironments and interact with neighboring cells to create a functional tissue. Tissue engineering usually adopts this biologically inspired approach, i.e., directs the organization of metabolically active cells into 3D arrangements (via scaffolds), parallelly setting the growth conditions so that the cells reconstruct a functional tissue (see the concept model of Fig. 6.1). As mentioned before, there are many challenges to be overcome in the tissue engineering field, and this section intends to show some applications of bioelectrical impedance technique as a helpful tool in facing some of these challenges.

The work of Giaever and Keese (1984) can be considered as one of the first papers to use electrical impedance to monitor tissue culture. They introduced a method of monitoring the growth pattern of human embryonic lung fibroblasts using the impedance of four small electrodes and one large electrode on the bottom of a sterile polystyrene petri dish employed to perform the cell culture. By observing the impedance changes, it was possible to follow the attachment and spreading of the cells, as well as, the cell response to various protein-covered electrodes. Posteriorly (e.g., Keese and Giaever 1994) they nominated the technique as ECIS (from Electric Cell–substrate Impedance Sensing) to refer to the use of electrode impedance of small gold film electrodes deposited on the bottom of cell culture dishes used to investigate cell cultures.

Evaluation of cell number and cell viability (for a better understanding about cell "viability" and "culturability," see Kell et al. 1998) is commonly performed to provide the information about the cell growth and is one of the important issues to be addressed in tissue engineering. For cell proliferation, which is traditionally achieved by counting cells microscopically; and for the assessment of cell viability, i.e., the ratio live/dead, the most common process involves fluorescent dye staining and microscopic observation. However, such classical approaches are not practical in a 3D cell culture environment. Discussing the use of the electric field in tissue engineering, Markx (2008) states that the electrical properties of biological tissues are mainly determined by their capacitance and indicates that an important potential application of dielectric measurements on tissue engineering is the online and continuous monitoring of the cell concentration. The author also indicates that by using a suitable experimental setup, bioelectrical impedance can be used to supply information on cell concentration and cell viability.

In a review about the potential uses of bioelectrical impedance technique to perform biomolecular screening and cell culture monitoring, K'Owino and Omowunmi (2005) state that bioelectrical impedance spectroscopy (BIS), unlike many other techniques used to characterize bio-events, provides information about the interior of a biological organism in their native environment in a noncontact and noninvasive experimental method. They argue the interaction of proteins with solid substrates holds an important role to improve applications in cell immobilization, protein purification, and tissue engineering. As an example, they mentioned that Wright et al. (2004) studied the adsorption of insulin on platinum electrodes as one of the processes for designing artificial pancreas and that Moulton et al. (2004) also used bioelectrical impedance spectroscopy to monitor the adsorption of HSA and IgG on gold electrodes. In both instances the researchers reported decreases in the double layer capacitance associated to the electrode-electrolyte interface and increases in charge transfer resistance consistent with addition of insulating layers of proteins on the electrode surface.

The assessment of viability and number of adipose stem cells were investigated by Jaatinen et al. (2009) by using a bioelectrical impedance setup that performed the measurements using a Biopac MP35 and Electrical Bioimpedance Amplifier EBI100C (Biopac Systems Inc., Goleta, CA, USA) at the frequency of 100 kHz. This work aimed to detect the changes in the impedance due to the different cell amounts in a single cell culture during the time the cells proliferate. The authors concluded that the magnitude of the observed bioelectrical impedance was related to the cell amounts and that the impedance measurements could offer a noninvasive method to validate standard tissue-engineered scaffolds with cells in the production phase.

In contrast to the classical fluorescence-based flow cytometer, some authors (Holmes et al. 2009) have proposed a so-called impedance flow cytometer (Fig. 6.4a). In such work, Homes and colleagues reported the implementation of a microfluidic single-cell impedance cytometer that counts white blood cells by measuring the impedance of single cells. The system is based on Maxwell's mixture theory to model dielectric properties of the cells that passes through the channel that contains the two pairs of electrodes, disturbing the electric field in the detection volume. The electrical model depicted in Fig. 6.4b was used to model what would be expected between a pair of electrodes with no cell (left) and with a cell between the electrodes (right). Figure 6.4c shows the expected impedance magnitude as a function of frequency for a bead used as a control particle and real cell. Two sinusoidal voltages at fixed frequencies (one at 503 kHz and a second varying from 0.5 to 30 MHz) were applied to both pairs of detection electrodes. A differential current was measured using custom-built electronics and two lock-in amplifiers (SR844, Stanford Research Instruments, USA), one for each frequency. The output from the lock-in consists of the real component (in phase) and imaginary component (90 out of phase) of the electrical impedance at each frequency. The outputs were sampled at 120 kHz and analyzed with a proper software. To analyze a mixture composed by blood cells and 5.62 mm diameter beads, the authors used an opacity index ($|Z|_{1.707\text{MHz}}/|Z|_{0.503\text{MHz}}$), defined as the ratio of high-frequency to low-frequency impedance and that reflected changes in the cell membrane vs. the impedance at 503 kHz (Fig. 6.4d).

Mesenchymal stem cells (MSCs) are multipotent stem cells that have the capacity to differentiate into some different lineage of cells, like osteoblasts, chondroblasts, and adipocytes (Robert et al. 2011). It is reasonable to argue that such differentiation would result in specific changes of the electrical properties of the cells, leading to a kind of signature in the bioelectrical impedance spectrum. Hildebrandt and Thielecke (2009) investigated the use of impedance spectroscopy for characterization of the osteogenic development of MSC in scaffold-free aggregates (spheroids) as an in vitro model for osteogenesis. They used a commercial impedance analyzer (4294, Agilent Technologies, Germany) to obtain the electrical impedance data in the range of 100 Hz to 1 MHz in a two-electrode arrangement of a microcapillary system containing the MSC aggregates held in a standard growth medium for control and the osteogenic-treated aggregates. The authors concluded that impedance spectroscopy has the potential to be a noninvasive method to characterize the osteogenic development.

The application of bioelectrical impedance technique in monitoring biological processes has already led to the development of some commercial devices. Diemert et al. (2012) used a commercial equipment, called xCELLigence system Real-Time Cell Analizer (RTCA) MP (Fig. 6.5), for real-time detection of cell death

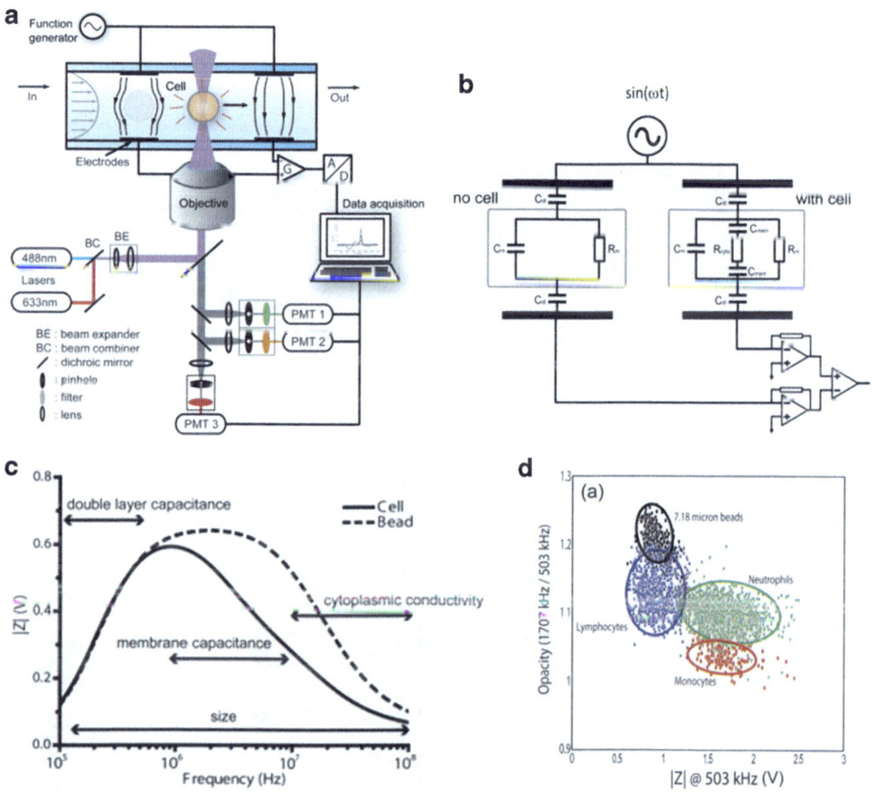

Fig. 6.4 (**a**) Diagram of an impedance flow cytometer with an optical detection system, showing the two sets of electrodes that are used for differential measurements. (**b**) The electric circuit used to model the impedance detection system, where the electrode pair on the left illustrates the equivalent circuit model without a cell between the electrodes, and the right electrode pair has the cell present between the electrodes. R_m and C_m are the equivalent resistance and capacitance of the medium, respectively. C_{mem} is the equivalent capacitance of the cell membrane. R_{cyto} is the equivalent resistance of the cell cytoplasm. C_{dl} represents the electrical double-layer capacitance at the liquid electrode interface. (**c**) Typical frequency-dependent impedance magnitude signal for a polymer bead and a cell of similar size. (**d**) Scatter plot of the impedance at 503 kHz vs. opacity (the ratio between the impedances at 1707 kHz and 503 kHz) for saponin-/formic acid-treated whole blood and 7.18 μm latex beads. The color in the plots corresponds to fluorescence also detected by the setup (Holmes et al. 2009)

in a neuronal cell line of immortalized hippocampal neurons (HT-22 cells), neural progenitor cells (NPC), and differentiated primary cortical neurons. Such analyzer uses an arbitrary cell index (CI) to monitor the cell growth, where CI = (impedance at time point n—impedance in the absence of cells)/nominal impedance value. Background impedance caused by the media was determined in each well before seeding the cells and subtracted automatically by the RTCA software. Using this device, the authors concluded that the impedance-based xCELLigence system

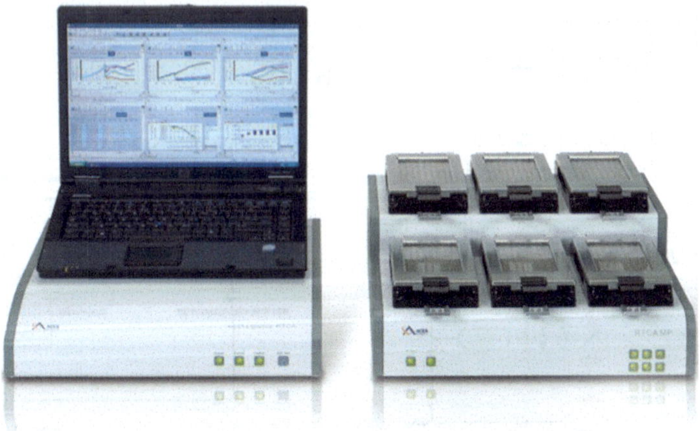

Fig. 6.5 xCELLigence Real-Time Cell Analizer (RTCA) (https://aceabio.com/products/xcelligence-rtca/)

Fig. 6.6 Interdigitated microsensor electrodes (IME) (http://www.abtechsci.com)

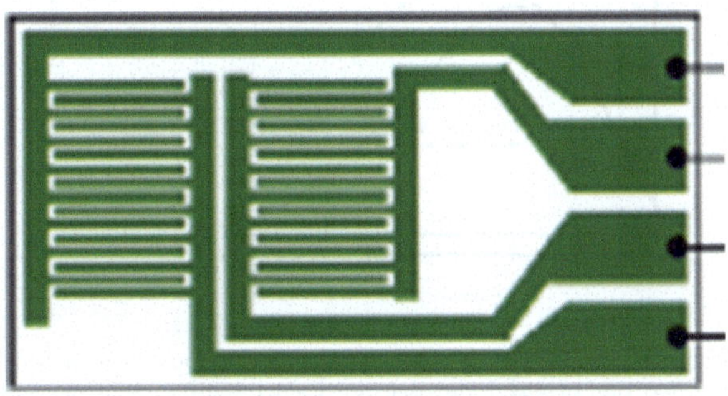

showed appropriate sensitivity for online monitoring of cell adhesion, proliferation, and detection of cell death in different neuronal cell lines.

Sarró et al. (2012) proposed an electrical impedance spectroscopy approach to improve online monitoring of cell concentration in adherent animal cell culture. The authors used an interdigitated microsensor electrodes (IME) (Fig. 6.6), purchased from Abtech Scientific (http://www.abtechsci.com), and a HP4192A impedance analyzer to perform measurements between 10 kHz and 10 MHz with a frequency resolution of 10 points per decade (31 total points for scan). However, only the frequencies of 1995 and 32 kHz were chosen as low and high frequencies used in a cell concentration estimator (E2) proposed by Bragós et al. (1999). Cell concentration was monitored using both four-electrode and two-electrode configurations in *Vero* cell and human mesenchymal stem cell cultures to analyze the improvement attained in two cell lines with opposite growth characteristics. The authors concluded that their measurements were linearly correlated to cell concentration through the cell concentration estimator E2, enabling its monitoring during cell growth.

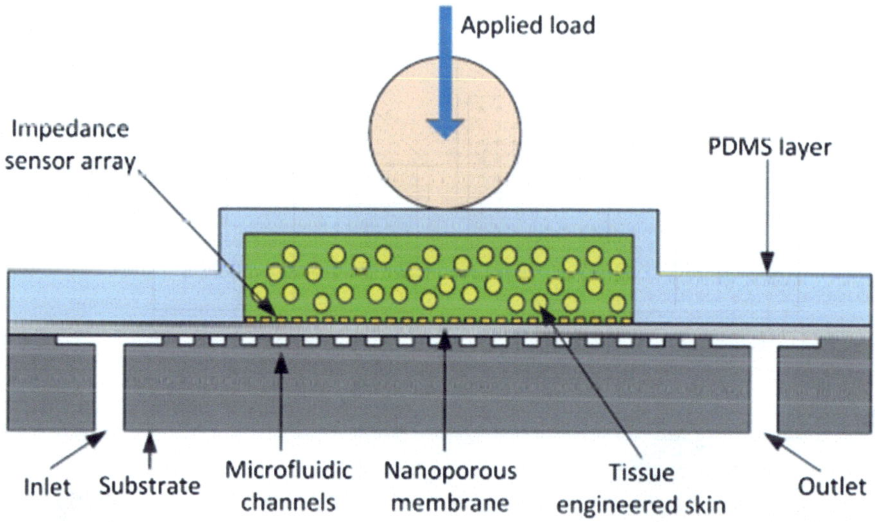

Fig. 6.7 Cross section of the bio-hybrid tactile sensor array (Cheneler et al. 2014)

In a very interesting approach, Cheneler et al. (2014) described the development of a bio-hybrid tactile sensor array composed by a tissue-engineered skin analogue in which the electrical properties are affected by mechanical stress induced during contact (Fig. 6.7). Such changes in electrical properties are detected by a ten-channel dual-electrode impedance sensing array. The skin analogue comprised of alginate encapsulated fibroblasts, which are known to be mechanosensitive, which are supported by a microfluidic system designed to maintain the viability of the encapsulated cells. By continuously monitoring the impedance of the sensor array at 250 Hz while normal and tangential loads are applied to the skin surface, the authors reported the existence of transient mechanotransduction. This type of tissue-engineered application can be used in neuroprosthetic artificial hands targeting the execution of human-like tasks such as grasping and manipulation, where the incorporation of sensory systems mimicking the human sense of touch is viewed as an important improvement.

Even the principles of the electrical impedance tomography (EIT) have been tested for tissue engineering applications. Ahn et al. (2014) applied some fixed boundary geometrical conditions and a special electrode configuration of a sample container (Fig. 6.8) to maximize the performance of Graz consensus reconstruction algorithm for EIT (GREIT). The authors concluded that the new micro-EIT system integrated with GREIT is robust and that, with some further refinement and scaling down to a microscale container, it may be a continuous, nondestructive monitoring tool for tissue engineering applications.

As previously mentioned, one of the technological challenges in the tissue engineering field is monitoring the spatial growth of the tissue which includes the scaffold characterization. Normally, polymer scaffolds are used because they

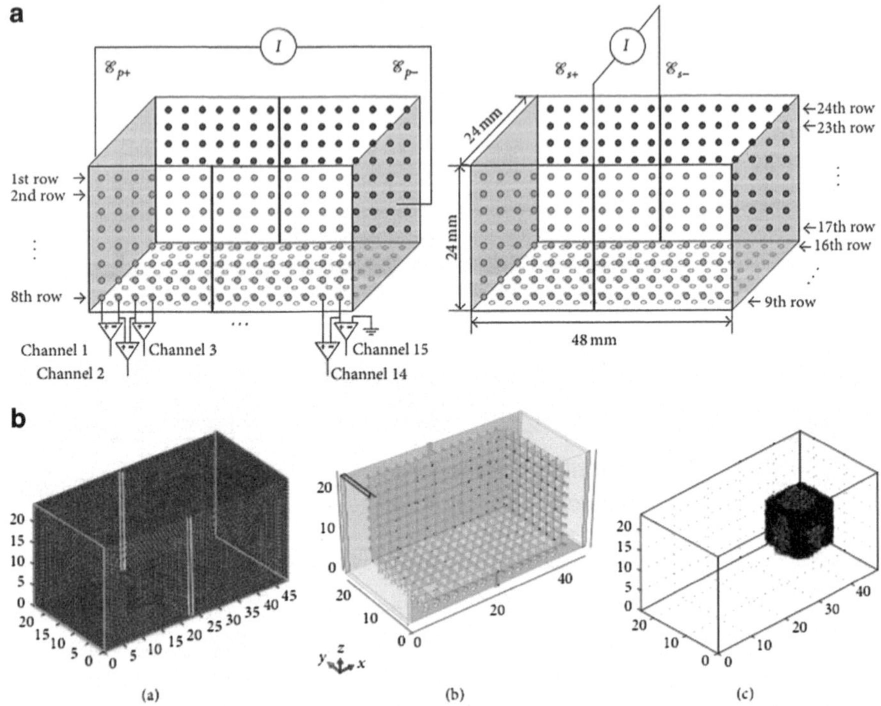

Fig. 6.8 (**a**) The structure of the sample container and the method used for the primary (E$p\pm$) and secondary injection currents (E$s\pm$). (**b**) Numerical simulation showing a sample reconstruction. For more details, please see Ahn et al. (2014)

present malleability, high porosity, and a surface that favors cell attachment, proliferation, and differentiation. Moreover, by monitoring a seeded scaffold, one can obtain information about the tissue growth. Classical methods for determining tissue constituent variation in these scaffolds are time consuming, laborious, and often destructive, encouraging the search for new approaches. In this sense, Canali et al. (2015a) used impedance spectroscopy to characterize the porosity of polydimethylsiloxane (PDMS) scaffolds. Scaffold parameters (pore size, geometry, orientation, interconnectivity) play an important role in the diffusion of nutrients and cell organization. The approach used was based on Archie's law that links the electrical resistivity of rock to its porosity, to the resistivity of the water that saturates its pores, and to the fractional saturation of the pore space with water. Using the scaffold characterization knowledge, Canali and her research group (2015b) also showed a bioelectrical impedance setup intended to monitor 3D cell cultures. The setup was composed by a special chamber with a set of four gold electrodes that, using the characteristics of the sensitivity field associated to tetrapolar bioimpedance measurements (Grimnes and Martinsen 2008), were switched to provide information associated with many directions of the cell growth

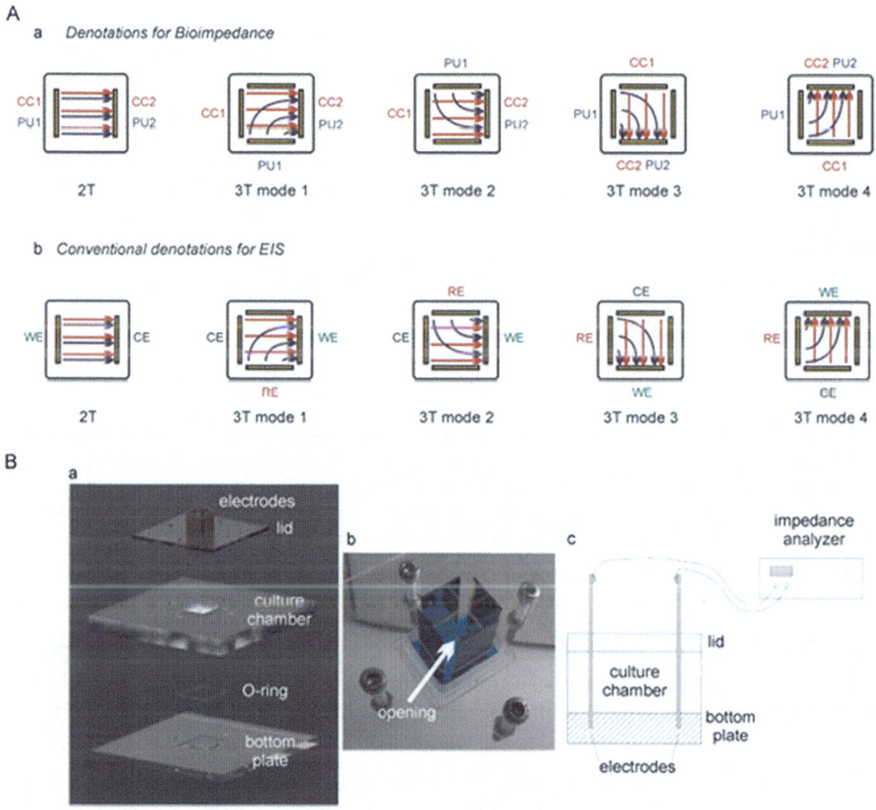

Fig. 6.9 Schematics of the 2T and 3T (modes 1–4) configurations tested. In bioimpedance measurements (**a**), CC1 and CC2 form the current-carrying couple and PU1 and PU2 form the pick-up couple. Red (CC) and blue (PU) arrows represent a top view of the equipotential surface directions. In EIS analysis (**b**): WE¼ working electrode, CE¼ counter-electrode, RE¼ reference electrode. (B) Design of the overall bioimpedance platform. (**a**) The system consists of three layers: bottom plate accommodating the electrodes, culture chamber, and lid. (**b**) Photograph of the system embedding four electrodes for 3T measurements. (**c**) Experimental setup for the 2T configuration (Canali et al. 2015b)

(Fig. 6.9). The authors monitored MSCs (mesenchymal stem cells) growth in a gelatin scaffold during 48 h as proof of the concept on the setup feasibility and concluded the proposed setup had been the first attempt of generating a flexible bioimpedance-based method for monitoring large 3D cell cultures with spatial resolution.

Sarró and collaborators (2016) argued that despite the existence of real-time monitoring systems capable of supplying parameters associated to the cell concentration and cell viability, there is still a lack of well-established systems for monitoring cell morphology and other biological parameters. BIS (10 kHz to 1 MHz with a resolution of 10 frequencies per decade) was carried out using a HP4192A

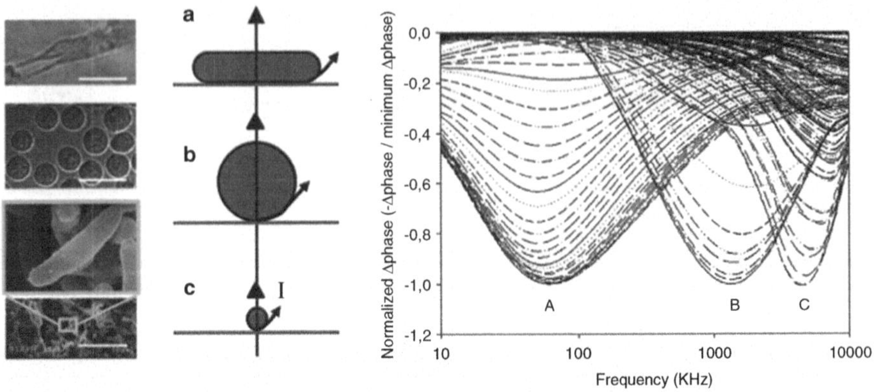

Fig. 6.10 Left: microelectronic images and diagrams representing shape and size of an adherent animal cell line (**a**), an animal cell line grown in suspension (**b**), and a bacteria (**c**). Possible paths of the injected current are marked with arrows depending on applied frequency (straight path for high-frequency current and surrounding path for low-frequency current). Right: impedance-phase profiles during an ample frequency range to show the β relaxation of all three cell line types (**a**, **b**, and **c**) (Sarró et al. 2016)

impedance analyzer (Hewlett–Packard, USA) and a four-electrode interdigitated microsensor setup in cultures of three cell lines (*Vero* cells, hybridoma cell, and *Escherichia coli* bacterium) with different shapes and sizes. The results (Fig. 6.10) confirmed the relationship between parameters of the Cole model obtained from BIS data and important morphological cell characteristics, i.e., changes on the central relaxation frequency and dispersion in frequency response (related to the α parameter of the Cole model).

6.4 Conclusion

This chapter did not intend to be an introduction to tissue engineering and what concerns its biological principles (for this purpose, see, e.g., Lanza Robert, Robert Langer, and Joseph P. Vacanti, Principles of tissue engineering, Academic Press, 2011). Instead of this, the chapter tried to show that the application of the bioelectrical impedance technique in tissue engineering has already produced remarkable advances, including the development of commercial devices.

The use of an electric field for monitoring tissue growth has some potential advantages over other approaches, for example, as a noninvasive and nondestructive method; does not use staining, labeling, or toxic chemicals; does not interfere with the tissue growth; is not expensive; and exhibits the possibility of investigating 3D opaque tissues (which cannot be observed by most optic techniques), among others.

As disadvantage one could say that bioelectrical impedance technique is not yet well known in the biology field, and many researchers still prefer to use classical gold-standard methods to study tissue growth.

Despite such promising perspectives, many aspects related the use of bioelectrical impedance in tissue engineering still need to be studied and developed because several challenges have not yet been solved.

References

Ahn, S., Wi, H., Oh, T. I., McEwan, A. L., Jun, S. C., & Woo, E. J. (2014). Continuous nondestructive monitoring method using the reconstructed three-dimensional conductivity images via GREIT for tissue engineering. *Journal of Applied Mathematics, 2014.*

Asami, K. (2002). Characterization of heterogeneous systems by dielectric spectroscopy. *Progress in Polymer Science, 27*(8), 1617–1659.

Asami, K., Takahashi, Y., & Takashima, S. (1989). Dielectric properties of mouse lymphocytes and erythrocytes. *Biochimica et Biophysica Acta (BBA)-Molecular Cell Research, 1010*(1), 49–55.

Bragós, R., Gámez, X., Cairó, J., Riu, P. J., & Gòdia, F. (1999). Biomass monitoring using impedance spectroscopy. *Annals of the New York Academy of Sciences, 873*(1), 299–305.

Canali, C., Mohanty, S., Heiskanen, A., Muhammad, H. B., Martinsen, Ø. G., Dufva, M., et al. (2015a). Impedance spectroscopic characterization of porosity in 3D cell culture scaffolds with different channel networks. *Electroanalysis, 27*(1), 193–199.

Canali, C., Heiskanen, A., Muhammad, H. B., Høyum, P., Pettersen, F. J., Hemmingsen, M., et al. (2015b). Bioimpedance monitoring of 3D cell culturing—Complementary electrode configurations for enhanced spatial sensitivity. *Biosensors and Bioelectronics, 63*, 72–79.

Cheneler, D., Buselli, E., Camboni, D., Anthony, C., Grover, L., Adams, M. J., et al. (2014). A bio-hybrid tactile sensor incorporating living artificial skin and an impedance sensing array. *Sensors, 14*(12), 23781–23802.

Diemert, S., Dolga, A. M., Tobaben, S., Grohm, J., Pfeifer, S., Oexler, E., et al. (2012). Impedance measurement for real time detection of neuronal cell death. *Journal of Neuroscience Methods, 203*(1), 69–77.

Dvir, T., Timko, B. P., Kohane, D. S., & Langer, R. (2011). Nanotechnological strategies for engineering complex tissues. *Nature Nanotechnology, 6*(1), 13–22.

Giaever, I., & Keese, C. R. (1984). Monitoring fibroblast behavior in tissue culture with an applied electric field. *Proceedings of the National Academy of Sciences, 81*(12), 3761–3764.

Grimnes, S., & Martinsen, Ø. G. (2008). *Bioimpedance and bioelectricity basics* (2nd ed.). Oxford, UK: Academic Press.

Heileman, K., Daoud, J., & Tabrizian, M. (2013). Dielectric spectroscopy as a viable biosensing tool for cell and tissue characterization and analysis. *Biosensors and Bioelectronics, 49*, 348–359.

Hildebrandt, C., & Thielecke, H. (2009). Non-invasive characterization of the osteogenic differentiation of hMSCs in 3D by impedance spectroscopy. In: *World Congress on Medical Physics and Biomedical Engineering*, September 7–12, 2009. Munich, Germany: Springer.

Holmes, D., Pettigrew, D., Reccius, C. H., Gwycr, J. D., van Berkel, C., Holloway, J., et al. (2009). Leukocyte analysis and differentiation using high speed microfluidic single cell impedance cytometry. *Lab on a Chip, 9*(20), 2881–2889.

Jaatinen, L., Sippola, L., Kellomäki, M., Miettinen, S., Suuronen, R., & Hyttinen, J. (2009). Bioimpedance measurement setup for the assessment of viability and number of human adipose stem cells cultured as monolayers. In: *World Congress on Medical Physics and Biomedical Engineering*, September 7–12, 2009. Munich, Germany: Springer.

Keese, C. R., & Giaever, I. (1994). A biosensor that monitors cell morphology with electrical fields. *IEEE Engineering in Medicine and Biology, 13*, 402–408.

Kell, D. B., Kaprelyants, A. S., Weichart, D. H., Harwood, C. R., & Barer, M. R. (1998). Viability and activity in readily culturable bacteria: A review and discussion of the practical issues. *Antonie Van Leeuwenhoek, 73*, 169.

K'Owino, I. O., & Sadik, O. A. (2005). Impedance spectroscopy: A powerful tool for rapid biomolecular screening and cell culture monitoring. *Electroanalysis, 17*(23), 2101–2113.

Markx, G. H. (2008). The use of electric fields in tissue engineering: A review. *Organogenesis, 4*(1), 11–17.

Moulton, S. E., Barisci, J. N., Bath, A., Stella, R., & Wallace, G. G. (2004). Studies of double layer capacitance and electron transfer at a gold electrode exposed to protein solutions. *Electrochimica Acta, 49*(24), 4223–4230.

Robert, L., Langer, R., & Vacanti, J. P. (Eds.). (2011). *Principles of tissue engineering*. Burlington, MA: Academic Press.

Sarró, E., Lecina, M., Fontova, A., Solà, C., Gòdia, F., Cairó, J. J., et al. (2012). Electrical impedance spectroscopy measurements using a four-electrode configuration improve on-line monitoring of cell concentration in adherent animal cell cultures. *Biosensors and Bioelectronics, 31*(1), 257–263.

Sarró, E., Lecina, M., Fontova, A., Gòdia, F., Bragós, R., & Cairó, J. J. (2016). Real-time and on-line monitoring of morphological cell parameters using electrical impedance spectroscopy measurements. *Journal of Chemical Technology and Biotechnology, 91*, 1755–1762.

Wright, J. E. I., Cosman, N. P., Fatih, K., Omanovic, S., & Roscoe, S. G. (2004). Electrochemical impedance spectroscopy and quartz crystal nanobalance (EQCN) studies of insulin adsorption on Pt. *Journal of Electroanalytical Chemistry, 564*, 185–197.

Chapter 7
Basics of Numerical Simulations of Bioimpedance Phenomena

Dejan Križaj

7.1 Introduction

This chapter describes and explains the use of numerical simulations of bioimpedance phenomena based on numerically solving differential equations derived from fundamental laws of electromagnetism on the domain of interest. Due to limited chapter space, we will concentrate on some important aspects of numerical simulations that should be understood in order to use and profit from numerical simulations. In particular, one should understand the basic equations that are solved as well as their limitations. These limitations can be due to several factors: simulation limited to two dimensions, accuracy of the solution (convergence), accuracy due to the limited number of discretized mesh points, limitations of the simulation area, knowledge of electrical properties of materials (most often tissues) or models describing them, frequency-dependent limitations, etc. All these limitations can lead to a disappointment to those expecting to find a perfect fit between the simulation results and the experimental work. The goal of numerical simulation is not necessarily limited to finding a good fit between the experiment and simulation-often this fit can be obtained by tuning of some geometrical or material parameters. The main goal of numerical simulation should be better understanding of the electrical phenomena influencing bioimpedance determination such as influence of geometrical changes (electrode dimensions, positioning, changes of tissue areas, volumes) as well as influences related to changes of electrical parameters that can arise from several factors (temperature, biochemical processes, etc.). Results of better understanding should lead to modifications of electrode positioning, analysis of detection sensitivity, improvements in instrumentation, etc.

D. Križaj (✉)
Laboratory for Bioelectromagnetics, Faculty of Electrical Engineering, University of Ljubljana, Ljubljana, Slovenia
e-mail: dejan.krizaj@fe.uni-lj.si; http://lbm.fe.uni-lj.si

F. Simini, P. Bertemes-Filho (eds.), *Bioimpedance in Biomedical Applications and Research*, https://doi.org/10.1007/978-3-319-74388-2_7

7.2 Current Conduction

In the analysis of current flow inside the tissues (or other materials), we use a concept of current density J which is defined as current per unit area of a cross section, so the total current through a cross section is obtained by the integration of a normal component of current density over the cross section: $I = \int_A \vec{J} \cdot d\vec{A}$. The current density is a product of the velocity of the charges multiplied by the volume charge density: $\vec{J} = \rho\vec{v}$. Free charges move in a media (tissues) driven by an applied electric field (voltage) with an average velocity which is proportional to the electric field $\vec{v} = \mu\vec{E}$, where μ is known as mobility. The concept of mobility is more often used in the analysis of conduction in electrolytes, while in solid materials (as well as tissues), it is more convenient to use specific conductivity σ which is a product of mobility and charge density. This gives the expression for conduction current

$$\vec{J}_c = \sigma\vec{E} \tag{7.1}$$

which is also known as Ohm's law. For those less familiar with field quantities, this law can be regarded as written in a differential form (valid in a "point") while a more well-known integral form (determined between two separate conductors—electrodes) would be $I = GU$. I is of course the electric current in amperes, G conductivity in Siemens ($S = 1/\Omega$), and U the voltage between the two electrodes. Even more frequently, it is written as $U = RI$, where R is the resistance between the electrodes ($R = 1/G$). This equation is in fact close to the essence of simulation of bioimpedance phenomena only that it should be extended to resistance at harmonic (sinusoidal shape) excitation signals. In a DC (direct current) case, we are interested in the evaluation of the resistance between two electrodes, while in AC (alternative current) case, we evaluate complex resistance—impedance. Maybe at this point, it should be noted that Ohm's law cannot be regarded (is not) a fundamental law of electromagnetics such as Gauss law, Ampere's law, etc. In analysis of bioimpedance phenomena (or even impedance, not necessary related to the biomedical field), we assume the relation between current and voltage to be linear, that is, twice larger current results in two times larger voltage, etc. It is important to note that if this is not the case, (bio)impedance concept is not valid. We may though find a range of magnitudes in which the relations could be regarded as linear.

Specific conductivity is often frequency-dependent. In particular the electrical parameters of tissues are strongly frequency-dependent as they represent an average effect of conduction through the tissues composed of various types of cells with very different properties.

Figure 7.1 presents a variation of specific conductivity of some selected tissues depending on the frequency of the stimulation signal. The values were obtained from the work of Gabriel et al. (1996) which is most often considered in numerical simulations. More practically one can use the data from the web page http://niremf.ifac.cnr.it/tissprop/ which enables use and/or download of frequency-dependent

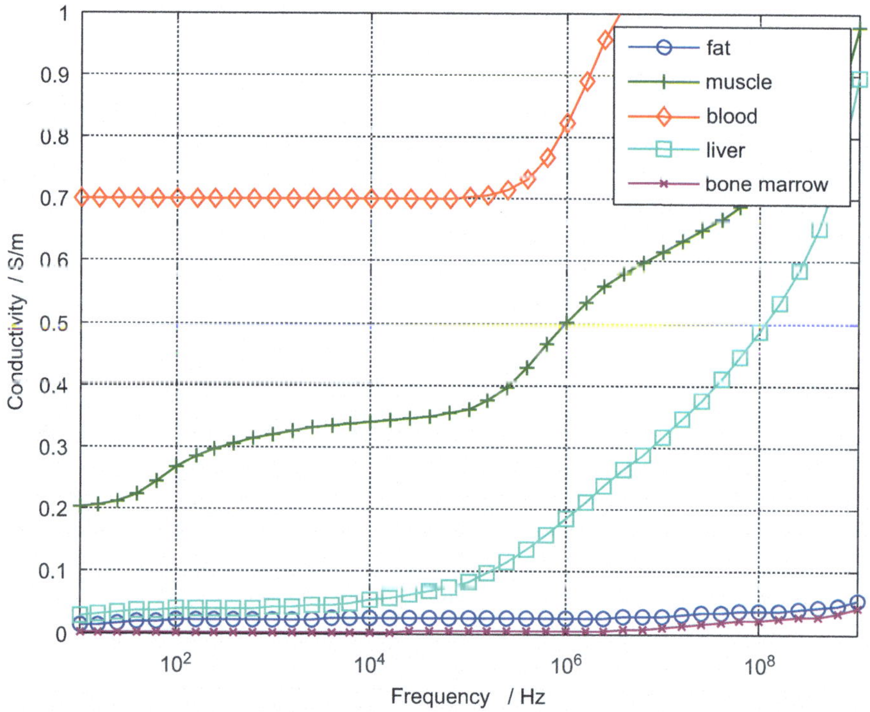

Fig. 7.1 Specific conductivities of selected tissues

material parameters for many tissues in various data formats. Typically, the values of specific conductivities of tissues range from around 1 mS/m for the dry skin up to 1 S/m for the body fluid. For comparison, good conductors (metals) have conductivities in the range of MS/m, drinking water in the range of mS/m, and deionized water in the range of uS/m). Dry skin is very weakly conductive and also in the range of mS/m which poses a problem in bioimpedance measurements. For this reason (and also reasons of safety), the bioimpedance stimulation signal is most often generated by a current source. Some tissues such as fat and bone have less frequency-dependent conductivity (see figure), while muscle, the liver, and blood express a more pronounced increase of conductivity with increasing frequency. This can be attributed to the fact that at higher frequencies the current flows more easily also through the cells due to the capacitive type current conduction explained later in the chapter. Some tissues (for instance, skeletal muscle) are known to have anisotropic properties, that is, their conductivity is direction-dependent. Hart et al. (1999) measured two to three times larger conductivity in longitudinal than transversal direction at 10 kHz, while some other studies (for instance, Epstein and Foster 1983) report even larger five to seven times differences. The anisotropic properties diminish with increasing frequency.

Beside conductive properties the tissues are characterized also by their dielectric properties. They are a consequence of the polarization phenomena. Some materials are intrinsically polarized even if they are not subjected to external electrical forces. For instance, water molecule is polar and has a so-called dipole moment characterized by intrinsic separation of charge inside the molecule. Other molecules get polarized when subjected to an external electric field. Usually it is assumed that their dipole moments increase linearly with the electric field by a factor χ called dielectric susceptibility. A differential volume average of dipole moments is called polarization vector P. In order to take into account the effect of free as well as polarized (bound) charges, a new quantity has been devised by James C. Maxwell called electric flux density (also electric displacement field) D: $\vec{D} = \varepsilon_0\vec{E} + \vec{P}$. If we assume polarization to be proportional to electric field $\vec{P} = \chi\varepsilon_0\vec{E}$ (which most often is assumed to be valid also for tissues), one has $\vec{D} = (1 + \chi)\,\varepsilon_0\vec{E} = \varepsilon_r\varepsilon_0\vec{E}$, where ε_r is the relative dielectric constant also called relative permittivity. In contrast to conductivities that vary in a relatively narrow frequency range, dielectric constants may vary for many orders of magnitude as shown in Fig. 7.2. In particular, very high dielectric constants are typically observed at very low frequencies. These very high values especially compared to common dielectrics such as plastic, glass, and

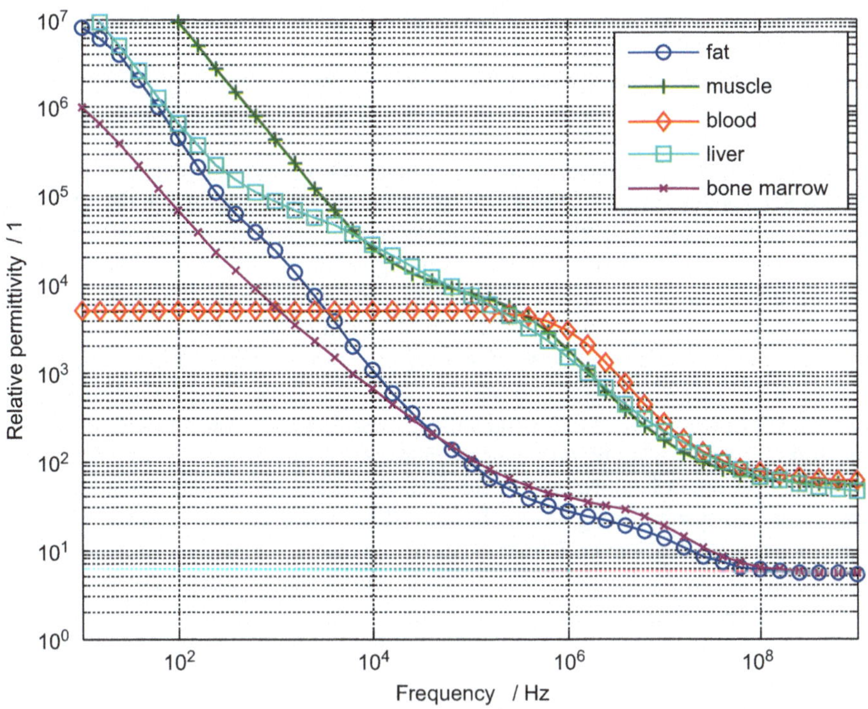

Fig. 7.2 Relative permittivity of selected tissues

silicon (which have relative dielectric constant around 10) are due to very large dipole moments that arise due to the diffusion processes of the ions which is often called alpha (α) dispersion. In the medium-frequency range (kHz and MHz region), the dispersion is due to the interaction of inter- and extracellular space with the cell membrane which is referred as Maxwell-Wagner or beta (β) dispersion. In GHz range, we have gamma (γ) dispersion related to the relaxation of water molecules. The terms α, β, and γ dispersion were coined by Schwan and Kay (1957). More detailed explanation can be found in Grimnes and Martinsen (2014).

Current density in dielectrics (displacement current J_{disp}) depends on the time change of electric flux density

$$\vec{J}_{\text{disp}} = \frac{\partial \vec{D}}{\partial t} = \varepsilon_r \varepsilon_0 \frac{\partial \vec{E}}{\partial t}. \tag{7.2}$$

For harmonic variation of electric field as $\vec{E}(t) = \vec{E}_o \cos(\omega t)$, the magnitude of time varying current density increases linearly with the increase of the excitation frequency as $\vec{J}_{\text{disp}} = \varepsilon_r \varepsilon_0 \omega \vec{E}_o \sin(\omega t)$. As we are interested in tissue response to harmonic signals, it is more customary to express this relation in a complex notation (with phasors). In this case the current density (as well as electric field) is a complex vector expressed as

$$\vec{\underline{J}} = \sigma \vec{\underline{E}} + j\varepsilon_r \varepsilon_0 \omega \vec{\underline{E}}. \tag{7.3}$$

We notice two important properties: (1) the ohmic and dielectric parts are phase shifted by a quarter of the period, and (2) the conductive part prevails if $\sigma > \varepsilon_r \varepsilon_0 \omega$. The ratio $\sigma / \omega \varepsilon_r \varepsilon_0$ also known as loss tangent for selected tissues is presented in Fig. 7.3 in linear as well as logarithmic scale of frequencies. It is interesting to note that this ratio is larger than 1 in a wide frequency range, meaning that ohmic conduction is usually prevailing over the dielectric one, in particular for frequencies below several MHz and above about 100 Hz.

In the literature, sometimes also the electrical parameters of tissues are described as phasors. Following Eq. (7.3), the current density can be written with complex conductivity as

$$\vec{\underline{J}} = (\sigma + j\varepsilon_r \varepsilon_0 \omega)\, \vec{\underline{E}} = \underline{\sigma} \vec{\underline{E}} \tag{7.4}$$

or equivalently with complex dielectric constant as

$$\vec{\underline{J}} = j\omega\varepsilon_0 \left(\varepsilon_r - j\frac{\sigma}{\omega \varepsilon_0} \right) \vec{\underline{E}} = j\omega\varepsilon_0 \left(\varepsilon' - j\varepsilon'' \right) \vec{\underline{E}} = j\omega\varepsilon_0 \underline{\varepsilon} \vec{\underline{E}}. \tag{7.5}$$

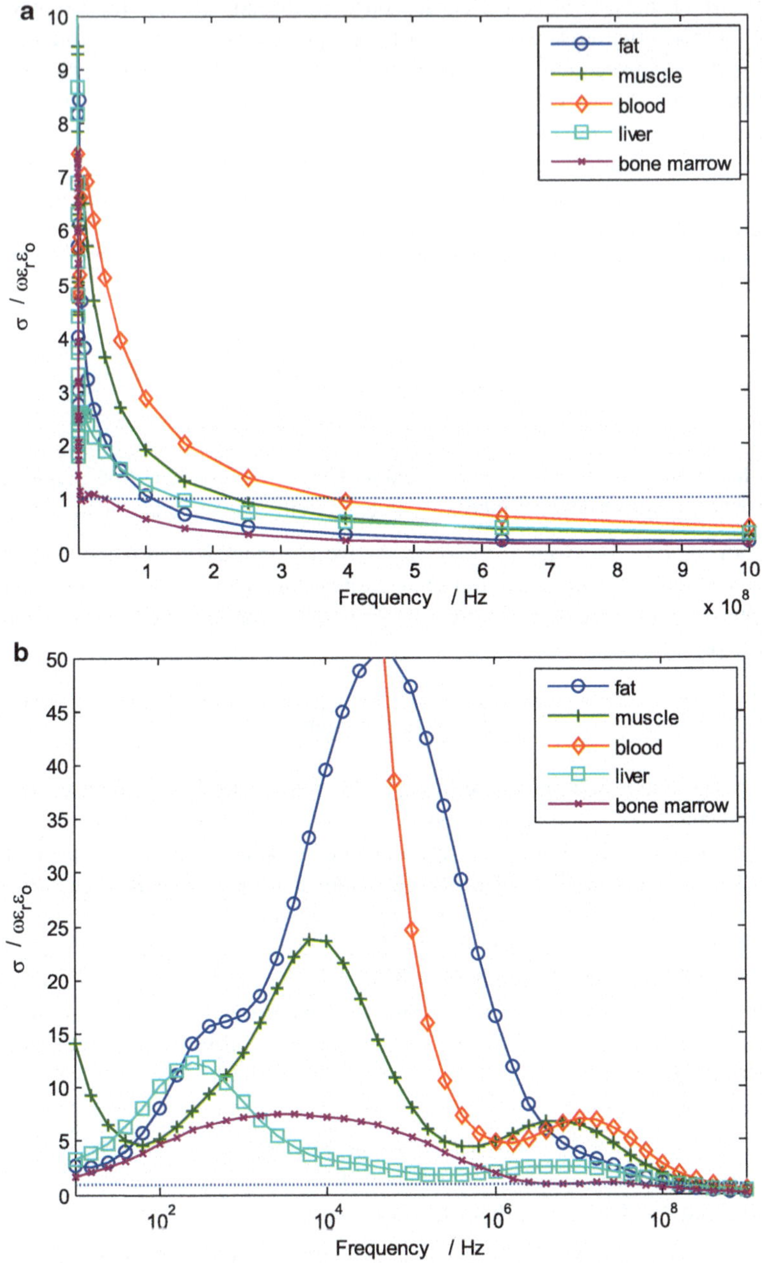

Fig. 7.3 Loss tangent—ratio of conductive to dielectric current (**a**) linear scale (**b**) logarithmic scale

7.3 Electrode-Tissue Interface

The electrode-tissue interface conduction phenomena are complex and require special attention. In general one needs to take into account the properties of the so-called Helmholtz double layer that can be modeled with a Helmholtz-Gouy-Chapman-Stern electrical lumped model presented in Fig. 7.4 (Yúfera et al. 2011). Interface capacitance C_I is modeled as a double layer capacitance composed of a series connection of a very thin Helmholtz layer and a Gouy-Chapman capacitance taking into account of the diffusive layer. Z_W is the Warburg impedance related to the mass diffusion process occurring in the electrode-electrolyte interface and can be modeled as $\underline{Z}_W = \frac{\omega^{-1/2} K_W}{A_e(1+j)}$, where A_e is the electrode area and K_w is a constant determined by the electrochemistry and mobility of the ions involved in the charge transfer reaction. Charge transfer resistance R_{ct} is determined by the electron transfer rate at the interface. R_s is the spreading resistance taking into account the spreading of the current from the electrode. If spreading resistance is not taken into account, the current density is much more densified (increased) to a limited area of the electrode where most current is escaping/entering the electrode. It is not straightforward to obtain the correct data for the specific simulation case. Yúfera et al. (2011) modeled double layer as a 0.5 μm layer with equation

$$\underline{\sigma}_{dl} = \sigma_{dl} + j2\pi f \varepsilon_{dl} = t\left[\frac{(2\pi f)^{0.5}}{K_w} + j\frac{(2\pi f)^{0.5}}{K_w} + jC_I 2\pi f\right] \qquad (7.6)$$

It should be noted, however, that whatever complex the model for electrode impedance may be, it is eventually described by two complex numbers (for real and imaginary part of the impedance) for a selected frequency.

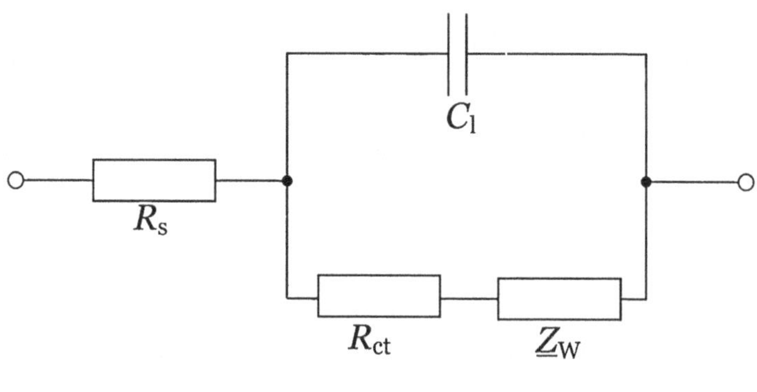

Fig. 7.4 A model of electrode/tissue interface composed of a capacitor, resistors, and a Warburg impedance

7.4 Numerical Simulations: Equations

Numerical simulation enables simulation of electrical (and in some cases also other physical) phenomena that are based on fundamental physical laws. These simulation models can range from a whole (human) body or just a part of a tissue or even some cell structures or a single cell. Bioimpedance is dealing with determination and evaluation of electrical impedance between two electrodes or more often between four electrodes where two electrodes are used for current injection and the other two for voltage determination. As the determined voltage is in this case not the same as the voltage between the current injection electrodes, the bioimpedance can be regarded as transfer impedance.

In numerical simulations of bioimpedance, we are most often interested in observations of current flow (current density) inside the simulated structure. We may also be interested in potential distribution (equipotential lines), electric field density, power density, etc. The most straightforward way to the development of suitable equations that need to be solved is by taking into account the current continuity equation stating that the total current flow through a closed region (volume) is zero unless the electric charges vary in time within the region:

$$\oint_A \vec{J} \cdot \mathrm{d}\vec{A} = -\int_V \frac{\partial \rho}{\partial t} \mathrm{d}V. \tag{7.7}$$

This equation is more conveniently expressed in a differential form as

$$\nabla \cdot \vec{J} = -\frac{\partial \rho}{\partial t}. \tag{7.8}$$

In case we neglect the time retardation effects, the right-hand term may be neglected (see Plonsey 1982)

$$\nabla \cdot \vec{J} = 0. \tag{7.9}$$

Such field is called quasi-static as it can be calculated by neglecting the time propagation of an electromagnetic field. By inserting the current density in complex notation (7.3) in (7.9), we obtain

$$\vec{\nabla} (\sigma + j\varepsilon_r\varepsilon_0) \underline{\vec{E}} = 0. \tag{7.10}$$

In a quasi-static case, the electric field can be related to the gradient of potential as $\vec{E} = -\vec{\nabla} V$ and the Eq. (7.10) may be written in a form that is most usually used for the numerical simulation of bioimpedance phenomena

$$\vec{\nabla} (\sigma + j\varepsilon_r\varepsilon_0) \vec{\nabla} \underline{V} = 0. \tag{7.11}$$

This equation needs to be solved in a specified region (domain) for complex potential $\underline{V}$ by taking into account appropriate electrical properties of the tissues and boundary conditions. Once the potential distribution in a domain is determined, all other quantities of interest (electric field, current densities, etc.) can be calculated. In case the electrode boundary conditions are fixed potential values, the current through the electrode is obtained by the integration of (a normal component) current density along the electrode interface.

7.4.1 *Influence of Magnetic Effects*

So far we have avoided the influence of magnetic effects on electric field and thus current conduction. We however know that electric current causes magnetic field or more exactly magnetic flux density B and also that the time-varying magnetic flux density induces an electric field which further influences the total current density. Due to a lack of space, we try to avoid description of a full set of Maxwell's equations but only discuss the relations that are relevant for practical considerations of numerical simulations of bioimpedance phenomena. There are several approaches to investigate the influence of magnetic fields. One of them is to consider the modification of electric field due to magnetic effects through the introduction of a magnetic vector potential A which is related to (rotor of) magnetic flux density through $\vec{B} = \vec{\nabla} \times \vec{A}$. By taking into account the magnetic (induction) effects, the electric field attains an additional term which is time-dependent

$$\vec{E} = -\vec{\nabla} V - \frac{\partial \vec{A}}{\partial t} \tag{7.12}$$

and is in complex notation for harmonic fields written as

$$\vec{\underline{E}} = -\vec{\nabla} \underline{V} - j\omega \vec{\underline{A}}. \tag{7.13}$$

Magnetic flux density and magnetic vector potential may be calculated directly by integrating current flux density as $\vec{A} = \frac{\mu_0}{4\pi} \int_v \frac{\vec{J}}{r} \mathrm{d}v$.

Now we have to make an estimate of the magnitude of time variation of magnetic vector potential with respect to other contributions to the electric field/current density. If we assume a case of a homogeneous current flow through a sphere of radius R (structure dimensions) and estimate the derivative of magnetic potential with the magnitude of a current density, we get

$$\left| \frac{\partial \vec{A}}{\partial t} \right| = \left| \frac{\partial}{\partial t} \left(\frac{\mu_0}{4\pi} \int_v \frac{\vec{J}}{r} \mathrm{d}v \right) \right| < \frac{\partial}{\partial t} \left(\frac{\mu_0 \left| \vec{J} \right|}{4\pi} \int_0^R \frac{4\pi r^2 \mathrm{d}r}{r} \right) = \mu_0 \frac{R^2}{2} \left| \frac{\partial \vec{J}}{\partial t} \right|. \tag{7.14}$$

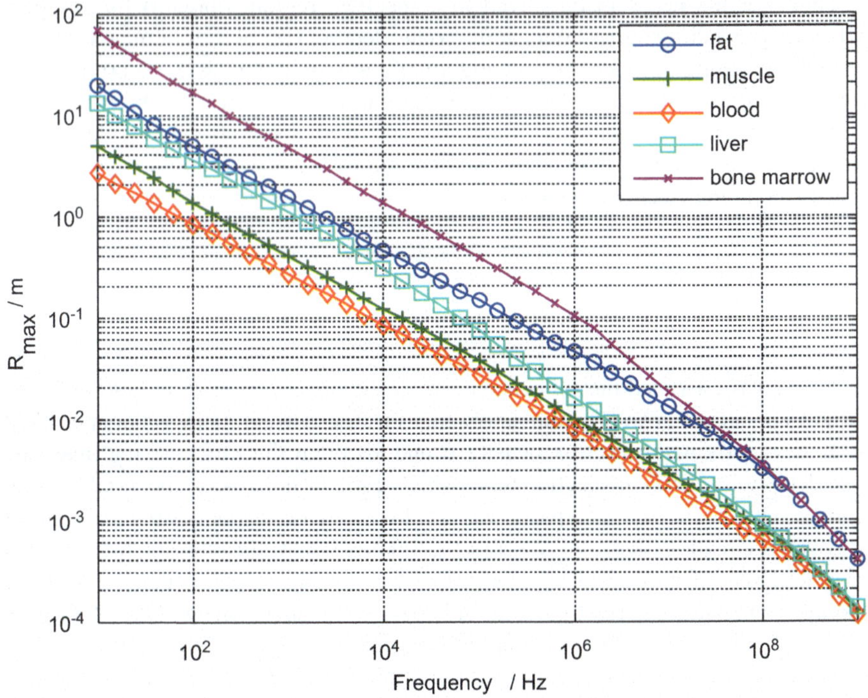

Fig. 7.5 Maximal permissible structure for validity of neglection of magnetic induction effects

Now we insert this result in Eq. (7.7) and then to Eq. (7.3), and we arrive to

$$\left|\underline{\vec{J}}\right| < (\sigma + j\omega\varepsilon_r\varepsilon_0)\left(|-\nabla\underline{V}| + \left|-\frac{j\omega\mu_0 R^2}{2}\underline{\vec{J}}\right|\right). \tag{7.15}$$

By rearrangement of the above equation, the last term (effect of the induced current density) can be neglected if $\left|(\sigma + j\omega\varepsilon_0\varepsilon_r)\, j\omega\mu_0\frac{R^2}{2}\right| \ll 1$. If we set the conservative value of 1%, then the maximal dimensions of the structure (R_{max}) in which the induction effects can be neglected is presented in Fig. 7.5 for some typical tissues.

Let us, for instance, assume we want to perform simulations up to a frequency of 1 MHz. In case our simulation includes tissues such as blood, the largest simulation domain in which inductive effects can be neglected is about 8 mm and less than 1 mm for frequencies higher than 100 MHz.

A similar expression can be obtained by considering the so-called skin effect known to occur in conductors at higher frequencies resulting in current conduction limited to the surface (skin) of the conductor. It should however be noted that the skin effect equation is often simplified by neglecting the dielectric term as specific conductivity of conductors is large and permittivity small. As shown in Fig. 7.4, in tissues the dielectric conduction is comparable to ohmic one in a wide frequency range and is even dominant at higher frequencies (>100 MHz), so it could (should) not be neglected.

7.4.2 Induced Current Electrical Impedance Tomography

In induced current electrical impedance tomography, the measured surface potentials are a consequence of the induced currents resulting from the currents in the current carrying coils placed outside the measured tissues. An advantage of this approach is that the method does not require injection electrodes as the currents develop in the tissues as a consequence of magnetic induction phenomenon. In this case, one needs to evaluate the magnitude of the induced electric field, taking into account the time derivative of magnetic vector potential. It is possible to extend Eq. (7.11) also to this type of problems by decoupling the magnetic and electric phenomena (Zlochiver et al. 2002, 2004). We do this by taking into account the influence of induced magnetic fields by the magnetic vector potential term in Eq. (7.13):

$$\vec{\nabla} \cdot (\sigma + j\omega\varepsilon)\left(\vec{\nabla}\underline{V} + j\omega\vec{\underline{A}}\right) = 0. \tag{7.16}$$

The magnetic vector potential distribution may be obtained directly from the known currents in the coil by the integral

$$\vec{A} = \frac{\mu_0 I}{4\pi}\oint_L \frac{\mathrm{d}\vec{l}}{R}, \tag{7.17}$$

where dl is the length element of the coil, I is the current in the coil, and R is the distance from the length element to a point where A is calculated. A more complex approach is to calculate A by solving another differential equation for A

$$\nabla^2\vec{\underline{A}} = \mu_0(\sigma + j\omega\varepsilon)\left(\nabla\underline{V} + j\omega\vec{\underline{A}}\right). \tag{7.18}$$

7.5 Simulations in Practice: A FEMM Case

There are several simulation packages that can be used for modeling bioimpedance phenomena. They are all based on the discretization of equations from the previous chapter but can use very different discretization and numerical solving approaches.

Professional software most often used in the simulation of bioimpedance phenomena are Comsol and Ansoft. In this chapter, we will however give a practical example by using the software FEMM which is freely available on the internet. FEMM stands for Finite Element Method Magnetics (www.femm.info). It was originally developed (by David Meeker) for simulation of magnetic structures, but it can be used also for simulation of electrostatic, current conduction and heat flow problems (in all cases, it solves a variant of Laplace/Poisson equation). An advantage of FEMM (besides it being free) is its simplicity of use, fast meshing and solving, and possibility to use scripting language LUA as well as Octave or Matlab (and even Mathematica). A disadvantage is that it is limited to two-dimensional (2D) problems and it does not enable multiphysics simulations—simulations of more physical phenomena such as current conduction with temperature dependence.

The starting point of every simulation is construction of a simplified model preferably in two dimensions. In this case we do not concentrate on geometrical difficulties, meshing problems, etc. but only on used physics and parameters. Figure 7.6 presents a starting point for FEMM simulation with Menus for problem definition (Problem), geometry setup (Operation), setting of material parameters (Properties), meshing (Mesh), and running simulations (Analysis). Also to notice is

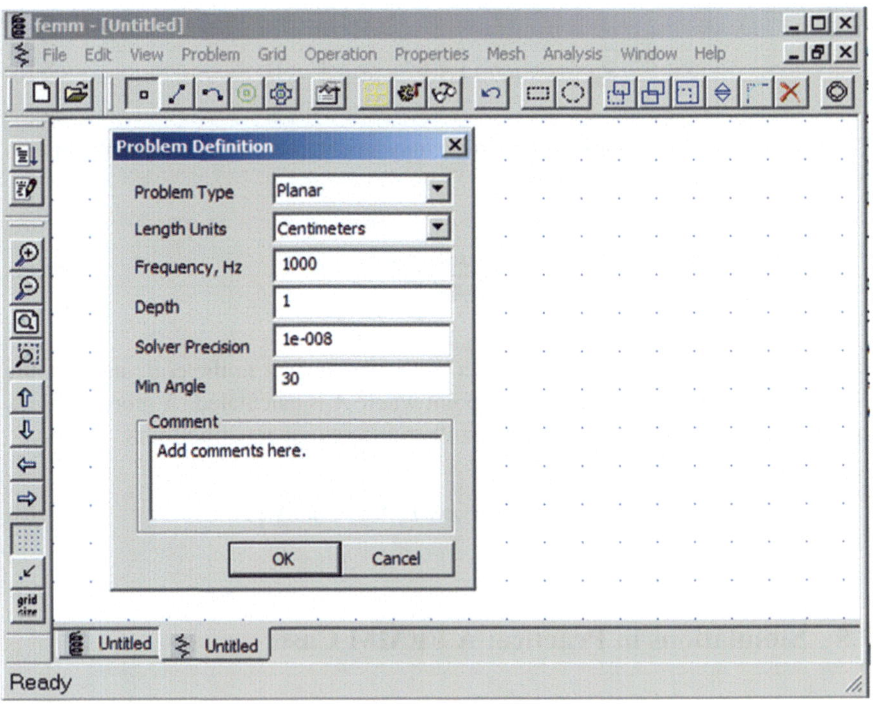

Fig. 7.6 Graphical interface of a FEMM language enables simple determination of geometry, material parameters, boundary conditions, meshing, and post-processing

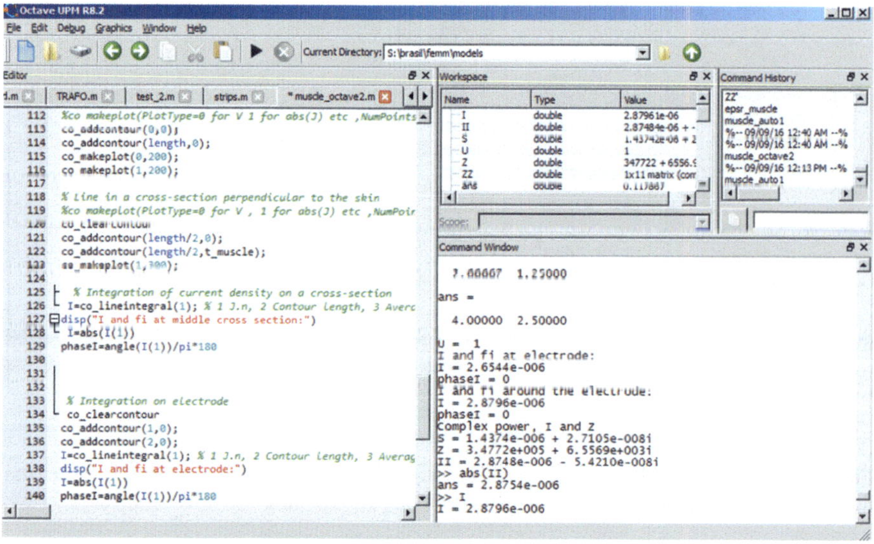

Fig. 7.7 Octave interface with FEMM scripting language for rapid parametric model evaluation

a possibility to import geometry in a DXF format which is a well-known format in CAD (Autocad/Autodesk). Furthermore, FEMM can be run directly from Matlab or Octave (Fig. 7.7) which gives the user another powerful tool for scripting the simulation runs, and particularly, it gives possibilities for advanced post-processing of the simulation results. LUA and Matlab/Octave syntax is very similar.

A simplified cross section of a (meshed) bioimpedance simulation case is presented in Fig. 7.8. It consists of three regions (skin, fat, muscle). The skin layer is in reality much thinner, but here it is made thicker for demonstration purposes. Tissue properties were obtained from the database http://niremf.ifac.cnr.it/tissprop/ by considering the parametrized properties of selected tissues. The boundary condition of fixed voltage (also called Dirichlet boundary condition) of 0 and 1 volt is set at the electrodes while a so-called Neumann boundary condition is automatically set on the rest of the boundaries. This condition assumes normal derivative of potential zero at boundary $\left(\frac{\partial V}{\partial n} = 0\right)$ which should be taken with care as this condition will result in zero normal component of electric field (and also current density) at these boundaries. This assumption may be valid if the currents (electric field) are much smaller at these boundaries compared to the inside region. As we usually have no better alternative, we design the simulation in such a way that this boundary condition does not result in too big errors. For simulations in two dimensions, this is not such a problem as the simulations usually run fast, while it can be more frustrating for 3D simulations where large simulation areas also require large meshes. This "problem" can be reduced to some extent by using a less dense mesh further away from the areas where most of the current is flowing.

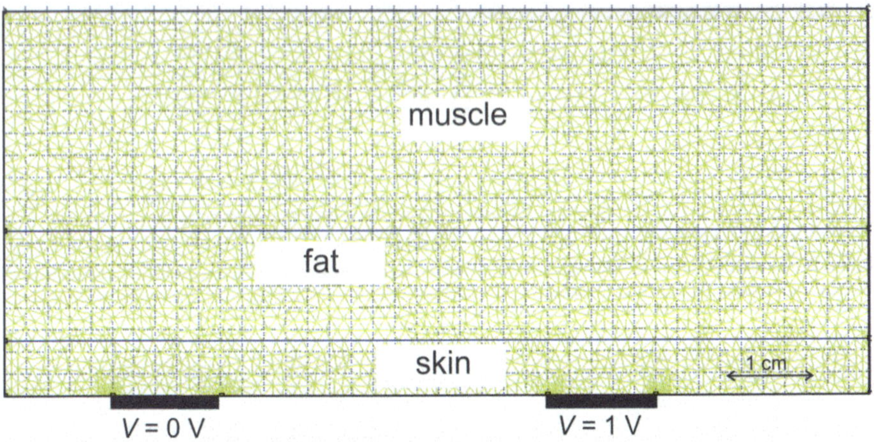

Fig. 7.8 Simplified simulation structure of a 2D cross section of a skin with muscle layer. The thickness of the skin is exaggerated for demonstration purposes only

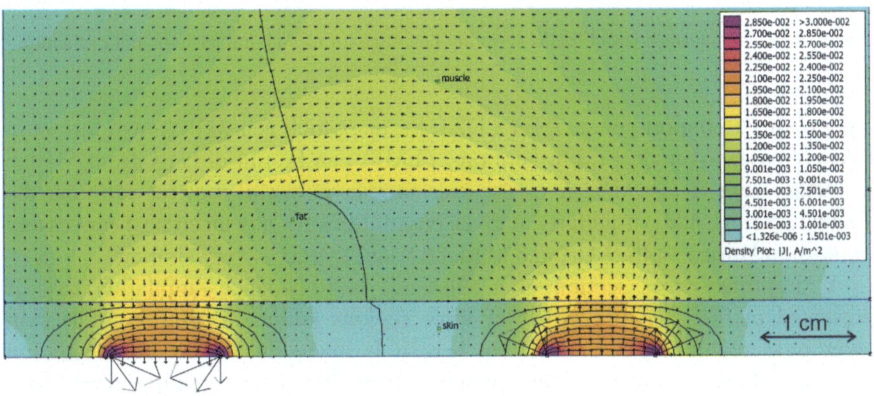

Fig. 7.9 Simulation result: equipotential lines (black lines) with absolute values of current densities (in colors) and current density vectors

Once the potential distribution is obtained by some sort of numerical methods, other quantities of interest may be calculated and visualized. Usually we plot equipotential lines as presented in Fig. 7.9 together with some other quantities such as magnitude of current density and/or current flow vectors. From Fig. 7.9 we notice that equipotential lines are very densely packed close to the electrodes indicating large fields and also current densities in this region. Furthermore, we notice that current flow is predominantly flowing through the fat and the muscle (layer) due to a much larger resistivity of the skin layer—in this model, we assumed dry skin electrical parameters. More exact simulation of skin bioimpedance is given, for instance, by Birgersson et al. (2011).

7.5.1 Calculation of (Bio)Impedance

Total current through the electrodes may be obtained by integration of a normal component of current density along the electrode $\underline{I} = \int_A \underline{\vec{J}} \cdot \mathrm{d}\vec{A}$. Bioimpedance is defined as a ratio of voltage and current complexors $\underline{Z} = \frac{\underline{U}}{\underline{I}}$, where $\underline{U}$ is the voltage between the voltage measurement electrodes. In two-electrode method, they are the same as the applied voltages, while in four-electrode bioimpedance simulation, this voltage is obtained by the difference of potentials at two points where voltage picking electrodes are put. In case we want to simulate also the effect of the area of the voltage electrode (and not only as point electrodes), they need to be specified as a boundary with a boundary condition of prescribed zero current.

The total current through the electrode can be calculated also over some other arbitrary cross section between the electrodes. This approach is even usually advantageous as at the electrode edges the current density is very high and the mesh may not be dense enough to assure high accuracy at numerical integration. For instance, in our case we obtain a current of 2.65 μA/m (per meter due to 2D simulation) by integration along the electrode and 2.879 μA/m through a cross section close to the electrode. Which value is more exact? Fortunately we have another possibility of determination of electrode current (and impedance)—through a concept of power. Power density may be calculated by multiplication of current density and electric field vectors as $\underline{p} = \underline{\vec{J}} \cdot \underline{\vec{E}}$. By integration of power density throughout the volume, we obtain apparent power $\underline{S} = \int_V \underline{p} \mathrm{d}V$. At the same time this power is (in complex notation) related to current and voltage phasors (and impedance) through

$$\underline{S} = \frac{1}{2}\underline{U}\underline{I}^* = \frac{1}{2}U^2/\underline{Z}^* \tag{7.19}$$

where * denotes conjugation. (If conjugation is not taken into account, one does not get the correct sign of the current and consequently the phase.) Calculation of current or impedance through calculation of apparent power should yield more exact results as it is "averaged" over the whole simulation domain. In our case by calculating current through (7.19), we get a current of 2.875 μA/m which is only a difference of 0.2% compared to the current in a cross section close to the electrode and about 12% difference to the surface integration. Eq. 7.19 can be though used to calculated (bio)impedance only for the two electrode system and not the three or four electrode system.

Interested reader can access the files for simulation at http://tinyurl.com/biabook-muscle.

7.6 Conclusion

Numerical simulation of bioimpedance phenomena is a powerful tool that can be used to verify or improve new concepts of bioimpedance measurement method, assessing the effects of electrode positioning, influence of material parameters, geometry, etc. Usually—for frequencies below several MHz—it is sufficient to solve complex Poisson equation on a suitably chosen domain. Two-dimensional simulations can be effectively performed using free software such as FEMM as presented briefly in this chapter.

References

Birgersson, U., Birgersson, E., Aberg, P., Nicander, I., & Ollmar, S. (2011). Non-invasive bioimpedance of intact skin: Mathematical modeling and experiments. *Physiological Measurement, 32*, 1–18.

Epstein, B. R., & Foster, K. R. (1983). Anisotropy in the dielectric properties of skeletal muscle. *Medical & Biological Engineering & Computing, 21*, 51–55.

Gabriel, C., Gabriel, S., & Corthout, E. (1996). The dielectric properties of biological tissues: I. Literature survey. *Technology, 41*, 2231–2249.

Grimnes, S., & Martinsen, O. G. (2014). *Bioimpedance and bioelectricity basics*. Cambridge: Academic Press.

Hart, F. X., Berner, N. J., & McMillen, R. L. (1999). Modelling the anisotropic electrical properties of skeletal muscle. *Physics in Medicine and Biology, 44*, 413–421.

Plonsey, R. (1982). The nature of sources of bioelectric and biomagnetic fields. *Biophysical Journal, 39*, 309–312.

Schwan, H. P., & Kay, C. F. (1957). The conductivity of living tissues. *Annals of the New York Academy of Sciences, 65*, 1007–1013.

Yúfera, A., Olmo, A., Daza, P., & Cañete, D. (2011). Cell biometrics based on bio-impedance measurements. In G. Chetty (Ed.), *Advanced biometric technologies*. Rijeka: InTech. https://doi.org/10.5772/21742. Available also from: http://www.intechopen.com/books/advanced-biometric-technologies/cell-biometrics-based-on-bio-impedance-measurements.

Zlochiver, S., Radai, M. M., Abboud, S., Rosenfeld, M., Dong, X. Z., Liu, R. G., et al. (2004). Induced current electrical impedance tomography system: Experimental results and numerical simulations. *Physiological Measurement, 25*, 239–255.

Zlochiver, S., Radai, M. M., Rosenfeld, M., & Abboud, S. (2002). Induced current impedance technique for monitoring brain cryosurgery in a two-dimensional model of the head. *Annals of Biomedical Engineering, 30*, 1172–1180.

Chapter 8
Numerical Basics of Bioimpedance Measurements

Alexander Danilov, Sergey Rudnev, and Yuri Vassilevski

8.1 Introduction

Bioimpedance measurements estimate passive electric properties of biological tissues in response to application of a low-level alternating electric current (see Grimnes and Martinsen 2014). The measurements form a basis for a number of safe, noninvasive, portable, and relatively low-cost techniques of health monitoring, such as impedance cardiography (ICG) (Cybulski 2011), impedance plethysmography (IPG) (Jindal et al. 2016), electrical impedance tomography (EIT) (Holder 2005), and bioimpedance analysis of body composition and spectroscopy (BIA, BIS) (Jaffrin and Morel 2008; Bera 2014).

In BIA and BIS, simple representation of the human body as a homogeneous isotropic cylindrical conductor is used to assess body composition and body fluids (Jaffrin and Morel 2008; Brantlov et al. 2016). A few contact electrodes (usually four or eight) are utilized to obtain whole body, segmental body, or local body characteristics from measured transfer impedance at single or multiple current frequencies. Similarly, a small number of electrodes (typically, four band or eight spot ones) are used in measurements of the transfer impedances for the assessment

A. Danilov (✉) · Y. Vassilevski
Institute of Numerical Mathematics, Russian Academy of Sciences, Moscow, Russia

Moscow Institute of Physics and Technology, Moscow, Russia

Sechenov University, Moscow, Russia

S. Rudnev
Institute of Numerical Mathematics, Russian Academy of Sciences, Moscow, Russia

Lomonosov Moscow State University, Moscow, Russia

Federal Research Institute for Health Organization and Informatics, Moscow, Russia

F. Simini, P. Bertemes-Filho (eds.), *Bioimpedance in Biomedical Applications and Research*, https://doi.org/10.1007/978-3-319-74388-2_8

of central hemodynamics in ICG (Cybulski 2011; Bera 2014). The same applies for the evaluation of vascular function in IPG (Jindal et al. 2016).

The human body is highly inhomogeneous anisotropic structure. The most promising way of substantiation and further development of the above methods, as well as accurate data interpretation, is accounting this inhomogeneity by using high-resolution mathematical models of the human body. Such models are used to identify the source of bioimpedance signal and proper electrode placement in ICG (Kauppinen et al. 1998; Belalcazar and Patterson 2004; Patterson 2010) and to find optimal electrode configuration and frequency range for the detection of lung edema in BIS (Beckmann et al. 2007; Ulbrich et al. 2013).

In a sense, EIT is an extension of BIA and BIS to a larger number of electrodes which are needed for reconstruction of conductivity distribution in the body from measured transfer impedance data (Holder 2005). Since EIT is an imaging technique, it exploits various software tools generating anatomically correct discrete models of the body that improve the quality of reconstructed images (Bayford et al. 2001; Adler et al. 2015). The discrete model of the body constituents allows to generate a computational mesh which is the basis for application of a finite element method (FEM). The latter is appealing in bioengineering simulations since, in contrast to finite difference methods, it can use very general meshes which discretize fine anatomical structures. An alternative and competitive discretization method, the finite volume method, has weaker theoretical basis and may be more sensitive to mesh properties.

In this chapter, we present essential elements and the workflow of the FEM-based computational technology (Danilov et al. 2017) in bioimpedance modeling (Danilov et al. 2012, 2013; Vassilevski et al. 2012). The cornerstone of the technology is an anatomically correct 3D model of the human body from the Visible Human Project (VHP) (Ackerman 2003). The outline of the chapter is as follows. In Sect. 8.2, we formulate the mathematical model of electrical fields generated during bioimpedance measurements. The 3D image segmentation, adaptive mesh generation process, and finite element discretization are considered, and the convergence of the proposed numerical method is analyzed in Sect. 8.3. The traditional approach to identification of regions of high measurement sensitivity is presented in Sect. 8.4. In Sect. 8.5, we introduce an online numerical simulator of bioimpedance measurements which models the conventional 4-electrode and 10-electrode placement schemes (Danilov et al. 2017). A summary of the proposed numerical technology, its advantages, limitations, and future prospects are discussed in Sect. 8.6. Section 8.7 collects conclusive remarks.

8.2 Mathematical Model

The mathematical model of bioimpedance measurements is based on the following stationary boundary value problem (see, e.g., Danilov et al. 2012; Grimnes and Martinsen 2014):

$$\mathrm{div}(\mathbf{C}\nabla U) = 0 \quad \text{in} \quad \Omega, \tag{8.1}$$

$$(\mathbf{J}, \mathbf{n}) = I_0/S_\pm \quad \text{on} \quad \Gamma_\pm, \tag{8.2}$$

$$(\mathbf{J}, \mathbf{n}) = 0 \quad \text{on} \quad \partial\Omega\backslash\Gamma_\pm, \tag{8.3}$$

$$U(x_0, y_0, z_0) = 0, \tag{8.4}$$

$$\mathbf{J} = \mathbf{C}\nabla U. \tag{8.5}$$

In these equations Ω is the computational domain, $\partial\Omega$ is its boundary, $\Gamma_\pm$ are the electrode contact surfaces, $\mathbf{n}$ is the outward unit normal vector, U is the electric potential, $\mathbf{C}$ is the conductivity tensor allowing to describe anisotropic materials, $\mathbf{J}$ is the current density, I_0 is the electric current, and $S_\pm$ are the areas of the electrode contacts. Equation (8.1) determines the distribution of electric field in the domain with heterogeneous conductivity $\mathbf{C}$. Equation (8.2) sets the constant current density on the electrode contact surfaces. Equation (8.3) defines the no-flow condition on the boundary. Uniqueness of the solution is guaranteed by Eq. (8.4), where (x_0, y_0, z_0) is a given point inside the domain Ω.

In order to analyze the response of materials to alternating electric fields, it is convenient to replace the conductivity with a complex tensor called the *admittivity*. Admittivity is the sum of a real component called the conductivity and an imaginary component called the susceptivity. Typical values of some materials and tissues admittivity are presented in Table 8.1 (Vassilevski et al. 2012).

The conventional measurement technique involves two pairs of electrodes attached to the skin. The first pair of *current carrying* (CC) electrodes is used for injecting electrical current into the body. The second pair of *pick-up* (PU) electrodes is used for measuring the electric potential difference. This difference divided by the value of injected current is called the *transfer impedance*.

Table 8.1 Admittivity of some tissues and materials at different frequencies, Sm/m (Vassilevski et al. 2012)

	Admittivity @ 5 kHz		Admittivity @ 50 kHz	
Material	Conductivity	Susceptivity	Conductivity	Susceptivity
Skin	0.0015	0.002	0.03	0.05
Heart	0.13665	0.0356	0.19543	0.047215
Lungs	0.23484	0.0183	0.26197	0.02372
Abdominal tissues	0.10–0.12	0.0045–0.0055	0.11–0.13	0.007–0.009
Contact electrode layer	0.9	0.01	0.9	0.01
External electrode layer	10	0	10	0

8.3 Computational Technology

The approximate solution of (8.1)–(8.5) is provided by the finite element method with P_1 (continuous piecewise linear) basis functions on unstructured tetrahedral meshes. We assume that each computational element has a constant admittivity coefficient related to one of the human tissues.

We can rewrite Eq. (8.1) with complex values as a system of two equations with real values:

$$\begin{aligned} \mathrm{div}(\mathbf{C_R}\nabla U_R) - \mathrm{div}(\mathbf{C_I}\nabla U_I) &= 0 \quad \text{in} \quad \Omega, \\ \mathrm{div}(\mathbf{C_R}\nabla U_I) + \mathrm{div}(\mathbf{C_I}\nabla U_R) &= 0 \quad \text{in} \quad \Omega, \end{aligned} \tag{8.6}$$

where $\mathbf{C} = \mathbf{C_R} + i\mathbf{C_I}$ and $U = U_R + iU_I$.

The FEM discretization is based on the weak formulation of (8.2)–(8.6). One seeks U_R, U_I from the Sobolev space $W_2^1(\Omega)$, that satisfy the identities

$$\begin{aligned} &\int_\Omega \mathbf{C_R}\nabla U_R \nabla V \,\mathrm{d}\Omega - \int_\Omega \mathbf{C_I}\nabla U_I \nabla V \,\mathrm{d}\Omega - \int_{\Gamma_\pm} VI_0/S_\pm \,\mathrm{d}\Gamma_\pm = 0, \\ &\int_\Omega \mathbf{C_R}\nabla U_I \nabla V \,\mathrm{d}\Omega + \int_\Omega \mathbf{C_I}\nabla U_R \nabla V \,\mathrm{d}\Omega = 0 \end{aligned} \tag{8.7}$$

for arbitrary function V from $W_2^1(\Omega)$. Let a conformal tetrahedral mesh be given in Ω. We construct the subspace $W_{2,h}^1(\Omega)$ of the Sobolev space $W_2^1(\Omega)$ composed of functions which are continuous in Ω, linear in each mesh tetrahedron and vanish at the point (x_0, y_0, z_0). Substituting $W_2^1(\Omega)$ by $W_{2,h}^1(\Omega)$ in the weak formulation (8.7) we arrive at a system of linear equations

$$\begin{aligned} A_R\mathbf{U}_R - A_I\mathbf{U}_I &= \mathbf{F}_R, \\ A_I\mathbf{U}_R + A_R\mathbf{U}_I &= \mathbf{F}_I, \end{aligned} \tag{8.8}$$

where $\mathbf{U}_R$ and $\mathbf{U}_I$ are vectors of coefficients in the expansion of the approximate FEM-solutions U_R^h, U_I^h with the FEM basis functions.

In order to substantiate the numerical scheme, we consider a simple geometrical model of the human torso (Vassilevski et al. 2012) (Fig. 8.1) and a series of unstructured tetrahedral meshes with variable element size. For each mesh we compute the numerical solution and compare it with the numerical solution obtained on the finest mesh by using L_2-norm to evaluate the difference between the numerical solutions.

We consider the series of hierarchical meshes. The initial coarse mesh contains 9359 tetrahedra. Starting from this mesh, we decompose each tetrahedron into 8 smaller tetrahedra by splitting each face into four triangles by the middle points on the edges. We call this operation a uniform refinement of the mesh. Now we can apply the uniform refinement to the new mesh, and so on. At each step the

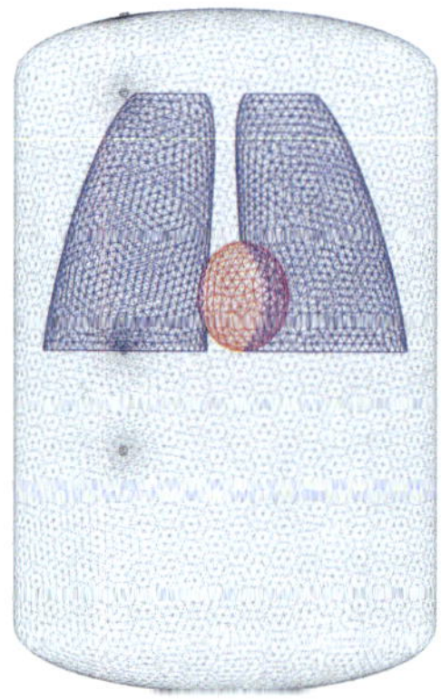

Fig. 8.1 Simplified geometrical model of the human torso used for the convergence study

Table 8.2 The results of the convergence study on hierarchical meshes

N_V	N_T	Memory (Mb)	N_{it}	Time (s)	L_2-norm
2032	9359	7.16	13	0.02	1.24E−03
14,221	74,872	37.3	23	0.18	9.31E−04
106,509	598,976	299.1	58	3.70	5.07E−04
824,777	4,791,808	2437.5	127	68.55	1.53E−04
6,492,497	38,334,464	20,015.3	353	2634.15	–

element size of the mesh is decreased in two times. The finest mesh in the series of five meshes has more than 38 millions tetrahedra and requires nearly 20 GB of the memory for computation. We used GMRES based iterative linear solver with the second-order ILU preconditioner (Kaporin 1998). All necessary routines were taken from the open source library Ani3D (Lipnikov and Vassilevski 2007).

The results of the convergence study on the series of hierarchical meshes are presented in Table 8.2. The first two columns show the number of vertices N_V and the number of tetrahedra N_T. The next three columns show the memory usage, the number of linear solver iterations N_{it} needed for the 10^{12}-fold reduction of an initial residual, and the overall time usage, respectively. The last column contains the relative L_2-norm that reduces asymptotically with roughly the second order convergence. If the meshes are not nested hierarchically, the error estimate requires extra interpolation of the finite element solution onto the finest grid. Application of this interpolation reduces the convergence order to one on non-nested meshes.

The numerical scheme verified on the simplified discrete model of the torso can be applied to a high-resolution human body geometrical model. The model was constructed in two steps. First, the geometrical model of the human torso was created for the VHP man data (Ackerman 2003). The data were clipped and downscaled to an array of 567×305×843 colored voxels with the resolution 1×1×1 mm. The initial segmented model of the VHP man torso was kindly provided by the Voxel-Man group (Höhne et al. 2000). This model has been produced primarily for visualization purposes, contained a significant amount of unclassified tissue and thus was not entirely suited for numerical purposes. Therefore, a further

Table 8.3 Admittivity of human body tissues at different electric current frequencies, Sm/m (Gabriel et al. 1996a,b; Vassilevski 2010)

	Admittivity @ 5 kHz		Admittivity @ 50 kHz	
Material	Conductivity	Susceptivity	Conductivity	Susceptivity
Blood vessels	0.50573	0.0031	0.50883	0.009496
Bones	0.051324	0.000532	0.052032	0.00122
Brain	0.088129	0.00881	0.102552	0.012525
Cartilage	0.17554	0.00169	0.17706	0.007679
Diaphragm	0.33669	0.0146	0.35182	0.028064
Esophagus	0.52811	0.00422	0.53369	0.009873
Eye	0.33369	0.005	0.33849	0.007302
Eye nerve	0.034567	0.0138	0.069315	0.026656
Fat	0.023589	0.000783	0.024246	0.000479
Gallbladder	0.90006	0.000064	0.90012	0.000317
Heart	0.13665	0.0356	0.19543	0.047215
Intestine	0.4393	0.011303	0.452783	0.023694
Kidney	0.12881	0.0193	0.15943	0.031776
Larynx	0.17554	0.00169	0.17706	0.007679
Liver	0.047666	0.0119	0.072042	0.029722
Lung	0.23484	0.0183	0.26197	0.02372
Muscle	0.33669	0.0146	0.35182	0.028064
Pancreas	0.52811	0.00435	0.53395	0.011185
Skin	0.0015	0.002	0.03	0.05
Spinal cord	0.034567	0.0138	0.069315	0.026656
Spleen	0.10829	0.00661	0.11789	0.015272
Stomach	0.52811	0.00422	0.53369	0.009873
Testis	0.37809	0.0049	0.385275	0.013202
Thyroid gland	0.52811	0.00435	0.53395	0.011185
Tongue	0.27812	0.00477	0.28422	0.015281
Trachea	0.30507	0.00984	0.32987	0.019219

processing of the segmented model was performed semi-automatically on the basis of ITK-SNAP segmentation software program (Yushkevich et al. 2006). At the final stage, we used several post-processing algorithms for filling remaining gaps between tissues and final segmented data smoothing (Serra 1984). Our segmented model of the human torso contains 26 labels describing tissues and major organs or their constituents (e.g., left/right kidney and lung, arteries/veins), with 22 admittivity properties presented in Table 8.3.

The segmented data are given on a very fine voxel grid; a coarser mesh with the same resolution of tissue interfaces should be used in computations. We tested several meshing techniques for the mesh generation of the segmented data. In our work we opted for the Delaunay triangulation algorithm from the CGAL-Mesh library (Rineau and Yvinec 2007). This algorithm enables defining a specific mesh size for each model material. In order to preserve geometrical features of the

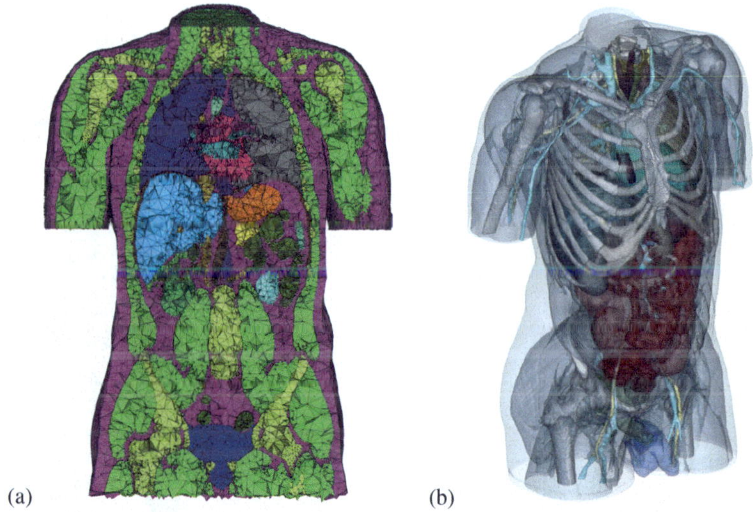

Fig. 8.2 Unstructured tetrahedral mesh (**a**) for geometrical model of the segmented image (**b**)

segmented model while keeping a feasible number of vertices, we assigned a smaller mesh size to blood vessels and a larger mesh size to fat and muscle tissues.

After the initial mesh generation, we applied mesh cosmetics from the Ani3D package (Lipnikov and Vassilevski 2007). This essential step reduces discretization errors and the matrix stiffness in the resulted systems of linear equations. The segmented model and the generated mesh containing 413,508 vertices and 2,315,329 tetrahedra are presented in Fig. 8.2. This mesh retains most anatomical features of the human torso.

In addition to the segmented torso model, we prepared a segmented model of the whole body. Missing parts were segmented using the ITK-SNAP software. The position of the arms was adapted manually by moving them away from the body (Danilov et al. 2013) as is the case with bioimpedance measurements. The final model is a $1100 \times 333 \times 1878$ voxels array with the resolution $1 \times 1 \times 1$ mm segmented with 30 labels associated with organs, their constituents and other tissues, altogether 26 materials with admittivites presented in Table 8.3. We used the same approach to construct the computational mesh for the whole body model based on the VHP data. The related segmented model and generated mesh containing 479,198 vertices and 2,725,980 tetrahedra are shown in Fig. 8.3.

After mesh generation for the torso and the whole body, we added a skin layer and multilayered electrodes to the surface of the constructed mesh. The original boundary triangulation was used to create a prismatic mesh on the surface, and then each prism was split into three tetrahedra resulting in a conformal tetrahedral mesh.

A series of computational meshes were generated in accordance with different positions of electrodes. Along with the conventional tetrapolar wrist-to-ankle measurement configuration (4-electrode scheme), a 10-electrode segmental BIA

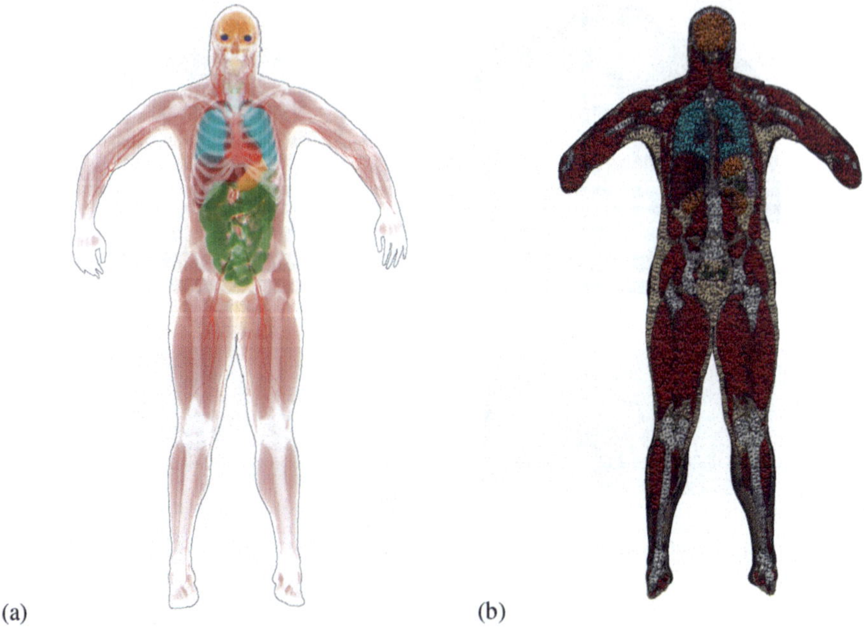

Fig. 8.3 Segmented whole body model of the VHP man with stretched arms (a) and a cut of the generated mesh by the frontal plane (b)

measurements scheme was considered with the placement of current and potential electrodes at a mutual distance of 5 cm on the back surfaces of the wrists and ankles, and also on the forehead (Fig. 8.4). Accordingly, five pairs of thin bilayer square objects 23×23 mm in size as electrodes were added on the forehead and distal parts of arms and legs of the segmented model.

8.4 Sensitivity Analysis

Sensitivity analysis is used to visualize the regions contributing most to the measured transfer impedance for the tetrapolar scheme measurements.

In order to perform the sensitivity analysis, we introduce the reciprocal lead field $\mathbf{J}'_{\text{reci}}$ which is equal to density vector field generated by a unit current excitation using the two PU electrodes. Field $\mathbf{J}'_{\text{reci}}$ is computed from (8.1)–(8.5), with electrode surfaces $\Gamma_{\pm}$ corresponding to PU electrodes and $I_0 = 1$ (Geselowitz 1971; Grimnes and Martinsen 2014) (Fig. 8.5b).

The lead field may be used for sensitivity distribution analysis of the PU electrodes for CC electrodes. We will use the following two equations: the general transfer signal equation

$$u = \int_{\Omega} \rho\, \mathbf{J}_{\text{cc}} \cdot \mathbf{J}'_{\text{reci}}\, \mathrm{d}x, \tag{8.9}$$

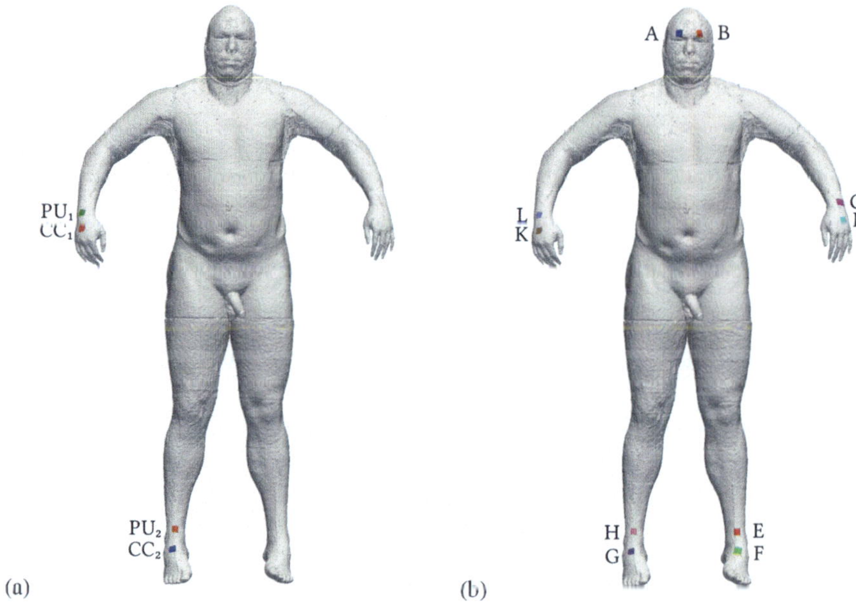

Fig. 8.4 Electrode positions for the conventional tetrapolar wrist-to-ankle measurement configuration (**a**) and the 10-electrode segmental BIA scheme (**b**)

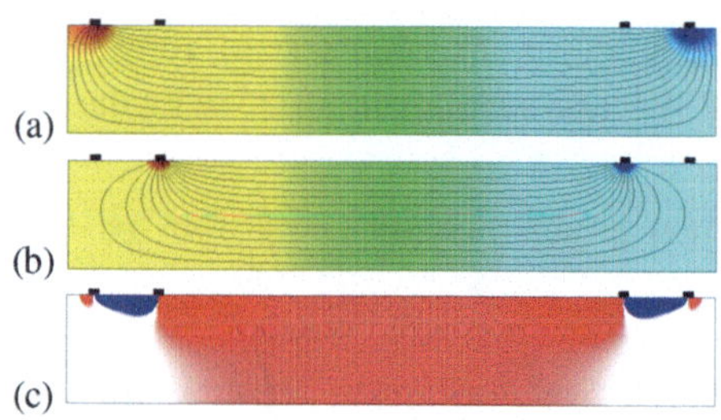

Fig. 8.5 Sensitivity analysis of the tetrapolar scheme on a homogeneous cylinder, cross section through the electrode plane: (**a**) electric potential and current lines for CC pair; (**b**) electric potential and current lines for PU pair; (**c**) sensitivity field: red, positive; blue, negative; white, nearly zero sensitivity

and the general transfer impedance equation:

$$Z_t = \int_{\Omega} \rho \, \mathbf{J}'_{cc} \cdot \mathbf{J}'_{reci} \, dx. \tag{8.10}$$

In these equations, u is the measured signal between PU electrodes, ρ is the resistivity, $\mathbf{J}_{cc}$ is computed from (8.1)–(8.5) with electrode surfaces $\Gamma_{\pm}$ corresponding to the current carrying electrodes (Fig. 8.5a), Z_t is the transfer impedance, and $\mathbf{J}'_{cc} = \mathbf{J}_{cc}/I_0$.

The sensitivity analysis is based on the distribution of the sensitivity field, which is computed as

$$S = \mathbf{J}'_{\mathrm{cc}} \cdot \mathbf{J}'_{\mathrm{reci}}. \tag{8.11}$$

Sensitivity multiplied by resistivity is called *volume impedance density*.

Using this notation, we have the following relations:

$$Z_{\mathrm{t}} = \int_{\Omega} \rho\, S\, \mathrm{d}x, \quad \Delta Z_{\mathrm{t}} = \int_{\Omega} \Delta\rho\, S\, \mathrm{d}x. \tag{8.12}$$

The last formula is applicable only to relatively small changes of ρ and moderate variations in S.

For sensitivity analysis purposes, we shall split the domain Ω in three parts according to the value of the sensitivity field (Fig. 8.5c):

$$\Omega^{-} = \{\mathbf{x} | S(\mathbf{x}) < 0\}, \quad \Omega^{+} = \{\mathbf{x} | S(\mathbf{x}) > 0\}, \quad \Omega^{0} = \{\mathbf{x} | S(\mathbf{x}) = 0\}. \tag{8.13}$$

We can introduce the positive and negative components of the transfer impedance:

$$Z_{\mathrm{t}}^{+} = \int_{\Omega^{+}} \rho\, S\, \mathrm{d}x, \quad Z_{\mathrm{t}}^{-} = \int_{\Omega^{-}} \rho\, S\, \mathrm{d}x, \quad Z_{\mathrm{t}} = Z_{\mathrm{t}}^{+} + Z_{\mathrm{t}}^{-}. \tag{8.14}$$

In addition, for a specific threshold value $t \in [0, 100]$, we shall define W_t^{-} as a subdomain of Ω^{-} and W_t^{+} as a subdomain of Ω^{+} with the following restrictions:

$$\inf_{\mathbf{x} \in \Omega \setminus W_t^{-}} \rho(\mathbf{x}) S(\mathbf{x}) \geqslant \sup_{\mathbf{x} \in W_t^{-}} \rho(\mathbf{x}) S(\mathbf{x}), \quad \int_{W_t^{-}} \rho\, S\, \mathrm{d}x = \frac{t}{100} Z_{\mathrm{t}}^{-}, \tag{8.15}$$

$$\sup_{\mathbf{x} \in \Omega \setminus W_t^{+}} \rho(\mathbf{x}) S(\mathbf{x}) \leqslant \inf_{\mathbf{x} \in W_t^{+}} \rho(\mathbf{x}) S(\mathbf{x}), \quad \int_{W_t^{+}} \rho\, S\, \mathrm{d}x = \frac{t}{100} Z_{\mathrm{t}}^{+}. \tag{8.16}$$

Thus, W_t^{+} is the region of high volume impedance density, which have the transfer impedance contribution equal to t percents of the total transfer impedance. The same applies to W_t^{-}, which is the region of the most negative volume impedance density.

We shall also define subdomains V_t^{+} and V_t^{-}:

$$\sup_{\mathbf{x} \in \Omega \setminus V_t^{\perp}} \pm S(\mathbf{x}) \leqslant \inf_{\mathbf{x} \in V_t^{+}} \pm S(\mathbf{x}), \quad \int_{V_t^{\pm}} S\, \mathrm{d}x = \frac{t}{100} \int_{\Omega^{\pm}} S\, \mathrm{d}x. \tag{8.17}$$

In our analysis we investigate the shape of the subdomains $W_t^{\pm}$ and $V_t^{\pm}$. These body regions represent the most sensitive parts of the human body for specified measuring scenario. The shape of $W_t^{\pm}$ describes the part of the body in which one measures the positive and negative components $Z_t^{\pm}$ of the transfer impedance. The shape of $V_t^{\pm}$ represents the part which is the most sensitive to local changes of resistivity. This approach may be applied for validation of empirically designed electrode schemes.

If the media is homogeneous, the shape of the respective subdomains $V_t^{\pm}$ and $W_t^{\pm}$ will be the same. Figure 8.5 represents the sensitivity analysis of the tetrapolar scheme on a homogeneous cylinder. We can see two negative sensitivity areas between corresponding CC and PU electrodes, a positive sensitivity area between PU electrodes with a higher sensitivity near the PU electrodes, and two nearly zero sensitivity areas on both ends of the cylinder.

8.5 The Online Numerical Simulator

Within the above-described numerical modeling framework combining segmentation, mesh generation, FEM discretization, and post-processing techniques, the online numerical simulator of bioimpedance measurements was developed with publicly available user interface (Danilov et al. 2017). It is based on a JavaScript 3D library and WebGL technology. The segmented model of the VHP man with stretched arms (Fig. 8.3) is preloaded in the simulator. At present, the user can select either a 4-electrode or a 10-electrode measurement scheme (Fig. 8.4) with a predefined list of electrode pairs (Table 8.4). Optionally, the user may select manually the current carrying (CC) and pick-up (PU) electrode pairs.

Once the electrode pairs are selected, the server computes the respective solution of the boundary value problem (8.1)–(8.5), calculates the sensitivity field (8.11), and constructs the $V_t^{\pm}$ and $W_t^{\pm}$ regions. The user controls the threshold parameter t and examines the shape of V_t^{+} or W_t^{+}.

Table 8.4 Electrode pairs for 10-electrode measurement scheme (rf. Fig. 8.4)

Assessed body region	CC pair	PU pair
Left arm	D, B	C, E
Left leg	F, D	E, H
Right leg	G, F	H, L
Right arm	K, G	L, A
Head	B, K	A, C
Torso (parallel)	D, F	L, H
Torso (cross)	D, G	L, E

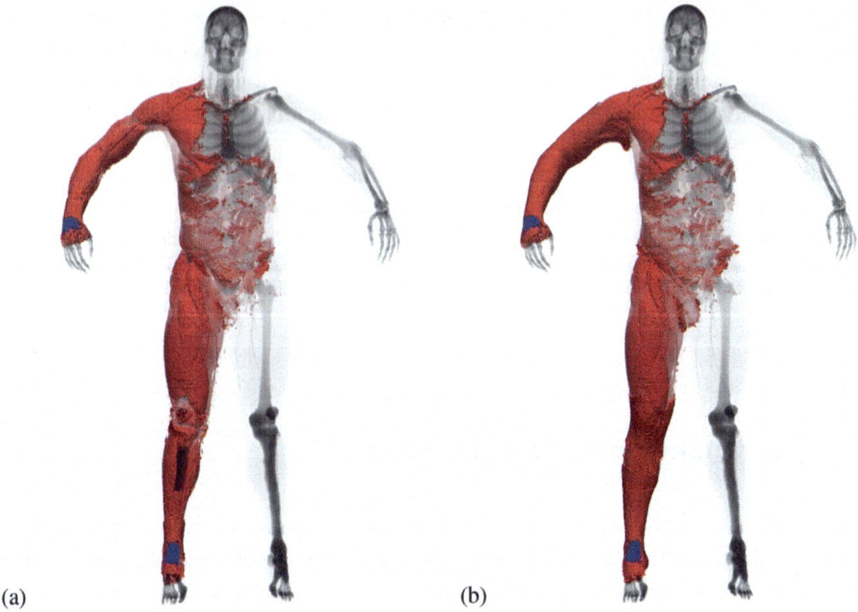

Fig. 8.6 Sensitivity analysis of the conventional 4-electrode scheme: (**a**) sensitivity distribution (red, V_{97}^{+}; blue, V_{97}^{-}); (**b**) volume impedance density distribution (red, W_{97}^{+}; blue, W_{97}^{-})

High-resolution images of $V_{97}^{\pm}$ and $W_{97}^{\pm}$ regions overlaid on translucent body structure were constructed for the conventional 4-electrode (Fig. 8.6) and 10-electrode (Figs. 8.7, 8.8, 8.9) measurement schemes. Since the body structure is heterogeneous, the shapes of $V_t^{\pm}$ and $W_t^{\pm}$ are slightly different, though their general shape is the same. One can see the negative sensitivity area between the adjacent CC and PU electrodes in all the figures. The conventional wrist-to-ankle 4-electrode scheme demonstrates high sensitivity in the right arm, the right leg, and partially in the torso; the sensitivity in the head and the left arm and leg is nearly zero. It can be seen that the positive sensitivity areas in the 10-electrode scheme for various combinations of CC and PU electrode pairs listed in Table 8.4 well correspond to the respective names of body regions. The noticeable negative sensitivity areas in the 10-electrode scheme actually have minor impact on the transfer impedance, since the absolute values of Z_t^{-} are much smaller than Z_t^{+} (Table 8.5).

8.6 Discussion

Computerized models of real human anatomy and digital human phantoms are extensively used in various areas of biomedicine such as radiography, nuclear medicine and radiation protection (Caon 2004; Xu 2014), as well as for educational

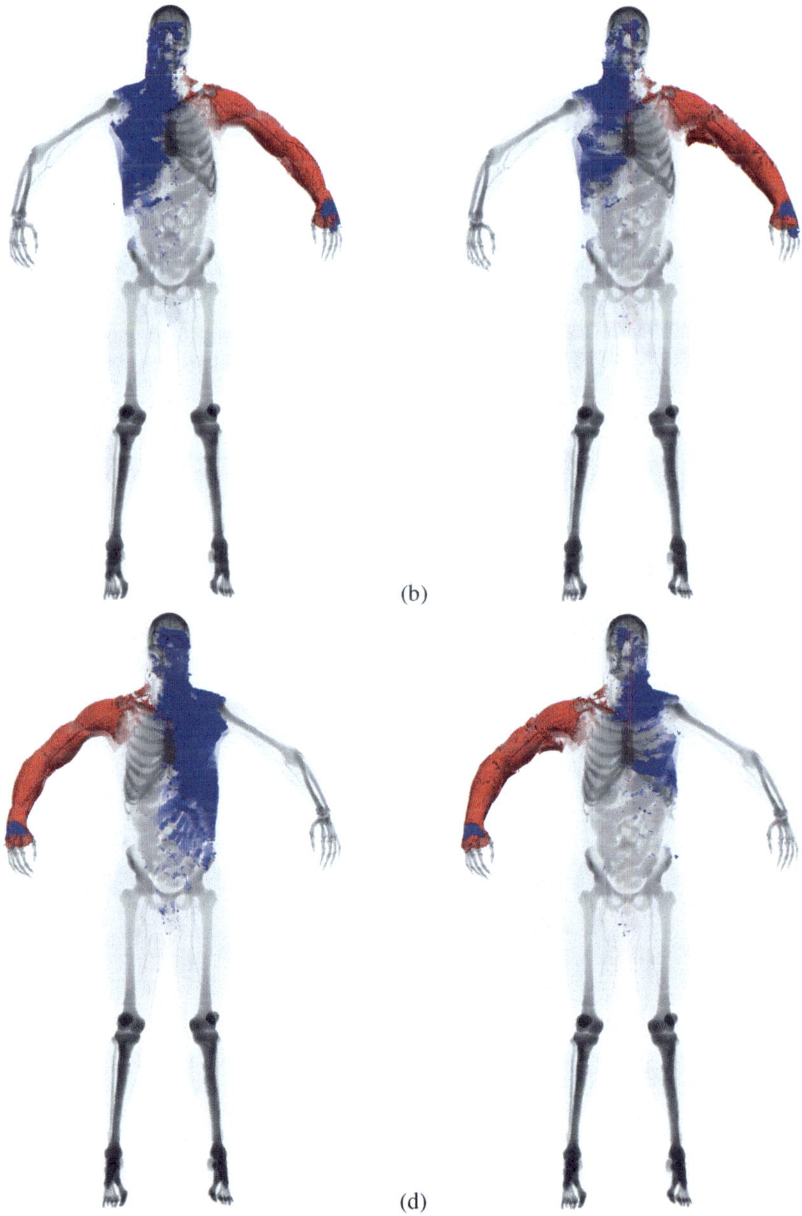

(a) (b) (c) (d)

Fig. 8.7 Sensitivity analysis of the left (**a,b**) and right (**c,d**) arms using the 10-electrode scheme: (**a,c**) sensitivity distribution (red—V_{97}^{+}, blue—V_{97}^{-}); (**b,d**) volume impedance density distribution (red—W_{97}^{+}, blue—W_{97}^{-})

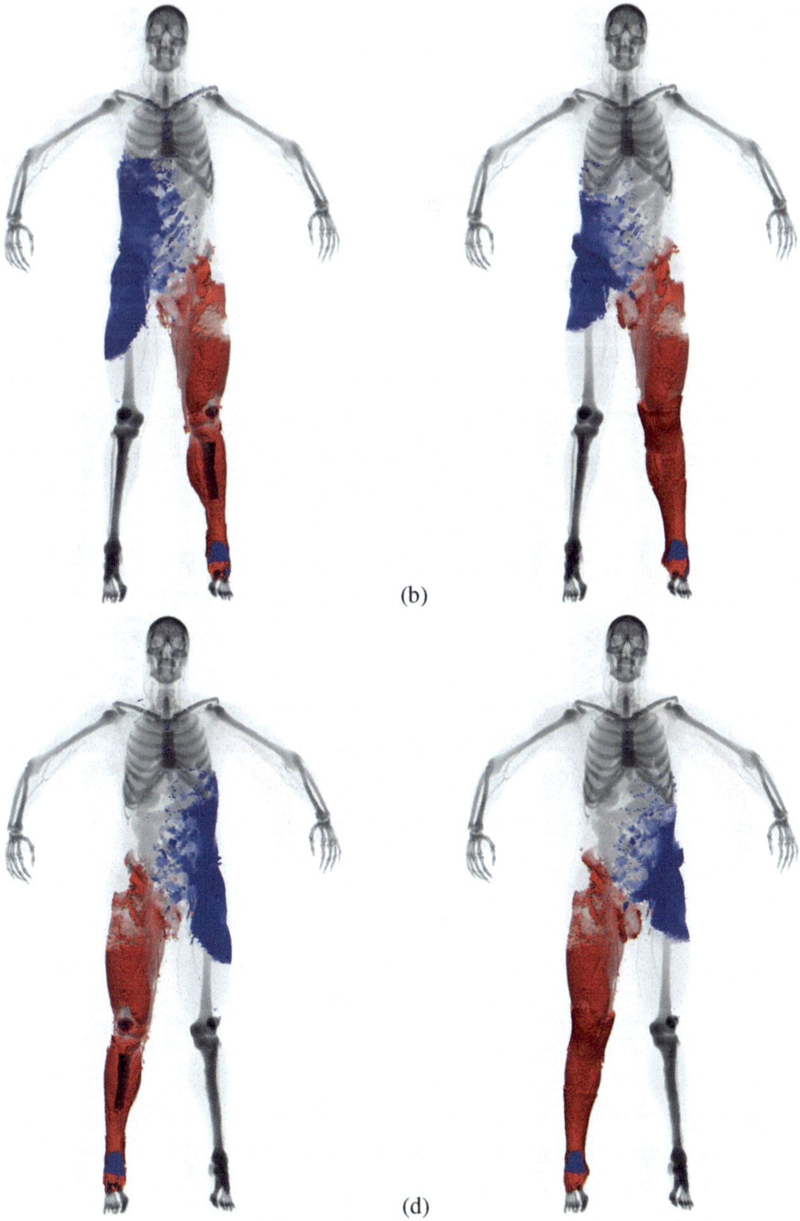
(a) (b) (c) (d)

Fig. 8.8 Sensitivity analysis of the left (**a,b**) and right (**c,d**) legs using the 10-electrode scheme: (**a,c**) sensitivity distribution (red, V_{97}^{+}; blue, V_{97}^{-}); (**b,d**) volume impedance density distribution (red, W_{97}^{+}; blue, W_{97}^{-})

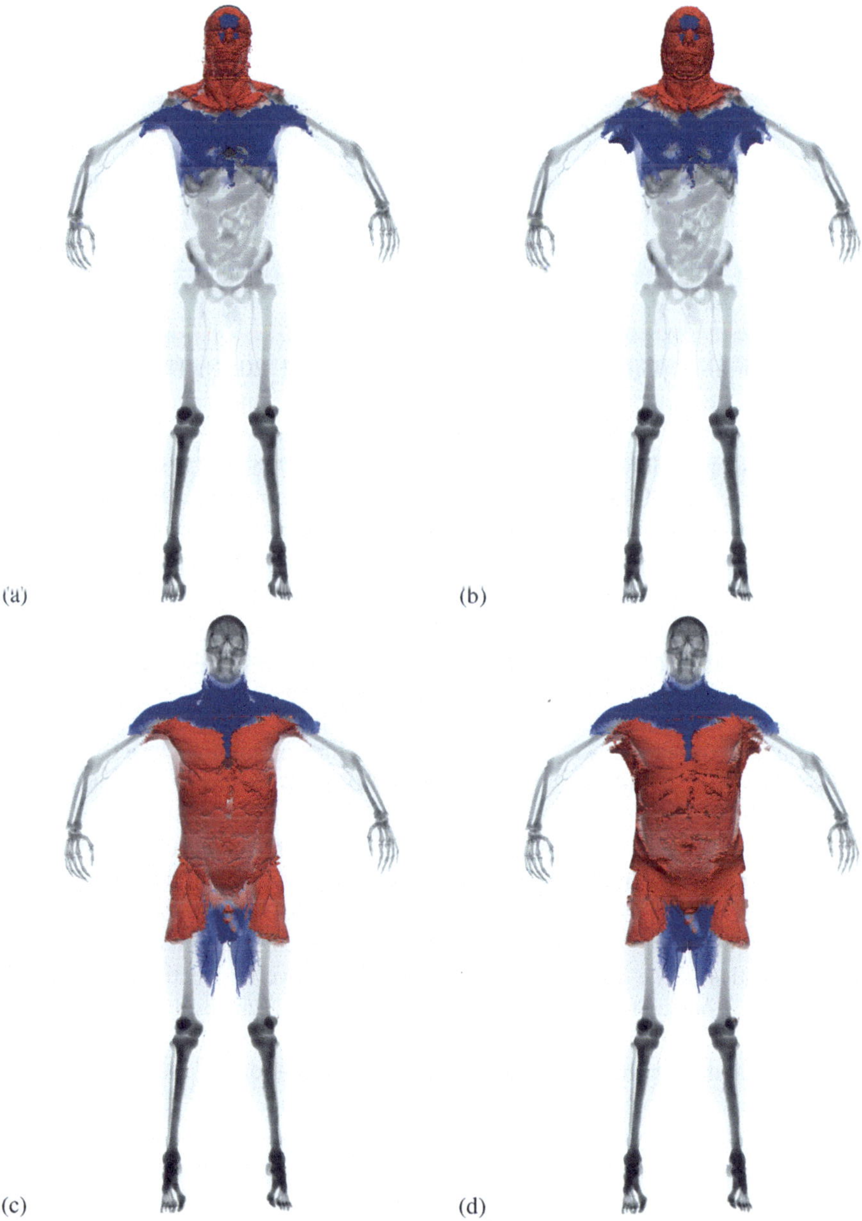

Fig. 8.9 Sensitivity analysis of the head (**a,b**) and torso (**c,d**) using the 10-electrode scheme: (**a,c**) sensitivity distribution (red, V_{97}^{+}; blue, V_{97}^{-}); (**b,d**) volume impedance density distribution (red, W_{97}^{+}; blue, W_{97}^{-})

Table 8.5 Contribution of positive and negative sensitivity areas to the transfer impedance

Measurement scheme	Z_t^+	Z_t^-	Z_t
Wrist-to-ankle	1004.84	−58.0518	946.79
Left arm	516.037	−35.8731	480.164
Right arm	558.69	−40.912	517.778
Left leg	413.024	−23.0604	389.963
Right leg	415.172	−23.418	391.754
Head	97.8496	−16.0922	81.7574
Torso (parallel)	38.1116	−5.62323	32.4884
Torso (cross)	38.1362	−5.7634	32.3728

and other purposes (Gosselin et al. 2014; Saltarelli et al. 2014). The same applies to bioimpedance methods of health monitoring, including impedance cardiography, electrical impedance tomography, and bioimpedance analysis of body composition and spectroscopy (Holder 2005; Jaffrin and Morel 2008; Cybulski 2011; Bera 2014; Jindal et al. 2016). High-resolution anatomically correct discrete models of the human body enable to study relative contribution of various organs and tissues to the bioimpedance signal, to find optimal electrode placement and electric current frequency range (Kauppinen et al. 1998; Belalcazar and Patterson 2004; Beckmann et al. 2007; Patterson 2010; Ulbrich et al. 2013; Orschulik et al. 2016). In electrical impedance tomography, such models are utilized to improve the quality of reconstructed images (Bayford et al. 2001; Adler et al. 2015).

In this chapter, the FEM-based computational technology and the online numerical simulator for high-resolution efficient modeling of bioimpedance measurements (Vassilevski et al. 2012; Danilov et al. 2012, 2013, 2017) are considered. The cornerstone of the technology is the anatomically correct 3D model of the VHP man (Ackerman 2003) and the finite element discretization of the stationary boundary value problem for the electric potential on tetrahedral meshes. FEM discretizations, in contrast to the conventional finite difference schemes on cubic meshes, account tissue anisotropy and provide more realistic description of electrical properties of the human body. The use of adaptive unstructured tetrahedral meshes makes our computational technology efficient: high-resolution models of the human body exploit significantly coarser meshes as compared to rectangular meshes thus reducing the computational cost. The limitations of the initial version of the online simulator are the use of the VHP man model only, fixed position of electrodes and, therefore, a limited number of measuring scenarios.

For other software and computational models used in the bioimpedance studies, we refer to Bayford et al. 2001, Pettersen and Høgetveit 2011, Adler et al. 2015, de Sitter et al. 2016. Validation of digital bioimpedance models can be performed by the MRI-based construction of an anatomically correct computational mesh of the measured subject and simulation of the bioimpedance measurements followed by the comparison with measured outcomes (Ulbrich et al. 2013). Bioimpedance applications utilizing finite element discretizations also include the development of new diagnostic tools (Ness et al. 2015), tissue engineering (Canali et al. 2015), and animal studies (Dowrick et al. 2016).

8.7 Conclusion

In this chapter, we presented essential elements and the workflow of the FEM-based high-resolution computational technology for modeling bioimpedance measurements. This technology is realized as the online numerical simulator with a user-friendly interface. The advantages of the technology are anatomically correct 3D model of the VHP man and the usage of FEM discretizations on the adaptive unstructured tetrahedral meshes. The technology allows to account tissue anisotropy and to adapt the mesh size and orientation of mesh faces for each organ/tissue geometry. Mesh adaptivity provides significant reduction of CPU time and computer memory requirements while saving the model accuracy. The simulator enables to calculate and visualize on a laptop sensitivity fields for the conventional tetrapolar and 10-electrode configurations at various measuring scenarios.

Acknowledgements The authors thank V.Yu. Salamatova, V.K. Kramarenko, and A.S. Yurova for segmentation of the VHP data and performing numerical experiments, G.V. Kopytov for the development of user's interface for the online numerical simulator, and D.V. Nikolaev and A.V. Smirnov for problem formulation, valuable discussion, and financial support of the initial part of this study. Our work was supported by the Russian Foundation for Basic Research (RFBR grants 17-01-00886 and 17-51-53160).

References

Ackerman, M. J. (2003). The National Library of Medicine's Visible Human Project. Accessed November 10, 2017. https://www.nlm.nih.gov/research/visible/

Adler, A., Lionheart, W. R. B., & Polydorides, N. (2015). EIDORS: electrical impedance tomography and diffuse optical tomography reconstruction software. Accessed November 10, 2017. http://eidors3d.sourceforge.net/

Bayford, R. H., Gibson, A., Tizzard, A., Tidswell, T., & Holder, D. S. (2001). Solving the forward problem in electrical impedance tomography for the human head using IDEAS (integrated design engineering analysis software), a finite element modelling tool. *Physiological Measurement, 22*(1), 55–64.

Beckmann, L., van Riesen, D., & Leonhardt, S. (2007). Optimal electrode placement and frequency range selection for the detection of lung water using bioimpedance spectroscopy. In *2007 29th Annual International Conference of the IEEE Engineering in Medicine and Biology Society* (pp. 2685–2688).

Belalcazar, A., & Patterson, R. P. (2004). Improved lung edema monitoring with coronary vein pacing leads: a simulation study. *Physiological Measurement, 25*(2), 475–487.

Bera, T. K. (2014). Bioelectrical impedance methods for noninvasive health monitoring: a review. *Journal of Medical Engineering, 2014*, 381251.

Brantlov, S., Andersen, T. B., Jødal, L., Rittig, S., & Lange, A. (2016). Bioimpedance spectroscopy in healthy children. *Journal of Clinical Engineering, 41*(1), 33–39.

Canali, C., Heiskanen, A., Muhammad, H. B., Hoym, P., Pettersen, F. J., Hemmingsen, M., Wolff, A., Dufva, M., Martinsen, O. G., & Emnéus, J. (2015). Bioimpedance monitoring of 3D cell culturing—Complementary electrode configurations for enhanced spatial sensitivity. *Biosensors and Bioelectronics, 63*, 72–79.

Caon, M. (2004). Voxel-based computational models of real human anatomy: a review. *Radiation and Environmental Biophysics, 42*(4), 229–235.

Cybulski, G. (2011). *Ambulatory impedance cardiography*. Berlin-Heidelberg: Springer.

Danilov, A. A., Kopytov, G. V., & Vassilevski, Y. V. (2017). Online simulator for bioimpedance measurements. Accessed November 10, 2017. http://dodo.inm.ras.ru/bia/

Danilov, A. A., Kramarenko, V. K., Nikolaev, D. V., & Yurova, A. S. (2013). Personalized model adaptation for bioimpedance measurements optimization. *Russian Journal of Numerical Analysis and Mathematical Modelling, 28*(5), 459–470.

Danilov, A. A., Nikolaev, D. V., Rudnev, S. G., Salamatova, V. Y., & Vassilevski, Y. V. (2012). Modelling of bioimpedance measurements: unstructured mesh application to real human anatomy. *Russian Journal of Numerical Analysis and Mathematical Modelling, 27*(5), 431–440.

de Sitter, A., Verdaasdonk, R. M., & Faes, T. J. C. (2016). Do mathematical model studies settle the controversy on the origin of cardiac synchronous trans-thoracic electrical impedance variations? A systematic review. *Physiological Measurement, 37*(9), R88–R108.

Dowrick, T., Blochet, C., & Holder, D. (2016). In vivo bioimpedance measurement of healthy and ischaemic rat brain: implications for stroke imaging using electrical impedance tomography. *Physiological Measurement, 36*(6), 1273–1282.

Gabriel, S., Lau, R. W., & Gabriel, C. (1996a). The dielectric properties of biological tissues: II. Measurements in the frequency range 10 Hz to 20 GHz. *Physics in Medicine and Biology, 41*(11), 2251–2269.

Gabriel, S., Lau, R. W., & Gabriel, C. (1996b). The dielectric properties of biological tissues: III. Parametric models for the dielectric spectrum of tissues. *Physics in Medicine and Biology, 41*(11), 2271–2293.

Geselowitz, D. B. (1971). An application of electrocardiographic lead theory to impedance plethysmography. *IEEE Transactions on Biomedical Engineering BME, 18*(1), 38–41.

Gosselin, M. C., Neufeld, E., Moder, H., Huber, E., Farcito, S., Gerber, L., Jedensjö, M., Hilber, I., Gennaro, F. D., Lloyd, B., Cherubini, E., Szczerba, D., Kainz, W., & Kuster, N. (2014). Development of a new generation of high-resolution anatomical models for medical device evaluation: the Virtual Population 3.0. *Physics in Medicine and Biology, 59*(18), 5287–5304.

Grimnes, S., & Martinsen, O. (2014). *Bioimpedance and bioelectricity basics* (3rd ed.) London: Academic.

Höhne, K. H., Pflesser, B., Pommert, A., Riemer, M., Schubert, R., Schiemann, T., Tiede, U., & Schumacher, U. (2000). A realistic model of the inner organs from the visible human data. In *Medical Image Computing and Computer-Assisted Intervention – MICCAI 2000* (pp. 776–785). Berlin: Springer Nature.

Holder, D. S. (2005). *Electrical impedance tomography*. Bristol: IOP.

Jaffrin, M. Y., & Morel, H. (2008). Body fluid volumes measurements by impedance: a review of bioimpedance spectroscopy (BIS) and bioimpedance analysis (BIA) methods. *Medical Engineering & Physics, 30*, 1257–1269.

Jindal, G. D., Sawant, M. S., Jain, R. K., Sinha, V., Bhat, S. N., & Deshpande, A. K. (2016). Seventy-five years of use of impedance plethysmography in physiological data acquisition and medical diagnostics. *MGM Journal of Medical Sciences, 3*, 84–90.

Kaporin, I. E. (1998). High quality preconditioning of a general symmetric positive definite matrix based on its $u^t u + u^t r + r^t u$-decomposition. *Numerical Linear Algebra with Applications, 5*(6), 483–509.

Kauppinen, P. K., Hyttinen, J. A., & Malmivuo, J. A. (1998). Sensitivity distributions of impedance cardiography using band and spot electrodes analyzed by a three-dimensional computer model. *Annals of Biomedical Engineering, 26*(4), 694–702.

Lipnikov, K. N., & Vassilevski, Y. V. (2007). Advanced numerical instruments 3D. Accessed November 10, 2017. https://sourceforge.net/projects/ani3d/

Ness, T. V., Chintaluri, C., Potworowski, J., Leski, S., Glabska, H., Wojcik, D. L., & Einevoll, G. T. (2015). Modelling and analysis of electrical potentials recorded in microelectrode arrays (MEA). *Neuroinformatics, 13*(4), 403–426.

Orschulik, J., Petkau, R., Wartzek, T., Hochhausen, N., Czaplik, M., Leonhardt, S., & Teichmann, D. (2016). Improved electrode positions for local impedance measurements in the lung—a simulation study. *Physiological Measurement, 37*(12), 2111–2129.

Patterson, R. P. (2010). Impedance cardiography: what is the source of the signal? *Journal of Physics: Conference Series, 224*, 012118.

Pettersen, F. J., & Høgetveit, J. O. (2011), From 3D tissue data to impedance using Simpleware ScanFE+IP and COMSOL Multiphysics a tutorial. *Journal of Electrical Bioimpedance, 2*, 13–32.

Rineau, L., & Yvinec, M. (2007). A generic software design for Delaunay refinement meshing. *Computational Geometry, 38*(1–2), 100–110.

Saltarelli, A. J., Roseth, C. J., & Saltarelli, W. A. (2014). Human cadavers vs multimedia simulation: a study of student learning in anatomy. *Anatomical Sciences Education, 7*(5), 331–339.

Serra, J. (1984). *Image analysis and mathematical morphology*. London: Academic.

Ulbrich, M., Marleaux, B., Mühlsteff, J., Schoth, F., Koos, R., Teichmann, D., & Leonhardt, S. (2013). High and temporal resolution 4D FEM simulation of the thoracic bioimpedance using MRI scans. *Journal of Physics: Conference Series, 434*, 012074.

Vassilevski, Y. V. (2010). Mathematical technologies for electroimpedance diagnostics and monitoring of cardiovascular and respiratory diseases, Report. Accessed November 10, 2017 (in Russian). http://dodo.inm.ras.ru/research/_media/fcp/14.740.11.0844-report-1.pdf

Vassilevski, Y. V., Danilov, A. A., Nikolaev, D. V., Rudnev, S. G., Salamatova, V. Y., & Smirnov, A. V. (2012). Finite-element analysis of bioimpedance measurements. *Zhurnal Vychislitelnoi Matematiki i Matematicheskoi Fiziki, 52*(4), 733–745 (in Russian).

Xu, X. G. (2014). An exponential growth of computational phantom research in radiation protection, imaging, and radiotherapy: a review of the fifty-year history. *Physics in Medicine and Biology, 59*(18), R233–R302.

Yushkevich, P. A., Piven, J., Hazlett, H. C., Smith, R. G., Ho, S., Gee, J. C., & Gerig, G. (2006). User-guided 3D active contour segmentation of anatomical structures: significantly improved efficiency and reliability. *NeuroImage, 31*(3), 1116–1128.

Chapter 9
Focused Impedance Method: Basics and Applications

Khondkar Siddique-e Rabbani

9.1 Introduction: How the Idea Evolved and a Brief Outline of the Possibilities

Any electrical impedance measurement involves driving a current and measuring the resulting potential using appropriate electrodes. The human body is a volume conductor of electricity in which injected electrical currents spread out in all possible directions, converging on areas with higher conductivity and bending around areas of lower conductivity. Therefore, it is very difficult to confine electrical currents to limited zones to target a specified region of the body or an organ. For this reason most of the earliest attempts at measuring electrical impedance in the human body involved measurement of whole body impedances connecting electrodes to different limbs. Again, in order to eliminate the effect of contact impedances on the measured impedance, the well-known four-electrode or tetrapolar impedance measurement (TPIM) was introduced in which current is driven through two electrodes while the resulting potential developed across the other two were measured, giving a "transfer impedance" (dividing the potential by the current) which could be related to the properties of the bulk of the volume conductor. TPIM can have the electrodes in any formation, a linear one being most popular as shown in Fig. 9.1 where the electrodes are arranged along a straight line. Here current is driven between the outer two electrodes, while the potential is measured across the inner pair. This basically measures the bulk impedance of the region between the inner potential electrodes. Another popular configuration is the square formation as shown in Fig. 9.2 in which the electrodes are placed at the corners of a square region. Current is driven through

K. S. Rabbani (✉)
Department of Biomedical Physics & Technology, University of Dhaka, Dhaka, Bangladesh
e-mail: rabbani@du.ac.bd

F. Simini, P. Bertemes-Filho (eds.), *Bioimpedance in Biomedical Applications and Research*, https://doi.org/10.1007/978-3-319-74388-2_9

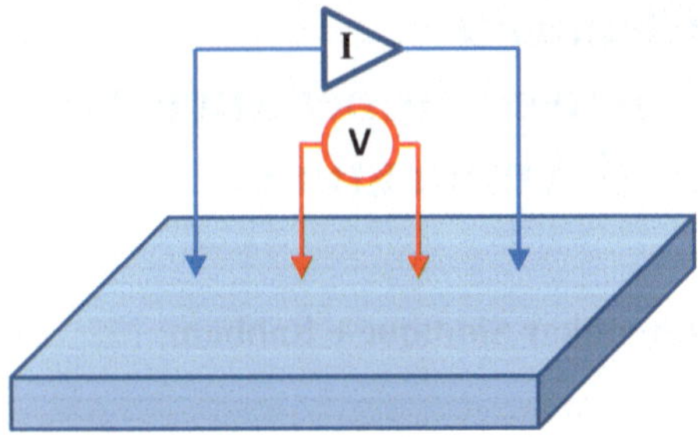

Fig. 9.1 Basic linear configuration of tetrapolar impedance measurement (TPIM)

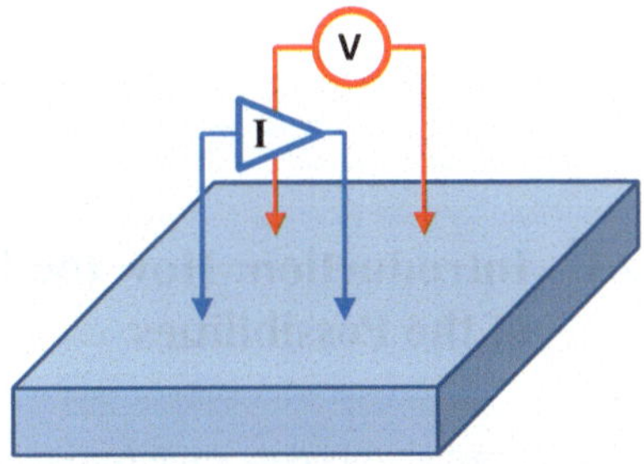

Fig. 9.2 Basic square configuration of tetrapolar impedance measurement (TPIM)

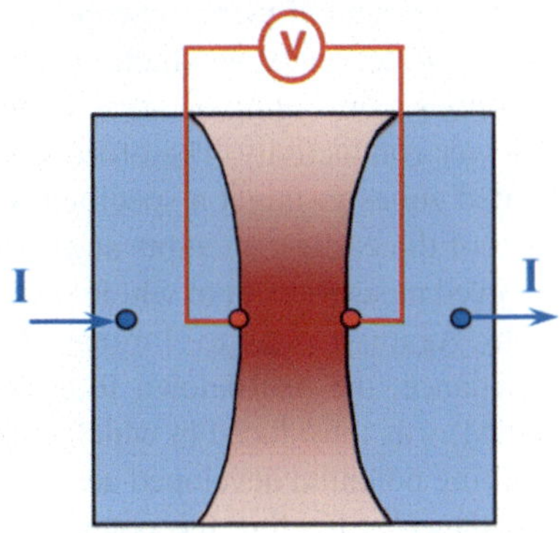

Fig. 9.3 Equipotential visualisation of sensitive zone for linear TPIM

two adjacent electrodes, while the potential is measured across the opposite pair. The bulk impedance within the square region mainly contributes to this measurement.

A simple and qualitative visualisation of the regions of the target object that contribute to the measurement may be made based on equipotential lines in 2D or surfaces in 3D. Figures 9.3 and 9.4 show such a visualisation for the two configurations of TPIM, respectively, in 2D. The red tinted inner zone, between the two equipotential lines that passes through the potential electrodes, represents the sensitive zone. This means the impedance of points within this zone influences the resulting measurement while those that are outside do not. Furthermore, this is the zone of positive sensitivity meaning an object with a higher impedance compared to the background and placed within this zone will cause the resulting transfer impedance to increase, and vice versa for a lower impedance or for a conductor. The shades of darkness in the figure are simply qualitative and try to present the fact

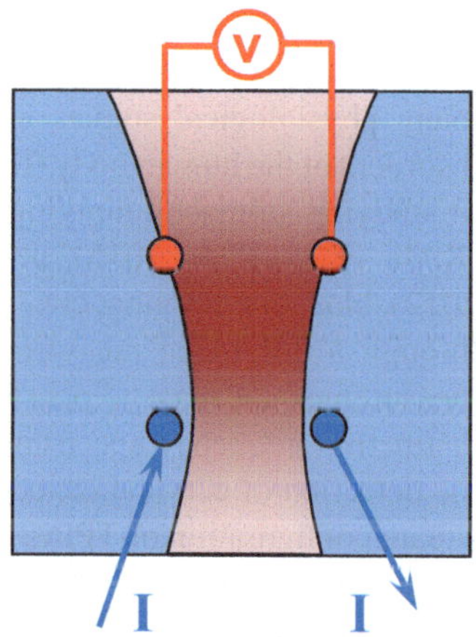

Fig. 9.4 Equipotential visualisation of sensitive zone for square TPIM

that the sensitivity decreases away from the centre. However, this representation is too simplistic and does not present an important aspect, i.e. the zones of negative sensitivity, which is associated with any TPIM system in the zone between a current and a potential electrode. By negative sensitivity it is meant that an object with higher impedance compared to the background and placed within this zone will cause the resulting transfer impedance to decrease, and vice versa for an object with impedance lower than the background. Sensitivity mapping of these measurement systems will be presented later in this chapter.

A significant enhancement in the field of electrical impedance came with applied potential tomography by a group in Sheffield, UK, which gives a 2D image (Barber et al. 1983; Brown et al. 1985). This was later named as electrical impedance tomography (EIT) by the bioimpedance community. The Sheffield system used 16 electrodes around a cross section of a body such as the thorax and performs many sets of TPIM combining all the electrodes in a specified way. Using a weighted back projection technique, these data are analysed to give a 2D image. Again, the current in EIT is not confined to 2D only, so this arrangement has contributions from the third dimension in a complex way (Rabbani and Kabir 1991). Therefore, pixel-level information given by the 2D EIT is not accurate. This naturally led to attempts in developing a volume image, a 3D EIT, by several groups (Wagenaar and Adler 2016). However, it is obvious that such a system would be immensely complex.

In many applications an image is not needed; rather, information of a localised object or of a localised region, separated from that of surrounding objects, is adequate. The idea of a focused impedance stemmed from this perception and the requirement of a simple, robust and low-cost system that could be of particular relevance to low-resource countries. Focused impedance method (FIM), a name coined by the author, is attractive because of its ability to localise a small region in a volume conductor using a few electrodes and simple instrumentation (Rabbani et al. 1998, 1999; Rabbani and Karal 2008). Utilising the 3D sensitivity of FIM, it is possible to detect objects at shallow depths using skin surface electrodes, typically of the order of the separation of the electrodes (Iquebal and Rabbani 2010; Kadir et al. 2013; Ahmed et al. 2014). FIM has potential applications in many physiological studies and in the detection and diagnosis of diseases and disorders of the human body (Rabbani and Kadir 2011). Particularly it may be useful in studying stomach emptying, stomach acid secretion, localised lung ventilation (Kadir et al. 2009, 2015), abdominal fat thickness measurement (Surovy et al. 2012), bladder emptying, etc., where temporal changes occur that may be discerned through a difference measurement. Again, through changes in the frequency of measurement, it may be possible to identify organs or localised tissues that have distinctive frequency spectrum from the surrounding tissues as has been done in the detection of cervical cancer by a group in Sheffield (Brown et al. 2000b) using square configuration of TPIM. This success led to the idea that it may be possible to characterise breast tumours, whether these are malignant or benign once these are felt by palpation or other means, which the author's group at University of Dhaka is trying using FIM and initial results looks promising (Al-Amin et al. 2014; Huda et al. 2016; Ahmed et al. 2016). It may even be possible to monitor tumour ablation at shallow depths using such multiple frequency measurements.

The next section will give the basics of the three different versions of FIM and their bulk sensitivities—obtained either using experimental findings or numerical simulations.

9.2 Focused Impedance Measurement (FIM): Basics

9.2.1 Basic Configuration and Visualisation

9.2.1.1 FIM-8, Eight-Electrode Version

This new method was conceived by the author and developed by his group at University of Dhaka, Bangladesh (Rabbani et al. 1998, 1999). The idea originated from a visualisation of sensitive zones in TPIM as shown in Fig. 9.2. It was envisaged that if two TPIM systems are placed orthogonally centred on the same region as shown in Fig. 9.5 and the two measured values of transfer impedances are summed, it should give enhanced sensitivity in the central region. This is because the central region will appear twice in the sum increasing its contribution compared to those of the outer regions. To illustrate the method, measurement using current I_1 and potential V_1 in Fig. 9.5 will give a transfer impedance Z_1 $(= V_1/I_1)$, while measurement using current I_2 and potential V_2 will give a transfer impedance Z_2 $(= V_2/I_2)$. The average transfer impedance is defined as the focused impedance,

$$\mathrm{FZ} = (Z_1 + Z_2)/2. \tag{9.1}$$

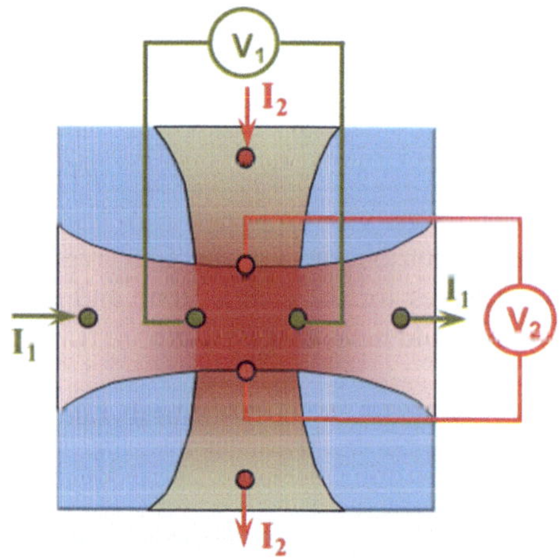

Fig. 9.5 Basic scheme of FIM-8, consisting of two orthogonal TPIM systems. The average of the two measurements gives the focused impedance

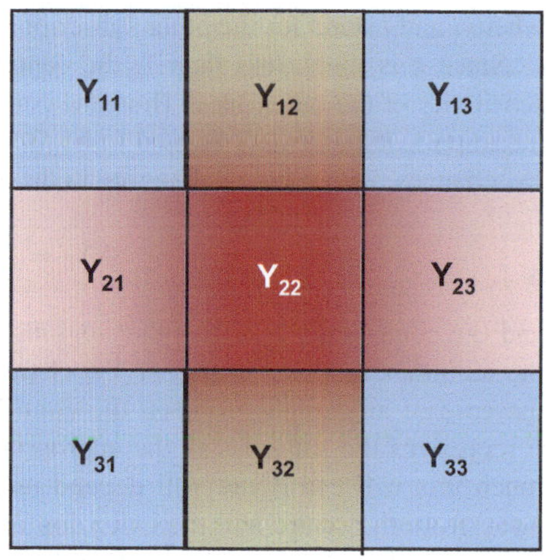

Fig. 9.6 Simplified admittance model of FIM-8 (Rabbani et al. 1999)

In order to obtain a simple expression for sensitivity with respect to localised zones around the electrodes, a simple admittance model as shown in Fig. 9.6 rather than an impedance model has been used where the curved equipotential lines in Fig. 9.5 have been replaced by straight lines for simplicity. The admittances of the nine zones are assumed isotropic and uniform within the matrix and named as shown. The relevant admittances add algebraically as the zones contribute to this measurement as parallel impedances as will be clear in the next paragraph.

Measurement using current I_1 and potential V_1 in Fig. 9.5 will give the admittance $(kY_{12} + Y_{22} + kY_{32})$ in Fig. 9.6, where k is a constant factor, usually much less than 1, which takes care of the sensitivity differences between the central zone and the outer adjacent ones. Similarly the orthogonal measurement using current I_2 and potential V_2 will give the admittance $(kY_{21} + Y_{22} + kY_{23})$. An algebraic addition of the two measurements gives a summed admittance as

$$Y_T = 2Y_{22} + k\,(Y_{21} + Y_{23} + Y_{12} + Y_{32}) \tag{9.2}$$

where contributions of the four regions in the outer corners are absent. The sensitivities for the central zone and each of the vertical and horizontal adjacent zones would be as follows:

$$\text{Central zone}: \quad \delta Y_T/\delta Y_{22} = 2 \tag{9.3}$$

$$\text{Adjacent zone}: \quad \delta Y_T/\delta Y_{ij} = k \tag{9.4}$$

where i and j stand for the other subscripts in Eq. (9.2).

Since k is much less than 1, the central zone has much more than twice the sensitivity of the outer ones. Thus, the central zone will dominate in the combined measurement and it may be said to be *focused*. In a physiological situation if only Y_{22} changes, with nothing changing in the adjacent regions, then

$$\Delta Y_T = 2\Delta Y_{22}, \tag{9.5}$$

and the contributions of the other regions would be eliminated. If one or more of the admittances Y_{21}, Y_{23}, Y_{12} or Y_{32} changes simultaneously with Y_{22}, such as in the case of an extended object, the contribution will be given as $\Sigma k\Delta Y_{ij}$, where ij represents the subscript of the appropriate admittance from the above list. How much this will contribute will depend on the degree of sensitivity with distance away from the centre and the extension of the object. However, as will be shown through findings presented soon that the contribution of areas beyond the focused zone is small, besides, negative sensitivities are also much less than that in TPIM.

Since impedance is just the inverse of the admittance, the sensitivities in terms of impedance values are expected to be the same.

9.2.1.2 FIM-6, Six-Electrode Version

Since the potentials are the same on a single equipotential line, replacing the four inner potential electrodes in Fig. 9.5 by two electrodes placed at the two points of intersection of the four equipotential lines on the electrode plane as shown in Fig. 9.7 will allow the number of electrodes to be reduced to six and will also allow a reduction in the number of potential measuring systems from two to one. This arrangement is expected to give essentially a similar result as given by FIM-8 but with simpler instrumentation. As shown in Fig. 9.7, the potential electrodes are common to both vertical and horizontal measurements of impedance; therefore, the

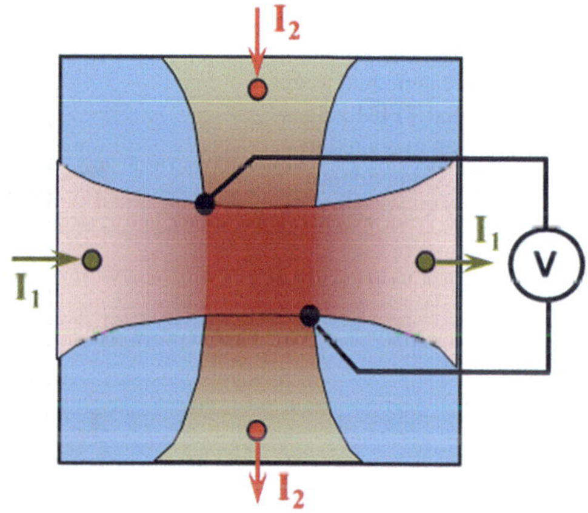

Fig. 9.7 Evolution of FIM-6 from FIM-8, replacing the four potential measuring electrodes by two, which are common to both TPIM measurements (Rabbani et al. 1999)

same voltage measurement system will give potential for both I_1 and I_2. Besides, it has the advantage that using two separate current drives that are in phase but electrically isolated from each other, a single potential measurement will give a summation of the two potentials, giving the focused impedance directly. Although the simple models described above suggest similar sensitivity in both FIM-8 and FIM-6 versions, some differences are expected because of the change in electrode placement geometry; besides, there will be some asymmetry in FIM-6.

9.2.1.3 FIM-4, Four-Electrode Version

The four-electrode version of FIM shown in Fig. 9.8 is derived from the square TPIM (Fig. 9.4) that uses four electrodes at the corners of a square region. Here current is driven between two adjacent electrodes, while the potential is measured across the opposite pair of electrodes. The whole arrangement is then switched to get an orthogonal measurement. Again, as before for FIM-8 and FIM-6, the average of the two measurements has an enhanced sensitivity at the central region giving the desired focused impedance. It has the least number of electrodes and has a symmetrical geometry which makes it attractive in many applications. However, since some of the electrodes are used for both current drive and potential measurement, the switching circuitry is slightly more complex.

9.2.2 *Instrumentation*

For FIM-8 and FIM-4, conventional instrumentation for TPIM can be used with switching of appropriate electrodes in sequence. This needs a computer- or microcontroller-based data acquisition system, and the data is then appropriately analysed to give the FIM. However, instrumentation for the FIM-6 can be designed

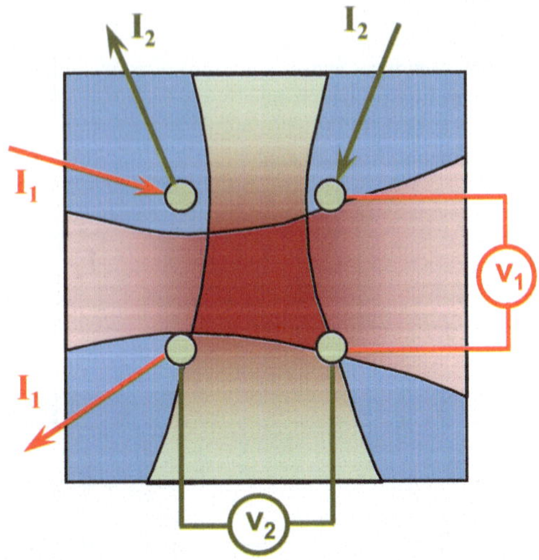

Fig. 9.8 Arrangement of FIM-4. Current and potential electrodes are switched to get two orthogonal TPIM values which are then averaged to get the focused impedance (Rabbani and Karal 2008)

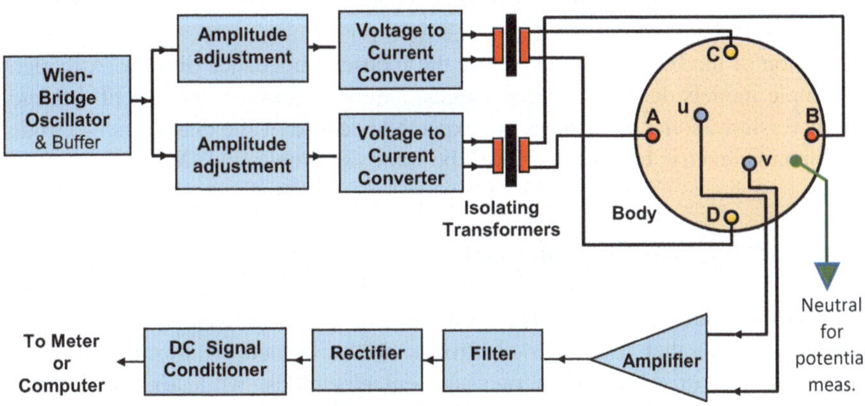

Fig. 9.9 Instrumentation for FIM-6 which uses a single potential measurement to give a summation of the two orthogonal TPIM Measurements (Rabbani et al. 1999)

without a computer or microcontroller (Rabbani et al. 1998, 1999), and Fig. 9.9 shows the schematics of such a circuitry developed by the Dhaka University group. Here output from a single sinusoidal oscillator is used to drive two separate voltages to current converters, the outputs of which are then electrically isolated using two separate current transformers. The two outputs from the transformers in turn drive two sets of orthogonal current electrode pairs of the FIM-6 system. Since the transformers may have slight differences in outputs, the amplitudes of the two current values are made the same using individual voltage adjustment circuitry placed before the voltage to current converters. Thus, the two orthogonal currents have constant amplitudes but are electrically isolated and are coherent, i.e. they have a constant phase relationship. Through appropriate choice of the polarity of the transformer terminals, the phases of the two current waveforms are chosen so that the potentials appearing at the two potential electrodes are in phase. Therefore, a single potential measurement between these electrodes gives a sum of the two orthogonal transfer impedances, i.e. twice the focused impedance. This single potential output is amplified, filtered and converted to dc signal through active rectification. This dc signal is then conditioned and either read off in a voltmeter or acquired in a computer for further processing and display. In this arrangement, the current sources are floating, and a seventh electrode is needed to act as the neutral or reference ground for the potential measuring circuitry, as shown.

Interestingly, reversing the outputs of the transformers will give a reversed phase leading to a subtraction of the two potentials, which may be combined with the focused impedance to give some specific information regarding the values of the matrix elements adjacent to the focused zone.

That the FIM-6 does not need a computer- or microcontroller-based system makes it essentially low cost and has an important bearing in view of its application in low-resource countries.

9.2.3 *Sensitivity Distribution in FIM Using Analytical and Numerical Methods*

The above sections give a basic idea of the electrode configuration and measurement techniques in all the three versions of FIM including visualisation of the focusing effect from the point of view of equipotential lines in 2D. The following two sections present some analytical and numerical approaches to the point sensitivity obtained using different versions of FIM as these will be important in the visualisation for their practical applications.

9.2.3.1 Sensitivity Based on Lead Fields

The first work on a sensitivity analysis of FIM was done by a group in Warwick, UK (Islam et al. 2010). This is based on the work of Geselowitz (1971) that showed that the sensitivity of a tetrapolar impedance measurement (TPIM) at a point can be defined as the scalar product of the vector current densities or the lead fields for unit current injection through both the pairs of electrodes—the current drive pair and the potential measuring pair. This method invokes the reciprocity theorem which essentially states that swapping the current injection and potential measuring electrode pairs in a TPIM gives the same transfer impedance. Since FIM essentially is a sum of two TPIMs, the sensitivities for two individual TPIMs are simply added together to obtain the sensitivity for FIM (strictly speaking, it should be the average, which is half the sum). To derive an expression of the lead fields, four electrodes a, b, c and d were considered that are positioned on the surface of a

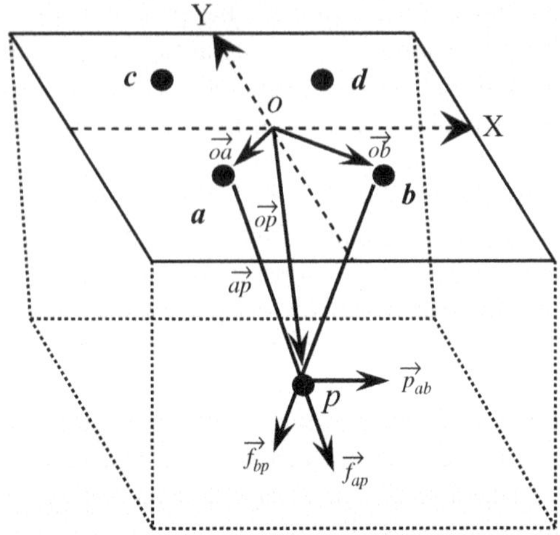

Fig. 9.10 Geometry used for sensitivity analysis based on Geselowitz's lead fields (Islam et al. 2010, slightly modified)

semi-infinite homogeneous medium, as shown in Fig. 9.10. For current injection through electrodes a and b, it was shown earlier that the vector lead fields at a point p within the volume are determined by the vector paths $\overrightarrow{ap}$ and $\overrightarrow{bp}$ from the respective electrodes (Brown et al. 2000a). Considering the origin of the axis system at o, $\overrightarrow{ap} = \overrightarrow{op} - \overrightarrow{oa}$ and $\overrightarrow{bp} = \overrightarrow{op} - \overrightarrow{ob}$. Therefore, for current injections through electrodes a and b, the respective vector lead fields $\vec{f}_{ap}$ and $\vec{f}_{bp}$ are given by

$$\vec{f}_{ap} = \frac{\overrightarrow{op} - \overrightarrow{oa}}{\left|\overrightarrow{op} - \overrightarrow{oa}\right|^3} \quad \text{and} \quad \vec{f}_{bp} = \frac{\overrightarrow{op} - \overrightarrow{ob}}{\left|\overrightarrow{op} - \overrightarrow{ob}\right|^3}. \tag{9.6}$$

Now the lead field at the point p due to the above current injection is given by $\vec{P}_{ab} = \vec{f}_{ap} + \vec{f}_{bp}$. Similarly, lead field at the point p due to current injections through electrodes c and d is given by $\vec{P}_{cd} = \vec{f}_{cp} + \vec{f}_{dp}$. The sensitivity S_p of the tetrapolar impedance measurement at point p is given by the scalar product of the two lead field vectors:

$$S_p = \vec{P}_{ab} \bullet \vec{P}_{cd}. \tag{9.7}$$

To find the sensitivity for FIM, the sensitivity for the two individual tetrapolar impedance measurements is simply summed (or averaged).

Using Matlab™, the sensitivity distributions were calculated for all three FIM electrode configurations for a 200 mm sided cube. Figure 9.11 shows the 2D sensitivity distribution at a depth of 2 mm from the electrode plane. For FIM-8 the drive electrode spacing was 48 mm, and the receive electrode spacing was 12 mm. For FIM-6 the drive electrode spacing was 48 mm and the receive electrode spacing was 25.5 mm measured across the diagonal, which corresponds to sides

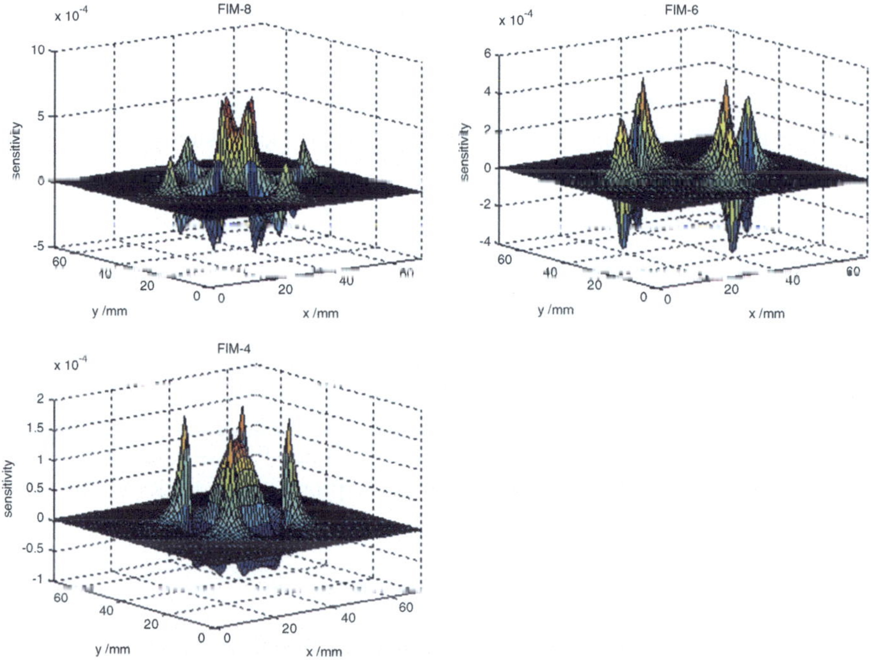

Fig. 9.11 2D sensitivity distribution of FIM-8, FIM-6 and FIM-4 at a depth of 2 mm from the electrode plane in a 200 mm cube. Electrode separations as given in the text. Focusing effect can be observed, together with high sensitivities at the electrodes. Negative sensitivities can also be seen. However, both of these decrease with depth (Islam et al. 2010)

of 18 mm. For FIM-4 the electrode spacing was 18 mm. Focusing effect can be observed, particularly for FIM-8 and FIM-4. High sensitivities can be observed at the electrodes and some negative sensitivity can also be seen. The focusing is not so evident for FIM-6 except for high values under the two potential electrodes. This is because of the relatively large size of the central square focusing region (18 mm side) while that shown for FIM-8 has 12 mm sides. Reducing the size of the zone would reduce the dip at the centre giving a better visual rendition of the focusing. All three figures show high sensitivities at the electrodes besides some negative sensitivity too. However, it was observed that at increased depths these values decrease and the focusing effect is enhanced.

9.2.3.2 Sensitivity Using Finite Element Method

For boundaries with irregular geometries, an analytical solution of the electrical impedance problem is very difficult if not impossible. Recent availability of simulation software packages using finite element method (FEM) like COMSOL has made the job of bioimpedance scientists much simpler and easier. Sensitivity of a tetrapolar impedance measurement at a point can be defined as the scalar product of the vector current densities for unit current injection between the two pairs of electrodes. This method also invokes the reciprocity theorem which essentially states that swapping the current injection and potential measuring electrodes gives the same transfer impedance. Based on this concept, the sensitivity equation for TPIM turns out to be

$$S_{\mathrm{TPIM}} = \frac{\vec{J_1}.\vec{J_2}}{I^2}, \tag{9.8}$$

where $\vec{J_1}$ is the current density vector at the concerned point due to current injection through the current drive electrode pair and $\vec{J_2}$ is the current density vectors at the same point due to current injection through the potential measuring pair. The magnitudes of the two injected currents I are the same, and its square in the denominator is used to normalise the sensitivity. Since FIM is essentially a sum of two orthogonal TPIM arrangements, through appropriate electrodes for the three versions, the sensitivity for FIM is defined as

$$S_{\mathrm{FIM}} = S_{\mathrm{TPIM1}} + S_{\mathrm{TPIM2}}, \tag{9.9}$$

where S_{TPIM1} and S_{TPIM2} refer to the two orthogonal TPIM configurations.

Based on the above mathematical analysis, Abir (2016, Personal communication) obtained sensitivity distributions of Square TPIM and FIM-4 using COMSOL for a comparative study which is shown in Fig. 9.12. The volume considered was a cube with sides of 500 mm, and electrodes for the two modalities were the same, assumed to be fixed centrally on one of the sides, having a separation of 20 mm between adjacent electrodes. 2D sensitivity distributions were obtained at depths of 5 mm, 10 mm, 15 mm and 20 mm, all measured from the electrode plane into the volume. The numerical values corresponding to the colour schemes are different for the different individual figures, which should be noted. The maximum and minimum values in each figure are mentioned above each which can be a guide in visualisation.

It can be seen that at 5 mm depth, the maximum negative sensitivity of TPIM is more than the maximum positive sensitivity. On the other hand, for FIM, the maximum negative sensitivity is less than half of the maximum positive sensitivity. At a depth of 10 mm, the TPIM has almost the same numerical values on both sides on the sensitivity scale, while the negative sensitivity value is 1/6 that of the positive sensitivity value for FIM. The negative sensitivity is still present for TPIM at 20 mm depth (one electrode spacing), while it disappears for FIM at 15 mm depth (3/4 electrode spacing) and beyond. Again, looking at the maximum positive sensitivity values, the fall of values with depth is sharp for TPIM but less so for FIM. This feature was also observed in the experimental sensitivity study shown in Figs. 9.15 and 9.16 for linear TPIM and FIM-6. This is an advantage for FIM when studying 3D objects with finite dimensions at shallow depths, as the contributions of different segments of the same object at different depths will have less variation among them.

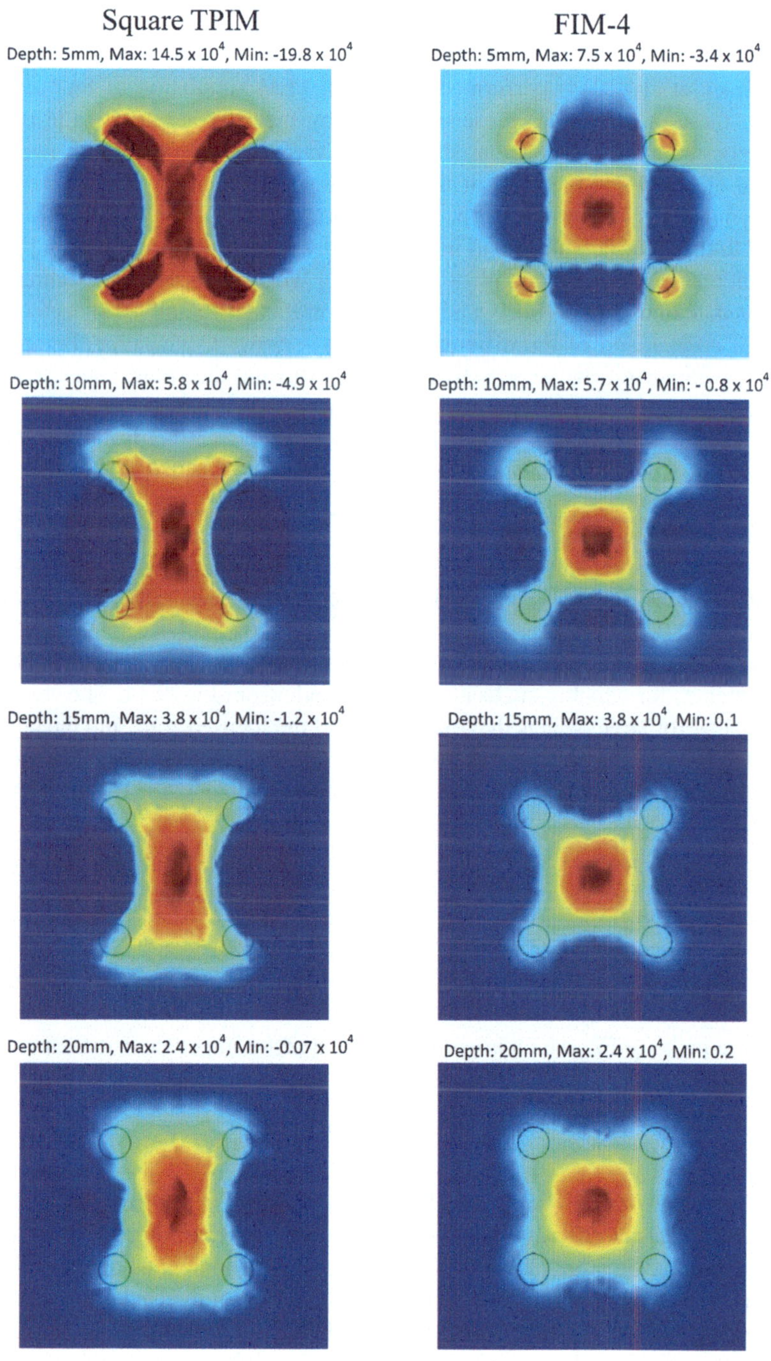

Fig. 9.12 Comparison of sensitivities square TPIM (left) and FIM-4 (right) at different depths using COMSOL finite element simulation. Electrode positions (separation = 20 mm) are indicated by black circles. The colour schemes for sensitivities are different for different figures, each plotted between their individual maximum (dark red) to minimum (dark blue). The sensitivity value is 0 for the background for each. The depth, maximum and minimum sensitivity values are indicated above each figure. Note, negative sensitivity disappears for FIM at 15 mm depth, while it is still present for TPIM at 20 mm depth (Abir, 2016, University of Dhaka, Bangladesh. Personal Communication)

Pettersen et al. (2014) performed COMSOL FEM simulation of four versions of FIM, three of which are those developed by the Dhaka group while the last one, suggested by them, is essentially a slight variation of FIM-6, with two more potential electrodes at the ends of the other diagonals, to make the focused zone more symmetric. They showed the sensitivities in three planes; that in a plane parallel to the electrode plane and in two suitably chosen planes orthogonal planes perpendicular to the electrode plane. From their work, the sensitivity figures have been separated out appropriately and presented in Figs. 9.13 and 9.14, respectively, so that a comparison of the different versions is easier to visualise. Figure 9.13 shows a comparison of the sensitivities parallel to the electrode plane at a very shallow depth. The focusing is evident, but there are individual variations in the pattern details. The negative sensitivity also varies between the different versions. Here also the numerical values of the colour schemes are different for individual figures.

Figure 9.14 shows that in FIM-4, the depth of positive sensitivity zone exceeds that of the negative sensitivity zone more compared to others. Besides, the extent of the negative sensitivity zone is also less compared to others in this version. Thus, Figs. 9.13 and 9.14 combined will help in choosing a particular version of FIM for any individual application. Figure 9.13 does not show the focusing effect for slightly deeper planes, for which visualisation can be made from Fig. 9.14. The study also suggests that for large objects that extend beyond the focused zone into the negative sensitivity zones, the results will have to be interpreted carefully. However, for objects that are limited to the dimensions of the focused zone, the results are easier to interpret.

9.2.3.3 Focused Impedance

When one makes a measurement of impedance, there will be contribution from the whole volume where the resistivity $\rho(x,y,z)$ at each of the points will contribute. The focused impedance FZ, which is essentially the transfer impedance, is then given as

$$\mathrm{FZ} = \iiint \rho\,(x, y, z)\; S_{\mathrm{FIM}} dxdydz. \tag{9.10}$$

The point sensitivities shown in the previous section help visualise the contributions of objects at different points of the volume. However, most of the practical applications involve large objects that have impedivity (analogous to resistivity but for frequency-dependent impedance) different from that of the background. In such situations, numerical solutions will involve solving Eq. (9.10). Figure 9.15 shows impedance variations due to a large object of finite dimension for FIM-4 (Abir and Rabbani 2014) obtained using COMSOL simulation. In this study, the object is moved through the centre of the square focused region, parallel to one of the sides, and is shown for different depths. The relevant parameters are given in the

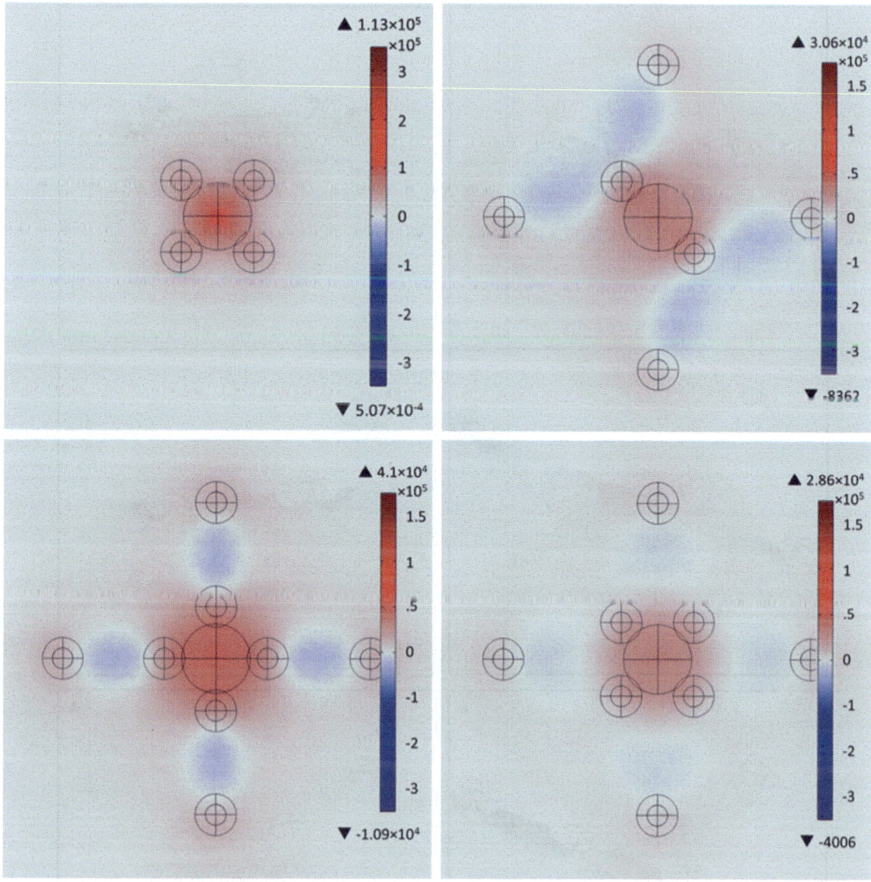

Fig. 9.13 2D sensitivities parallel to the electrode plane at a shallow depth of four versions of FIM: FIM-4 (top-left), FIM-6 (top-right), FIM-8a (original version, bottom left) and FIM-8b (new version, bottom right). The last one is a slightly different eight-electrode version based on FIM-6, suggested by Pettersen et al. (2014). Shades of red indicate positive sensitivity, while shades of blue indicate negative sensitivity. Sensitivity figure produced by Pettersen et al. (2014) using COMSOL finite element simulation. Note, the numerical values of the colour scales are not the same in the different figures

figure caption. The focusing effect is clearly reproduced, and there is no negative sensitivity in this direction. Discernible values are obtained to a depth almost equal to the electrode separation. Figure 9.16 shows the variation of the impedance value with depth for objects with different radii. The relevant parameters are given in the figure caption. Here objects with diameters equal or larger than the focused zone have been used. If the depth of an object is known, such curves may be used to obtain the sizes of objects embedded in a volume conductor at shallow depths.

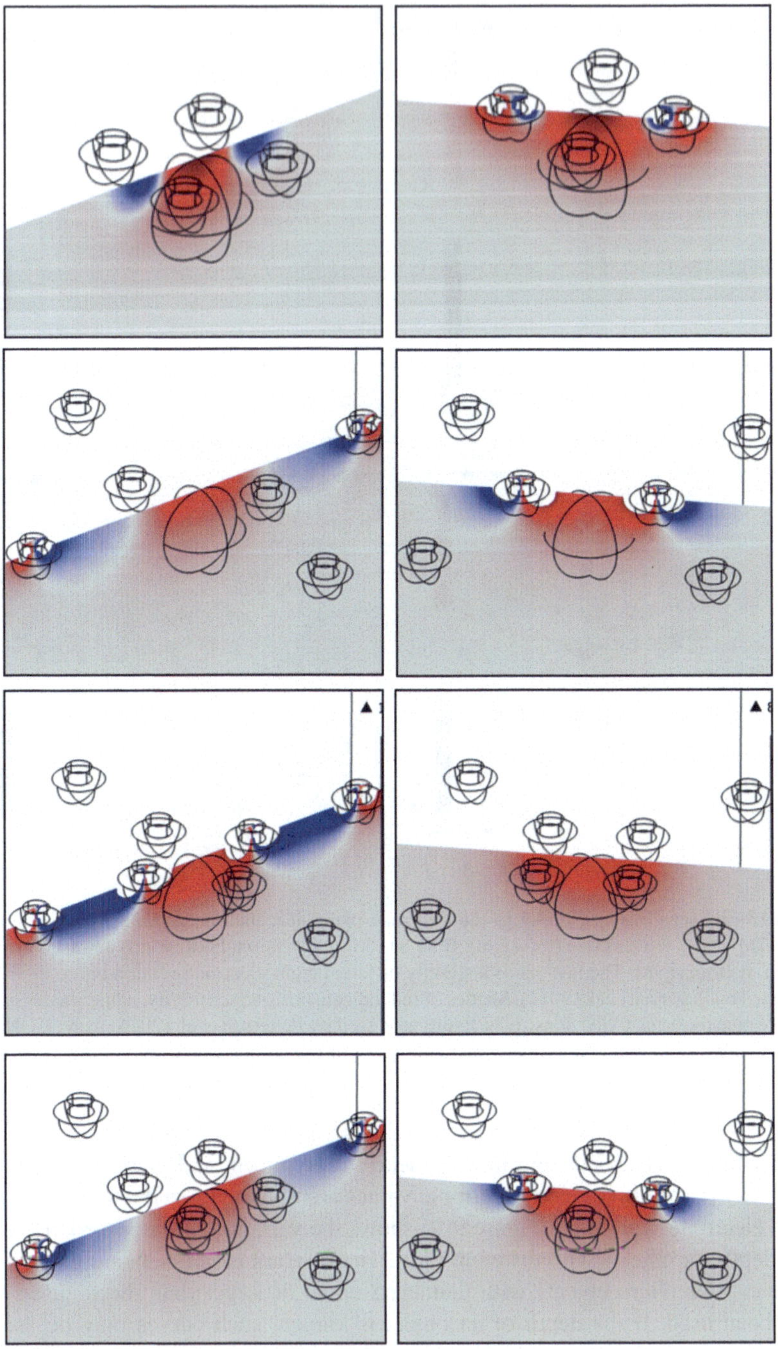

Fig. 9.14 2D positive (red) and negative (blue) sensitivity in two orthogonal planes perpendicular to the electrode plane of four versions of FIM, showing the relative depth of sensitivity in the third dimension (Pettersen et al. 2014). FIM-4 (top), FIM-6 (second from top), FIM-8a (third from top) and FIM-8b (bottom most)

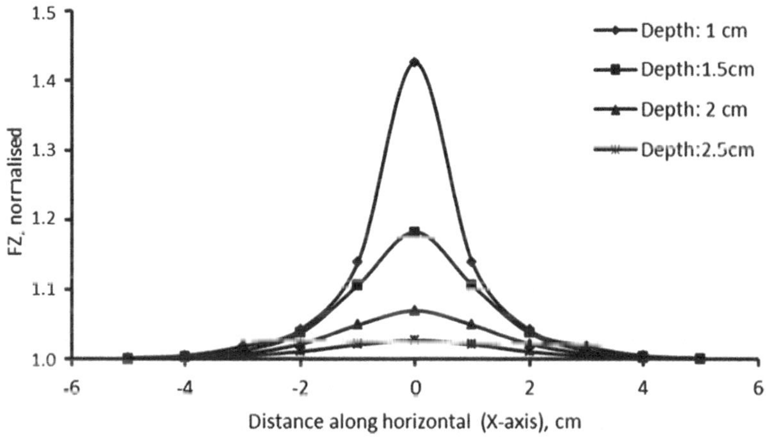

Fig. 9.15 Variation of normalised impedance FZ for FIM-4 with an object moved through the centre of the square focused region, parallel to one of the sides, for different depths. Background resistivity, 1 Ωm; object resistivity, 10 Ωm; electrode separation, 2 cm; object diameter, 2 cm (Abir and Rabbani 2014)

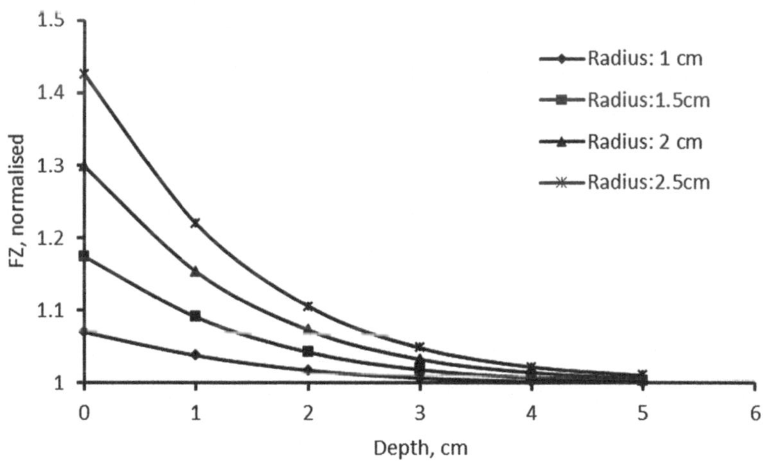

Fig. 9.16 Variation of normalised impedance FZ for FIM-4, with depth, for objects with different radii, placed at the centre of the focused region. Background resistivity, 1 Ωm; object resistivity, 10 Ωm; electrode separation, 2 cm (Abir and Rabbani 2014)

9.2.4 Sensitivity Distribution in FIM Using Experimental Methods

The first verification of the newly proposed FIM in fact came from experimental observation rather an analytical solution (Rabbani et al. 1998, 1999). At that time a numerical package like COMSOL was not available to the Dhaka University group, and the innovation mainly originated from a visualisation based on equipotential surfaces followed by phantom measurements using homegrown instrumentation for FIM-6 shown in Fig. 9.9 earlier. With the availability of the COMSOL package, the job of verification of a concept has been simplified, but it still has to go through a phantom measurement before any application on human subjects. The following sub-sections describe some of these phantom experiments and the results.

9.2.4.1 Sensitivity in 2D

Firstly sensitivity studies were performed on a 2D phantom made of a plastic-wood tray containing saline as shown in Fig. 9.17. Metallic pins were inserted vertically into the base to make the different electrode arrangements for TPIM and FIM, centrally within the tray. Saline depth was about 10 mm so that the pin heads were above the water from which electrical connections were taken out. Carbon rods glued to the base were also used in some experiments instead of the metallic pins as electrodes. The tray was kept horizontal as much as possible so that the water depth is uniform throughout. Cylinders made of metals and insulators of appropriate diameters were placed at different positions with respect to the electrodes. Impedance measurements were taken with and without the cylinder, and the differences gave the sensitivity. Figure 9.18 shows the results of a typical measurement of the change in impedance, in arbitrary units, using an insulating cylinder and presented in a matrix. The red- and blue-circled matrix positions indicate the positions of the current electrodes, while the yellow-circled matrix positions indicate the positions of the potential electrodes. The degree of darkness of the square matrix elements indicates approximately the resulting change in impedance in the corresponding matrix positions. There appear some matrix elements with small negative values. These are the points with negative sensitivity.

Sensitivities were also measured in a volume phantom as shown in Fig. 9.19. It is basically a plastic water tank with a square horizontal cross section measuring 30 cm × 30 cm inside. Water can be poured to a height of 30 cm, making a

Fig. 9.17 Phantom to study 2D sensitivity distributions for TPIM and FIM

0	0	0	0.17	0.3	2.6	2.1	1.7	0
0	0	0	- 1.7	XA	3.4	2.6	1.7	0
0	0	- 1.7	- 4.3	- 0.3	6.6	4.8	2.6	2.1
0	- 0.9	- 4.3	XU	14.7	10.7	6.4	3.4	2.6
1.3	XC	- 0.3	14.7	20.3	15.2	- 0.3	XD	0.3
2.6	3.4	6.4	10.7	14.7	XV	- 4.3	- 0.9	0
2.1	2.6	4.8	6.6	- 0.3	- 4.3	- 1.7	0	0
0	1.7	2.6	3.4	XB	- 1.7	0	0	0
0	1.7	2.1	2.6	0.3	0.17	0	0	0

Fig. 9.18 Sensitivity for FIM-6 measured using cylindrical objects in a 2D saline-filled phantom (Rabbani et al. 1999)

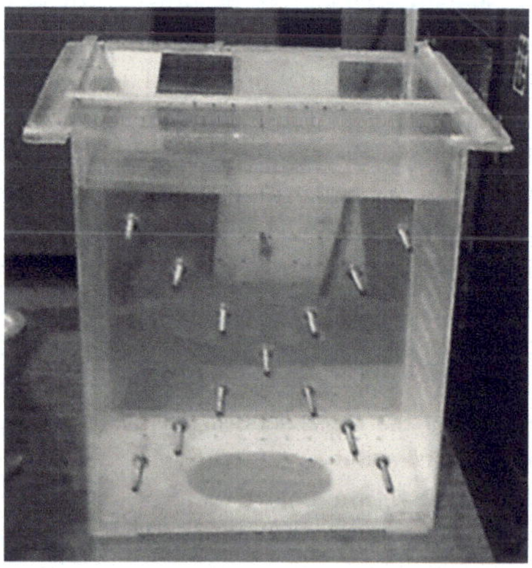

Fig. 9.19 Deep volume phantom for both 2D and 3D sensitivity studies. Electrodes were fixed onto a wall, while an object (cylindrical or spherical) was suspended from the top using a fine thread at desired positions

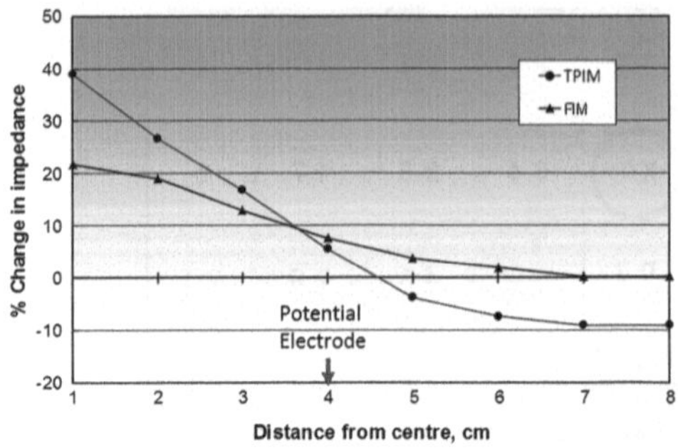

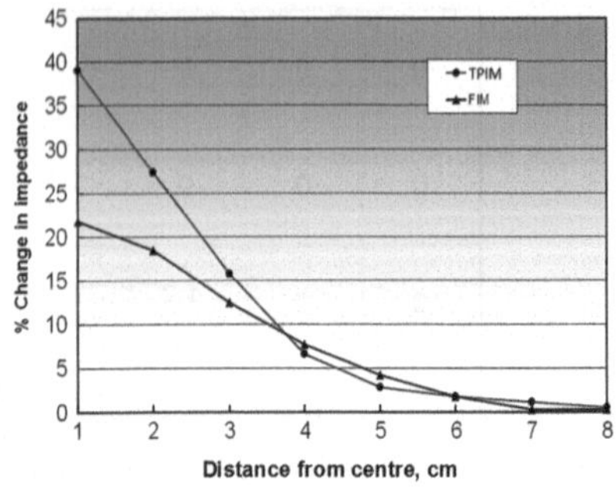

Fig. 9.20 Comparison of sensitivities for linear TPIM and FIM-6. The graphs in the left-hand picture shows values along the electrodes, measured from the centre outwards, while graphs in the right-hand picture shows values along a perpendicular direction to the above, again, measured from the centre outwards. There is a considerable negative sensitivity in TPIM after the position of the potential electrode on the left-hand graph. There is no negative sensitivity for TPIM in the perpendicular direction, and none for FIM in any of the two directions shown (Rabbani et al. 1999)

cubic phantom. Electrodes with the desired geometry were fixed onto a side wall so that 3D sensitivity can be studied hanging an object at different positions with respect to the electrodes using a fine thread. Using a similar phantom, a comparative study between the simple TPIM and FIM-6 showed the improvement obtained in terms of the negative sensitivity in the latter as shown in Fig. 9.20. It shows sensitivities along the direction of electrodes of the linear TPIM (left-hand figure) and in a direction perpendicular to the above (right-hand figure), measured from the centre of the system for both TPIM and FIM-6. It can be seen that FIM-6 has no negative sensitivity in either of these directions, while TPIM has a significant negative sensitivity beyond the position of the potential electrode (shown by a red arrow at 4 cm on the horizontal axis on the left-hand picture) along the direction of the electrodes and no negative sensitivity in the perpendicular direction. Of course FIM-6 has some negative sensitivity along the diagonals as can be seen in Fig. 9.18 but is much smaller in magnitude.

A comparative visualisation of the 2D sensitivities of FIM-6 and FIM-4 measured using a 3D phantom similar to that shown in Fig. 9.19 are presented in Fig. 9.21 where a spherical object was moved closed to the electrode plane. Here the matrix elements are larger than that in Fig. 9.18, and one element covers the whole of the central 'focused region'. The spherical object size was such that it almost covers the whole of the cross section of the focused zone. FIM values were normalised with reference to the respective values at the centre. In both cases, it can be seen that the values at the centre were at least four times as great compared to those at the adjacent matrices. Negative values are present in some matrix elements but are less than 10% of the value at the central focused zone.

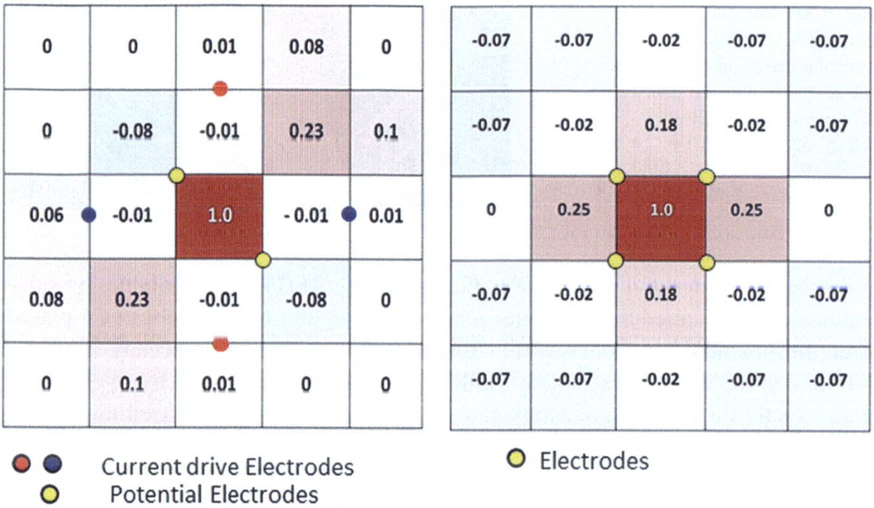

Fig. 9.21 Comparison of 2D sensitivities for FIM-6 and FIM-4. The values in the adjacent matrix elements are at best 25% of that at the central focused zone, but their positions are different in the two systems. Negative sensitivity values are less than 10% in both cases, again, at different locations (Rabbani and Karal 2008)

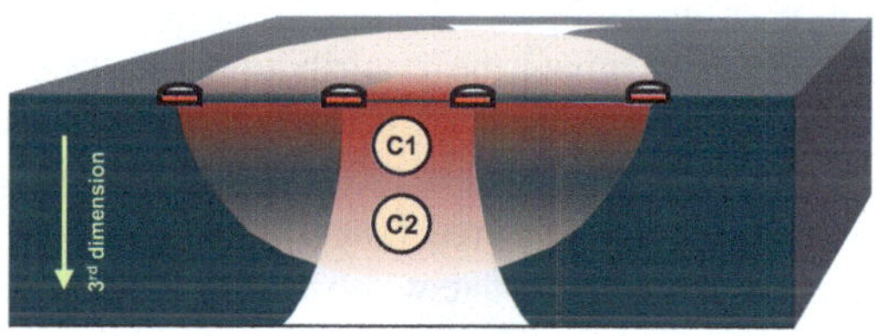

Fig. 9.22 Visualisation of sensitivity of TPIM in the third dimension. The zone of overlap between the two shaded zones is the sensitive zone. Details given in the text

9.2.4.2 Visualisation of Sensitivity in 3D

To visualise the sensitivity of FIM in 3D, Fig. 9.22 presents a simplified schematic diagram of a linear TPIM. Sensitivities for other configurations of TPIM and for FIM may be visualised in an analogous way. Since the TPIM measurement system depends on two parameters, the current driven through the two outer electrodes and the potential measured across the inner pair of electrodes, both will have effects on the sensitivity at any point of the volume conductor. Firstly, the distributed current density through the volume is highest near the line of electrodes (because of the shortest route length) and decreases with distance into the volume, as represented by the approximate hemispherical vignetted zones in the figure. Therefore, the sensitivity of the system will be high near the line of electrodes and decrease with distance into the volume.

Secondly, since it is the potential difference measured between the potential electrodes that gives the overall transfer impedance of the measurement, the volume of material bound by the two equipotential surfaces (in 3D) passing through the

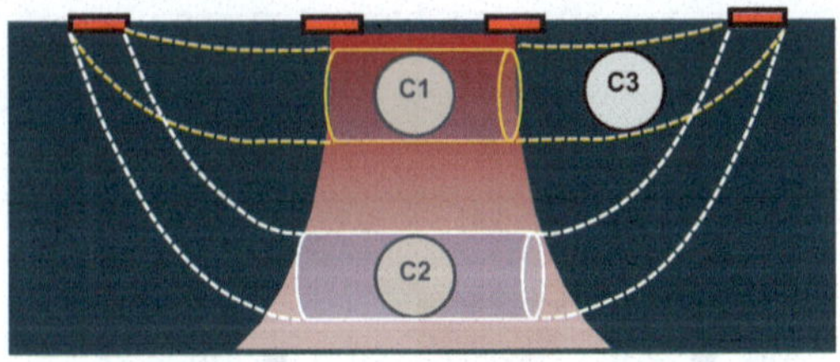

Fig. 9.23 Current pipes at two depths and the contribution of an object of the same size to each pathway as a means to visualise sensitivity

potential electrodes will contribute to the sensitivity. This is essentially the sensitive volume as represented by the lighter region in the figure. Now, an object C1 placed near the potential electrodes within this sensitive volume will occupy a greater fraction of the respective sensitive width compared to that occupied by an object C2 with exactly the same dimensions and electrical properties but placed away from the line of electrodes. Therefore, the contribution of object C1 will have a greater effect on the measurement compared to an object at C2, i.e. the sensitivity will again decrease away from the line of electrodes within the sensitive volume. These two factors combined will contribute to an overall reduction in sensitivity of the system away from the line of electrodes.

The latter concept is better visualised through Fig. 9.23 in which a concept of current pipe is introduced. The product of the impedance of the yellow cylindrical region around object C1 and the current through this cylinder (considered as a current pipe) gives the contribution of the cylinder to the measured potential. Similarly, the product of the impedance of the white cylindrical region around object C2 and the current through this cylinder gives the contribution of the white cylinder to the measured potential. The yellow and the white cylinders equal the respective sensitive zone widths between the equipotential surfaces passing through the potential electrodes. Clearly, length of the yellow cylinder is smaller than that of the white cylinder. Therefore, contribution of the impedance of the object C1 to the total impedance across the yellow cylinder will be more than the contribution of the impedance of the object C2 to the total impedance across the white cylinder. Or in other words, the sensitivity of C1 will be more than that of C2.

If the current pipes have the same cross section, it is clear that the total current flowing through the yellow cylinder will be more than that through the white cylinder since the former is closer to the current electrodes. This will also reduce the contribution of C2 compared to C1. Therefore, both the above factors will combine to further reduce the contribution of C2 compared to C1. This is stated in a slightly altered way that the sensitivity to the impedance measurement at the geometric position of C2 is less than that at the geometric position of C1.

However, this visualisation does not explain the negative sensitivity observed for which a bending of equipotential surfaces by objects needs to be considered, which is basic to TPIM. An object placed at C1 with resistivity higher than that of a uniform background bends the equipotential surfaces inwards (towards the object) so that the potential measuring electrodes measure a greater potential difference compared to the situation without the object at C1 (i.e. only for the background).

On the other hand, if the same object is placed at a position C3, between the current and potential measuring electrodes, it will bend the equipotential surfaces towards it which will result in a lower potential difference measured by the potential electrodes. If the position of the object is unknown, this measurement may be interpreted as being caused by an object with resistivity less than that of the background and placed in between the potential electrodes (such as C1). Therefore, it is said that a position between the potential and current electrode (such as C3) has a negative sensitivity.

9.2.4.3 Sensitivity in 3D

Sensitivity measurements in 3D for six-electrode FIM using a saline-filled 3D phantom are shown in Fig. 9.24 (Iquebal and Rabbani 2010). The dimension of the phantom was 50 cm × 50 cm × 50 cm, spacing of the current drive electrodes in each pair being 18 cm, while the spacing between the diagonal potential measuring electrodes was 5 cm. The latter corresponds to a square focused zone with about 3.5 cm sides at the surface. Insulating and conducting spheres of various diameters were used for the sensitivity measurements in 3D with the centres of these spheres kept concentric with that of the focused zone. Figure 9.24 shows the measured potential, which is proportional to the transfer impedance. It can be seen that for a given object, the impedance falls sharply with depth till it levels off to the background value. The behaviour is similar to that in Fig. 9.16, obtained through FEM simulation using COMSOL. The impedance is lower for smaller objects, which is expected. The results for conducting spheres were similar, but the curves were in the opposite direction, having potentials less than that for the background.

At a certain depth (here, 5 cm), the variation of the measured impedance (represented by potential here) with volume of the spherical objects is shown in Fig. 9.25. It appears to increase almost linearly with volume. This is an important result as this allows measurement of volumes of objects of known conductivity if the depth is known. Such relationships are also useful in designing applications of FIM to measure reasonably large organs at shallow depths of the human body

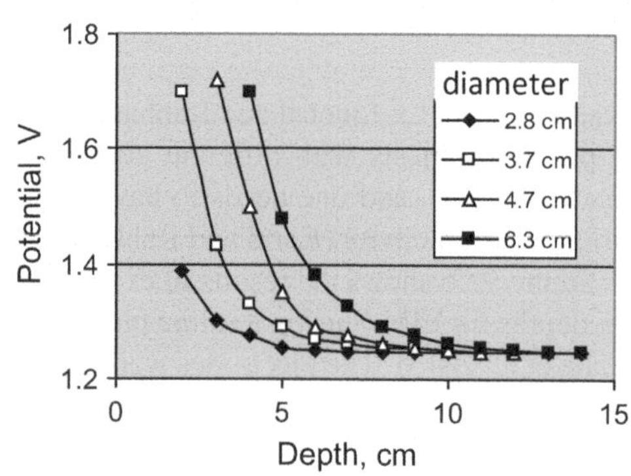

Fig. 9.24 Depth sensitivity of focused impedance (represented by potential) for insulating spherical objects of different diameters (Iquebal and Rabbani 2010)

Fig. 9.25 Variation of focused impedance (represented by potential) with volume of insulating objects, at a certain depth (Iquebal and Rabbani 2010)

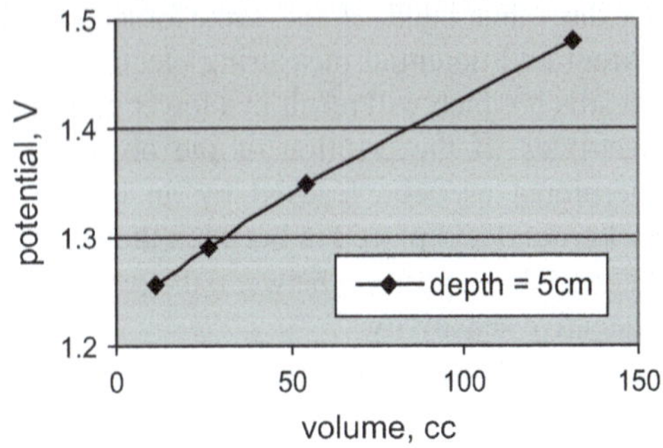

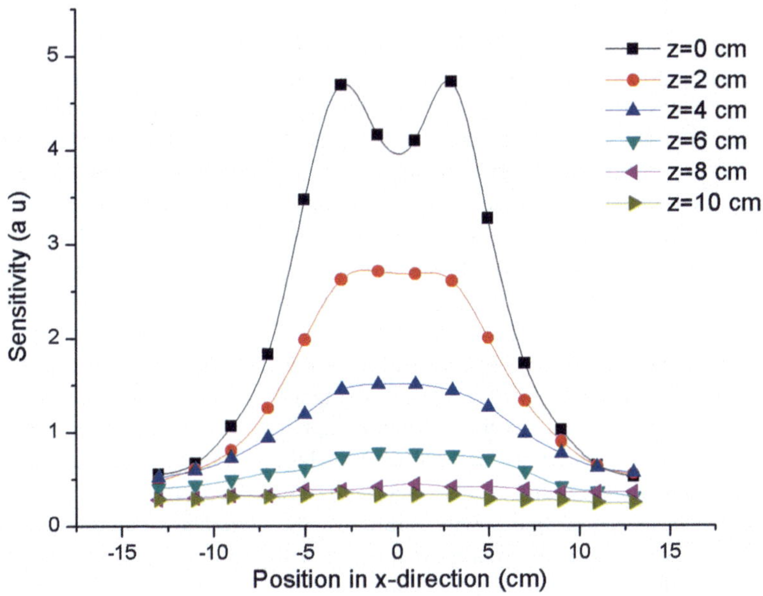

Fig. 9.26 Sensitivity variation of eight-electrode FIM along a line through the centre (x-direction), as measured using a spherical insulator of diameter 3.8 cm in a phantom. The central 7 cm range ($\pm$3.5 cm) is the focused zone. The variation of sensitivity (in arbitrary units—au) at different depths (z-values) are also shown (Saha et al. 2013)

(Kadir et al. 2013; Iquebal and Rabbani 2013; Abir et al. 2013; Ahmed et al. 2014). In practice, objects with different conductivities compared to a background will be encountered, and one needs to have an insight for sensitivities for objects with different conductivity (Karal and Rabbani 2009).

Figure 9.26 shows the results of experimental sensitivity measurements at different depths for FIM-8 using a saline phantom of dimension 30 cm × 30 cm × 30 cm (Saha et al. 2013). The electrodes were fixed on a side wall. The separation of each current electrode pair for the respective TPIM component was about 20 cm, while the separation of each electrode pair was about 10 cm. Since the latter corresponds to the diagonal of a square focused zone in FIM-8, the sides of the focused zone were about 7 cm. The sides were arranged to lie along the horizontal and vertical directions. An insulating plastic ball of diameter 3.8 cm was hung at different positions along the focused zone and beyond, in both directions, with $x = 0$ at the centre of the focused zone. Its centre was kept at a horizontal level of the centre of the focused zone. The depth, given by the z-direction, corresponded to centre of the spherical object, and it was defined as zero when the object just touched the electrode plane, i.e. when the centre of the ball was at a depth of 1.9 cm with respect to the electrode plane. The focusing effect is evident. In the above, the object covers the full cross section of the focused zone, and for the lowest depth, small spikes are seen in the sensitivity curves at the edges of the focused zone, which however disappears with increased depth. Discernible measurements are obtained to a depth of about 6 cm, almost equal to the separation of the potential electrodes.

9.2.5 *Advantages and Disadvantages of Different Versions of FIM*

From the sensitivity studies presented before, it is clear that out of the three versions, FIM-4 gives the best overall focusing, particularly beyond a certain depth when the large spike in sensitivity under the electrodes and the negative sensitivities are almost eliminated. It also needs the least number of electrodes among the three versions. However, since the same electrodes are used both for current drive and potential measurements through programmed switching and since there are issues of noise shielding and isolation between the current drive side and potential measurement side, the circuitry is more complex.

Between FIM-8 and FIM-6, the former has more geometrical symmetry of the focused region. However, the FIM-6 has the advantage that it allows the summed FIM to be obtained through a single potential measurement, which makes the instrument simpler and of low cost, as mentioned before. It can be designed without the need of computerised data acquisition and analysis which are needed for the other two versions of FIM.

Another point needs to be considered while considering applications for the different versions of FIM, which is the effect of the separation of electrodes. In the cross-sectional schematics of Fig. 9.27, two sets of outer current electrodes (red and dark red) and two sets of inner potential electrodes (blue and dark blue) are shown which are relevant to both FIM-8 and FIM-6. The current driven through any pair of outer electrodes will have the highest component near the surface and will gradually fade away in an approximately semi-circular fashion in the third dimension as represented by the vignetted approximate semicircles. Although the current will spread out throughout the volume, an effective limit can be visualised that will contribute significantly down to the noise limit of the instrumentation. This effective limit is shown schematically here, which will have a greater depth for the larger

Fig. 9.27 In eight or six electrodes, FIM it is possible to change the depth of sensitivity and the size of the focused zone independently changing the current drive and potential electrode separations separately as visualised by the semicircles through the outer current electrodes and equipotentials through the central potential electrodes

separation of the electrodes. Therefore, for fixed potential electrode separation, significant contribution from an object in the third dimension may be obtained to a greater depth with a larger separation between the current electrodes. As for the inner potential electrodes, the equipotential lines passing through these electrodes give a boundary to the inner volume that has more sensitivity (here we consider only positive sensitivity), obviously fading away from the surface (indicated by the vignetted zones). This implies that for a fixed separation of current electrodes, a smaller separation between the potential electrodes gives a smaller focused zone. Thus, the focused zone and the depth of sensitive zone can be independently varied to some extent in FIM-8 and FIM-6.

On the other hand, in FIM-4, increasing the electrode separation to achieve sensitivity to a greater depth automatically increases the size of the focused zone. These cannot be independently varied.

9.2.6 Visualisation in Terms of Equipotentials or Point Sensitivity?

In this article different aspects of FIM have been visualised in two ways—from a viewpoint of equipotentials and from a viewpoint of point sensitivity. Although some authors prefer the latter, it sometimes leads to mistaken visualisation, particularly if the electrodes are far apart. The treatment of point sensitivity shows high sensitivity under all the four electrodes of TPIM and low sensitivity at other places, even in between the potential electrodes. This is more pronounced if the potential electrodes are far apart. Now the total impedance measured is an integral over the whole volume as given by Eq. (9.10). Therefore, regions with non-zero but low sensitivity may contribute greatly to the total impedance if the volume is large, which is not apparent from a picture of point sensitivity. Therefore, it would be useful if both viewpoints are used in combination for visualisation, as this would suggest appropriate methods and electrode configurations for a particular geometry quickly, without going into elaborate numerical simulation first, which can be taken up later for details after a few initial choices are made up as above.

9.3 Application of FIM

The applications primarily suggested and tried for FIM are those that can be measured non-invasively, utilising the 3D sensitivity of FIM at shallow depths. In fact in several applications of EIT where large regions comprising of many pixels are marked out and the pixel values are summed to look for variations of large regions of interest, FIM may be applicable. In biomedical applications several attempts have been taken so far in the applications of FIM, mostly by our group at University of Dhaka. There have been some successes already, and several others look promising. The following sections present an account of some of these efforts.

9.3.1 Lungs Ventilation Measurement

9.3.1.1 Localised Measurement

This is a very promising application area of FIM. In fact this was one of the first applications in the early period of FIM development (Kadir et al. 2009). Figure 9.28 shows the electrode connections made for a study of the lower right lung using FIM-6. For this measurement, the subject inhaled air as much as possible and then breathed out a little at a time to a bellows-type "Vitalograph" spirometer and held breath in between for brief periods when FIM measurements were taken. Figure 9.29 shows the change in relative impedance against the volume of exhaled air as measured using the spirometer. This has a good linear fit indicating the success of the ventilation measurement. Of course if the lower boundary of the lung falls within the sensitive zone, this may contribute to a change in the measured impedance which is not truly related to ventilation, and one has to be careful since the lung moves up and down, through an excursion of a few cm. Further on, percentage change between full expiration and full inspiration was measured at the four quadrants of the lungs, and the results obtained on the same subject are shown in Fig. 9.30. It can be seen that the lower right lung has the greatest variation which is expected since the heart takes up a considerable space around the lower left lung.

The above shows measurement on a single individual, but there are large person-to-person variations, even if they have normal lungs. This could be due to various factors such as thickness of fat and muscle tissue between the skin and lungs, which can change the fraction of the sensitive volume of FIM within the lungs itself. Besides placement of electrodes may not be exactly the same for different persons. Furthermore, placement of an electrode directly on a rib or in an intercostal space may have an effect.

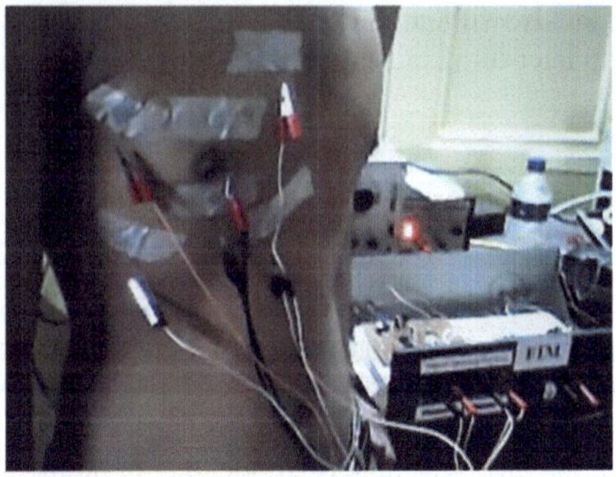

Fig. 9.28 Electrode connection for localised lung ventilation study using FIM-6 (Kadir et al. 2009)

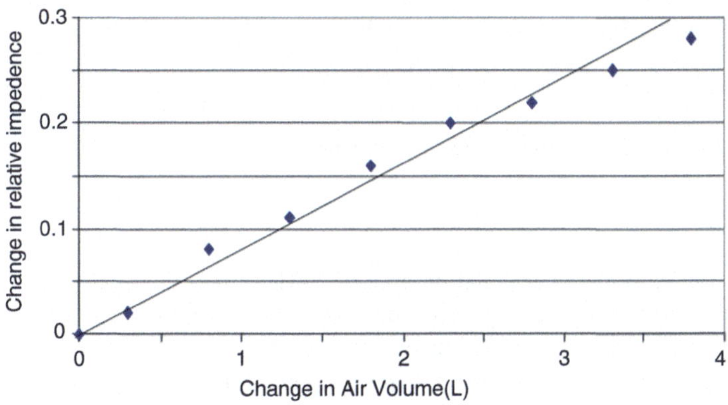

Fig. 9.29 Change in relative impedance (FIM-6) of lower right lung of a subject against exhaled air volume, measured using a bellows-type spirometer (Kadir et al. 2009)

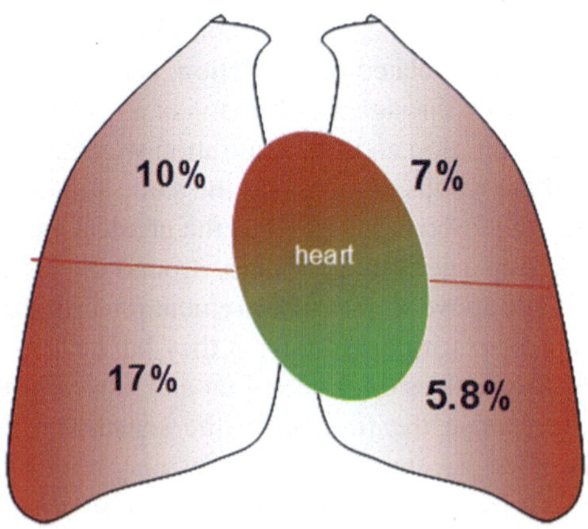

Fig. 9.30 Percentage change in relative impedance (FIM-6) of four quadrants of the lung of the same subject measured between full inspiration and full expiration (Kadir et al. 2009)

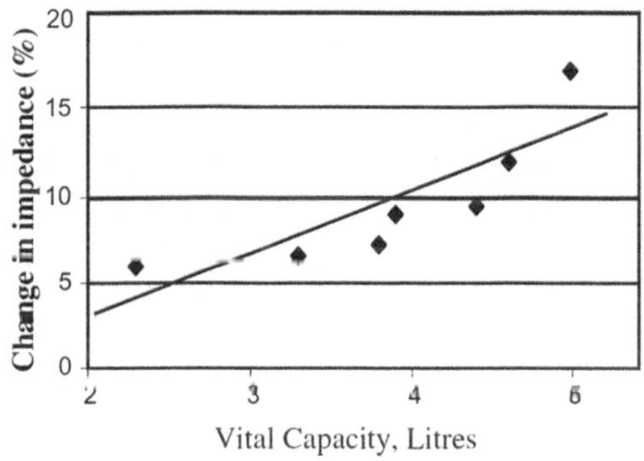

Fig. 9.31 Percentage change in focused impedance in lower right lung against vital capacity for seven normal subjects (Kadir et al. 2009)

However, to see how a localised measurement relates to the whole air volume transferred in a full inspiration-full expiration manoeuvre (vital capacity) for normal subjects, a small study was carried out on seven normal subjects, all in the 25–30 age group. Percentage change of FIM impedance of the lower right lung of these subjects was compared to the corresponding vital capacity, measured using a bellows-type spirometer (Vitalograph), and the results are shown in Fig. 9.31. There appears to be a linear dependence; however, the factors mentioned above possibly have contributed to the deviations, and if corrected for through other measurements, FIM may be applied in most of the lung-related applications attempted by EIT, which includes monitoring of patients in hospitals under artificial ventilation.

9.3.1.2 Mapping of the Upper Thorax for Lungs Ventilation

An attempt was made to map the whole of the upper thorax, both from the front and back, using a hand-held six-electrode FIM probe as shown in the left and middle pictures in Fig. 9.32 (Kadir et al. 2010; Ferdous et al. 2013). The electrodes were spring loaded to conform to body shapes. Wet cotton pads were used at the tips of the electrodes for electrical connection to the skin. The seventh electrode at the centre acted as the common terminal to the potential measuring circuitry. The probe was held diagonally so that the focused zones were small squares as shown in the rightmost schematic diagram. The probe was shifted laterally and vertically such that the focused square zones lie adjacent to each other as shown in the rightmost diagram. The percentage changes in FIM impedances between full inhalation and full expiration were measured and mapped into a matrix as shown in Fig. 9.33, for measurements both from the front and the back, for a typical normal subject (details given in the figure caption). That from the back was rearranged in the matrix so that the right and left sense of the body remains the same for both the maps in the figure. It can be seen that magnitude of changes are more from the front than from the back for the corresponding matrix positions. This is expected as the interposing muscle and other tissues between the skin and the lungs have more thickness in the back

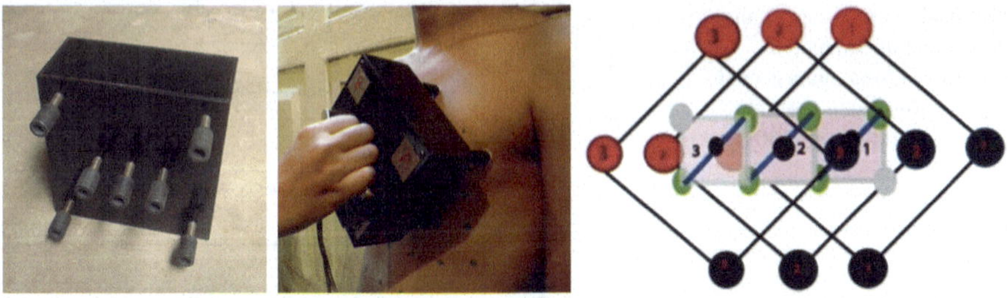

Fig. 9.32 Hand-held six-electrode probe for FIM-6 with spring-loaded electrode terminals to conform to body curvature (left and middle). Diagonal placement of probe giving square focused zones. Horizontal and vertical movement of probe produces measurement in a matrix. This is shown in the right-hand diagram for horizontal movement only (Ferdous et al. 2013)

Measurement from Front						Measurement from Back					
Right				Left		Right				Left	
7.7	5.3	15.1	6.8	4.8	2.4	3.8	7.8	4.8	2.6	4.6	7.6
8.1	11.7	14.3	6.7	4.5	3.8	1.5	1.6	4.4	7.2	4.4	7.6
11.3	15	16.7	12.4	13.6	6.3	3.3	10.7	4.8	5	10.5	6
.5	12.1	10.6	4.8	12.7	20.4	9.4	7.1	3.3	4.3	3.6	9.5
19.1	12.9	4.1	8.2	12.4	16	14.1	6	4.3	3.3	3.7	5.8

Fig. 9.33 Percentage change in localised impedance in each individual matrix, as measured from the front and from back. The right-left orientation are indicated and matched in columns, to be the same on both views. Subject details: male; age, 18 years; height, 174 cm; weight, 70 kg; vital capacity, 4.4 l; non-smoker (Ferdous et al. 2013)

than in the front of a human body. Therefore, with the same electrodes placed both in the front and the back, because of the limited depth of sensitivity, more volume of the lung will contribute when measured from the front of the body.

Recently a probe has been made for FIM-4 for lung ventilation study, waiting for a field trial of normal subjects and patients (Fig. 9.34). This FIM device is also intended for use in a telemedicine programme being run by the Dhaka University department. The probe has detachable electrodes to accommodate two or three different sizes, for persons with different chest sizes. The electrodes, made of stainless steel strips, also provide a springiness to conform to body shapes, and a provision has been made to make the body connection through a wet cotton or tissue pad inserted into a plastic tube at the end, which can be changed to eliminate cross infection between subjects.

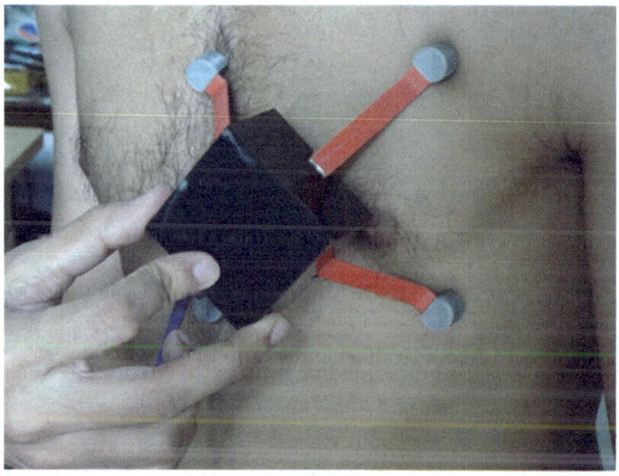

Fig. 9.34 New FIM-4 probe for localised lungs ventilation measurement

9.3.2 *Pneumonia Detection in Babies for Autodiagnosis*

A doctor working for low-resource countries came up with an interesting problem. He wanted to develop an automated pneumonia detection system for rural areas of low-resource countries, where the diagnosis is to be given through artificial intelligence. Since X-ray equipment is not available in rural areas, the method should depend on other diagnostic information, and the respiration rate plays a major role in such diagnosis. For the automated system, the device has to have a machine input, in electronic form. The problem that the doctor faced was that as soon as any sensor is attached to babies, they start crying, and crying changes respiration rate, making the whole procedure useless. The challenge was to have a machine output for respiration rate that does not upset the baby.

As a response to this problem, the Dhaka group developed a palm-worn electrode probe, made on a flexible rubber sheet as shown in Fig. 9.35, which was attached to the hand using a Velcro strap. The FIM-4 system was used, and the central fifth electrode was for a common terminal of the potential measuring circuitry. The electrode pad was covered using a cotton sheet to give more comfort to the baby, particularly during winter month when the rubber sheet may feel cold to the touch. Within the cotton sheet, folded cotton pieces were sewed in to make raised buttons at the position of the electrodes which were wetted using drops of tap water, to provide an electrical conducting path. The cotton sheet was wrapped around the rubber pad and fixed using pieces of Velcro fasteners. The idea was that when a mother touches, a baby or a child usually does not cry. The probe was used with a home-made FIM instrument, and the data was logged using a computer. Figure 9.36 shows a mother using this probe to acquire lungs ventilation data from a child (left) and the acquired data (right). The child did not cry! The group also performed similar studies on several (more than 5) child patients in a hospital through their respective mothers. None of the subjects cried, which is success of this innovation.

Fig. 9.35 Palm worn four-electrode FIM probe for on a flexible rubber sheet (left), covered by a cotton cloth with wet cotton buttons at the electrodes (middle) for lungs rate measurement on babies using the home made FIM machine (right)

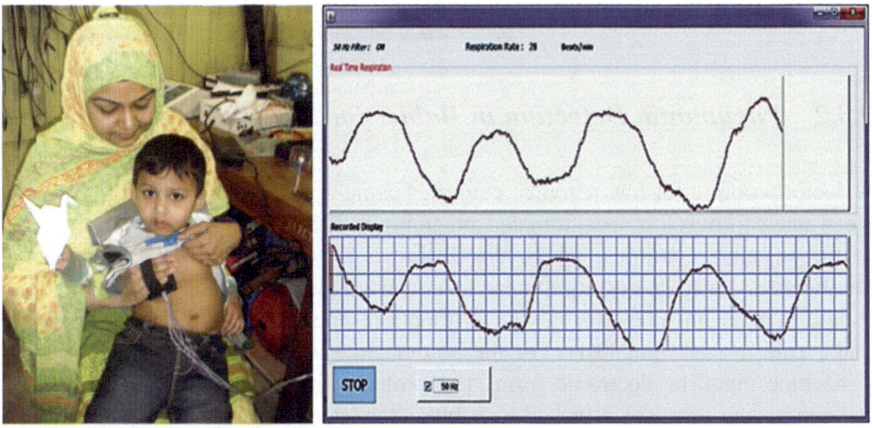

Fig. 9.36 Palm-worn FIM-4 probe applied on a 2-year-old child by her mother (left). The child did not cry during measurement. The tidal breathing output is shown on the right in which the beat rate is also shown. Similar success was achieved with five more babies in a hospital

This particular application just needs the respiratory rate; the magnitude of the variation is not important. Therefore, FIM, even the conventional TPIM, may be employed for this application straightway, with the present state of development. However, to assess the magnitudes of localised ventilation and to exploit the potential benefits of FIM, other contributing factors to the measurement will need to be isolated and compensated for, which needs further work. Furthermore, through measurements at different frequencies, it may be possible to distinguish between different types of lung disorders, whether there is simply edema or pneumonia or other types of lung disorders, etc.

9.3.3 *Gastric (Stomach) Emptying*

EIT has been used for gastric emptying studies with success (Sarker et al. 1997). Stomach is a large organ, so FIM was expected to give a useful output as well. The electrode placement for a six-electrode FIM is shown in Fig. 9.37 (Rabbani et al. 1999). The subject had a drink of saline (about 500 ml). The average FIM measurement before the drink was considered as the reference level. Readings were taken at 1 min interval just after having the drink, and the percentage change in FIM impedance is shown in Fig. 9.38. It can be seen that the impedance goes down sharply, and then it gradually goes back to the reference level in about 20 min. A similar experiment was carried out using EIT on another person. Here a ring electrode was strapped to the stomach region from the front, and images were taken at 1 min interval. The stomach region was identified and marked out as a region of interest. The sum of the pixel values within this region was then plotted with time. The changes in pixel sum are shown in Fig. 9.39. It shows a behaviour similar to Fig. 9.38 obtained using FIM. The EIT can be taken as the gold standard in this experiment as the stomach region is outlined and separated from duodenum where a concentration of saline appears as it is expelled from the stomach. For another comparison, a linear TPIM was used for a measurement on the stomach zone with a similar drink of saline on another day. The results are shown in Fig. 9.40. This clearly has a different pattern indicating that this measurement

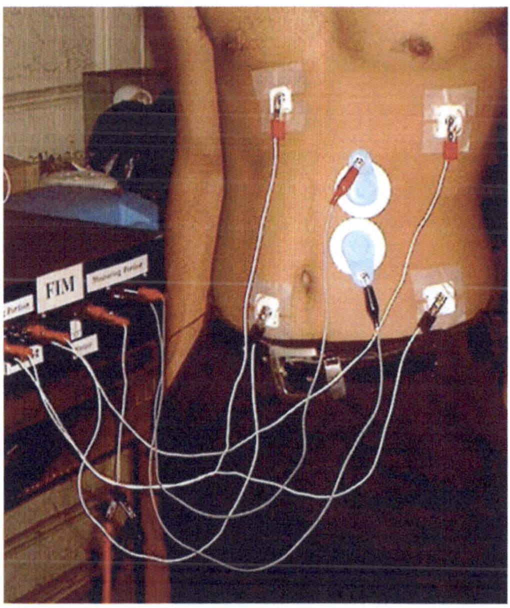

Fig. 9.37 Electrode placement for gastric (stomach) emptying study using FIM-6

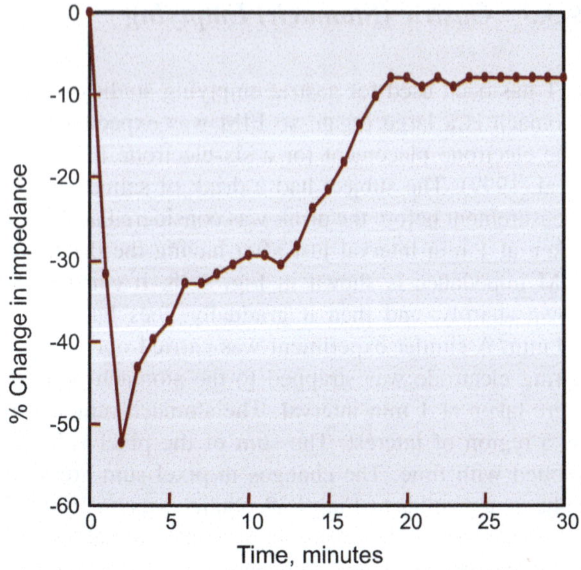

Fig. 9.38 Gastric emptying study using FIM-6 showing percentage change in impedance after a drink of saline (Rabbani et al. 1999)

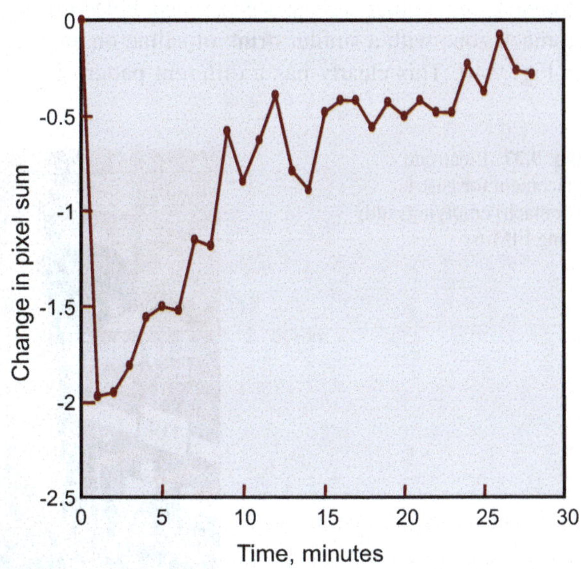

Fig. 9.39 Gastric emptying study using EIT showing change in pixel sum in the stomach region after a drink of saline. This can be used as a reference since the stomach region is clearly outlined (Rabbani et al. 1999)

possibly had contributions from both the stomach and the duodenum, and, therefore, TPIM could not isolate the stomach region that FIM could. Being much simpler in instrumentation and ease of measurement procedure than EIT, FIM could be the measurement tool of choice for such stomach studies.

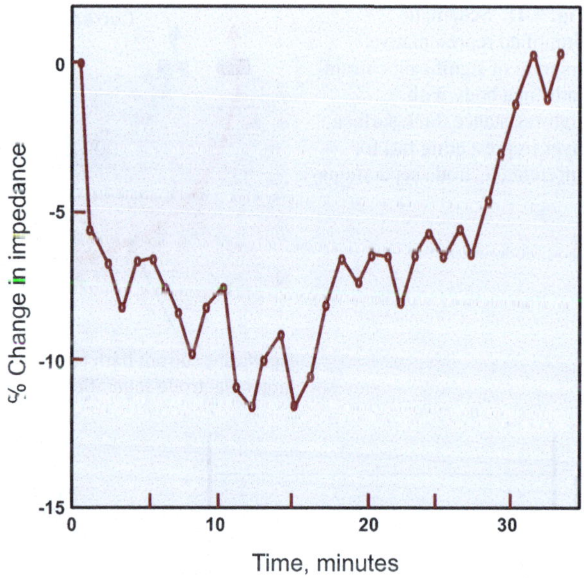

Fig. 9.40 Gastric emptying study using TPIM showing after a drink of saline. This does not have the correct behaviour. Presumably it cannot isolate the stomach from the duodenum (Rabbani et al. 1999)

9.3.4 Determination of Subcutaneous Abdominal Fat Thickness

Subcutaneous abdominal fat thickness can give information on the fitness of a person and is a risk indicator of several diseases and disorders like diabetes, heart problem, etc. Scharfetter et al. (2001) employed TPIM in obtaining the abdominal fat thickness and obtained good correlations with MRI results. The Dhaka University group used a concept of multiple electrode separations of TPIM which was expected to give a better result (Haowlader et al. 2010). However, since TPIM has a wide area of sensitivity and abdominal fat is extended over a large area that includes negative sensitivity zones, the results may have errors, particularly for variation in electrode placement. Therefore, the same concept of multiple electrode spacing was used with four-electrode FIM to give a more localised measurement (Surovy et al. 2012).

How multiple electrode spacing may help in getting fat thickness is explained briefly with the help of Fig. 9.41. It shows a high-resistive surface layer beneath the electrodes representing the fat layer and a lower-resistive volume underneath representing the bulk of the body. For a smaller electrode separation, most of the current has to traverse the high-resistive layer, so the measured impedance will be high. As the electrode separation is increased, a significant part of the current will pass through the low-resistance bulk, so the measured impedance will be lower. It was hypothesised that measurements using two or more sets of current electrodes with different separations will contain information about the surface layer thickness.

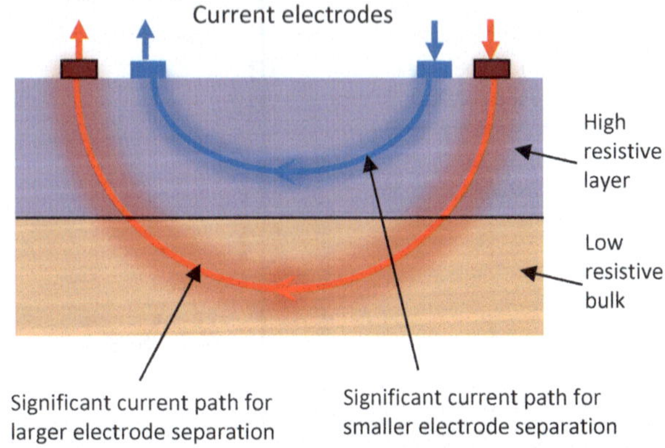

Fig. 9.41 Schematic simplified representative diagram of significant current paths in a body with high-resistance thick surface layer (representing fat) for different electrode separations

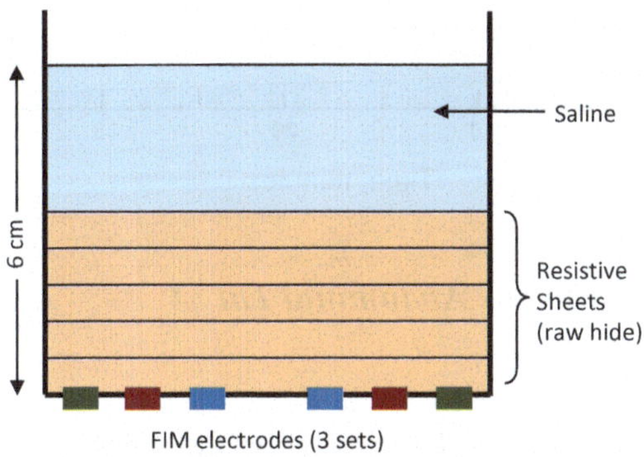

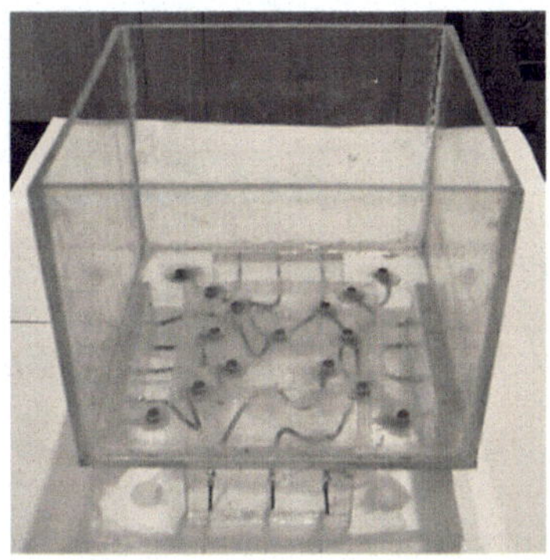

Fig. 9.42 Schematic of the saline phantom with multiple sets of FIM electrodes (four-electrode version) at different separations and multiple layers of raw cow hide as resistive sheets (left). Photo of the phantom, without inner structures to press down on the hide layers (right)

For FIM-4 measurement using multiple electrode separation, a saline phantom was designed to provide different electrode separations and different layers of resistive materials adjacent to the electrodes as shown in Fig. 9.42. The left-hand diagram is a schematic one, while a photograph of the phantom is shown on the right. Three sets of FIM electrodes (FIM-4 version) with different separations were fixed to the bottom of a plastic container with a 12 cm × 12 cm square base to make the phantom. The background saline simulated the internal organs having low impedance, while raw cowhide, soaked overnight in saline having the same concentration as used in the phantom, was used to make the resistive layers. To keep the hide layers together near the electrodes, a plastic framework was made to push these down without blocking the current distribution in the phantom significantly (not shown in the photo). Saline was added or removed (with removal or addition of rawhide layers as appropriate) to keep the total height of the effective phantom at 6 cm.

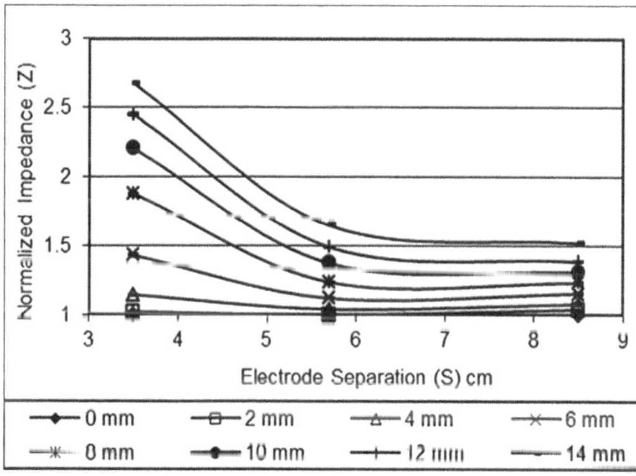

Fig. 9.43 Plots of normalised impedance (Z) vs. electrode separation (S) for different thicknesses of rawhide 0-14mm; symbols shown below the graph (Surovy et al. 2012)

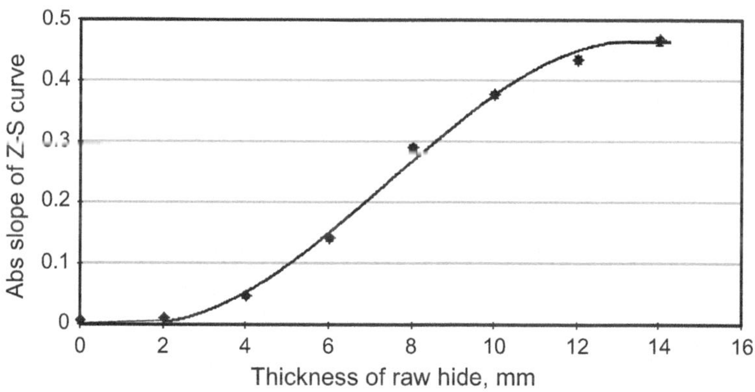

Fig. 9.44 Absolute slope of impedance vs. electrode separation curve for different thicknesses of rawhide (Surovy et al. 2012)

The measured impedances (at 10 kHz with 0.5 mA constant current) with different thicknesses of rawhide were normalised to the impedances with no hide (only saline) at the respective electrode separation. Plots of normalised impedance with electrode separation are shown in Fig. 9.43. The curves have different "slopes" for different thicknesses of resistive layers at low separations, which offered a method to obtain an unknown thickness of subcutaneous fat layer. The absolute value of slopes of all the curves at the lower end of electrode separation was plotted against the thickness of rawhide as shown in Fig. 9.44. The thickness of the rawhide can now be determined from the absolute slope of the normalised impedance vs. electrode separation curve. This is the basis which has been used to determine abdominal thickness of subcutaneous fat in the human body.

As an initial study of the method, FIM-4 measurements were performed on seven human male subjects using two electrode separations, 4 cm and 6 cm. Fat layer thickness was measured using mechanical callipers. A plot of the slope of the

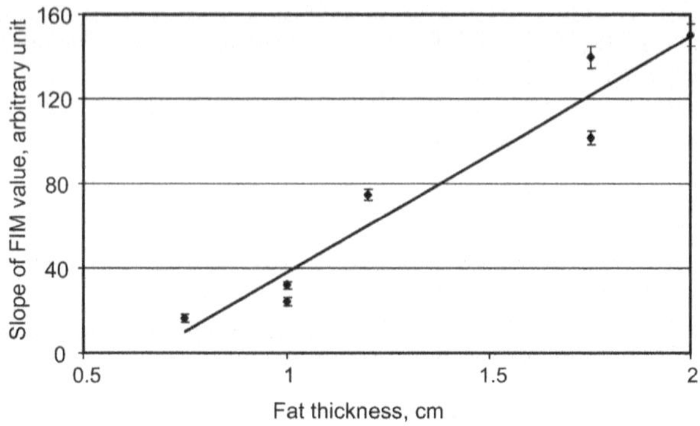

Fig. 9.45 Slope of FIM values with 4 and 6 cm electrode separation (in FIM-4) against abdominal fat thickness, measured using calipers of seven male subjects (Surovy et al. 2012)

FIM values due to these two separations are shown plotted against fat thickness in Fig. 9.45. The points could be fitted using a straight line. This appears to bear out the success of the proposed method. Further work using more accurate CT and MRI measurements may be taken up to study the feasibility of this method for human subjects of different age groups and gender.

An alternative technique was suggested through a simulation study by Al-Quaderi et al. (2014), where a point of inflexion of a curve of the focused impedance against electrode separation was utilised to get an estimate of the fat thickness. However, this may be difficult to apply practically as it needs measurements with many values of electrode separations.

9.3.5 Non-invasive Breast Tumour Characterisation Using FIM

In many cases patients with breast tumour report to a clinic only after they can feel a lump by hand. Usually the tumour is reasonably large by this time, typically more than 1 cm. The step for a medical intervention then requires a diagnosis—if the tumour is malignant (cancerous) or benign (non-cancerous). The most reliable technique for this is biopsy, which needs breast tissues to be collected from a few regions in and around the tumour. This is a semi-invasive procedure and has associated hazards, besides taking time to obtain an expert report contributing to a period of anxiety of the patient. The feasibility of using FIM for such characterisation of breast tumour was taken up by the Dhaka group, and the initial results indicate a potential of FIM in this application (Al-Amin et al. 2014).

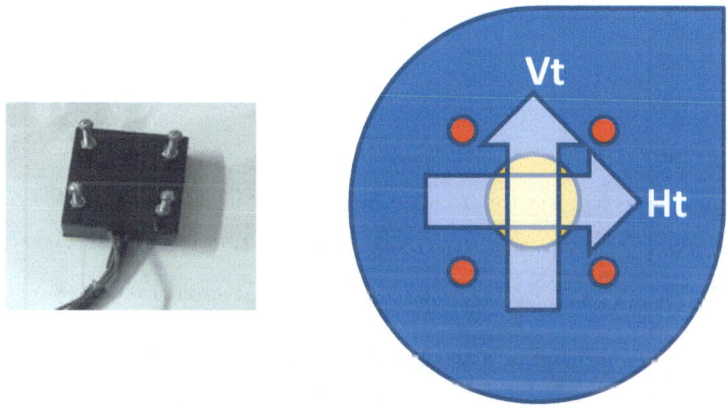

Fig. 9.46 Four-electrode probe with 5 cm separation used for measurements on breast (left). Scheme for TPIM measurements in vertical (Vt) and horizontal (Ht) directions (right)

A four-electrode probe as shown in Fig. 9.46 using long metallic screws was used to measure TPIM in the vertical and horizontal directions choosing the current and potential electrodes appropriately. Then the two measurements were summed which essentially gave four-electrode FIM. The measurements were carried out using a MALTRON Bioscan 920-II equipment at 5 kHz and 200 kHz on 19 female patients in the age range 17–55 years at a breast clinic. Four of the patients were identified to have malignancy through core biopsy tests later. However, core biopsy was not done on all patients; the clinician declared many of the tumours as benign based on her experience with other clinical tests. So it does not rule out the possibility that some of the tumours identified as benign may in fact be malignant.

The patients chosen had tumours on one side only, and measurements were taken on both sides—affected and unaffected. The electrode probe was placed keeping the tumour, where present, at its centre. A feature plot was drawn with all the parameters obtained, which included individual TPIM values, combined FIM values and slopes between points at 5 kHz and 200 kHz. There was significant overlap between the plots of the malignant and the benign cases. On close observation, it appeared that there was some contribution of age in the dispersion of data. So all the data were divided by the respective patient age when some distinction appeared and is shown in Fig. 9.47. The data were tested using the K-NN classifier (Haykin 2009) using "one-leave-out cross validation" method with K taking the values 1 to 3, i.e. up to the three nearest neighbours. All of these three tests produced the same result. Subsequently, treatment of the results for classification of malignancy detection gave a result which looks promising. Table 9.1 presents the results of this analysis.

However, there was still considerable overlap between the feature plots for the two categories of tumour. In view of the large person-to-person variation, an attempt was made to further improve the result through a variation in the measurement technique so that location of the same breast of a patient that does not have a

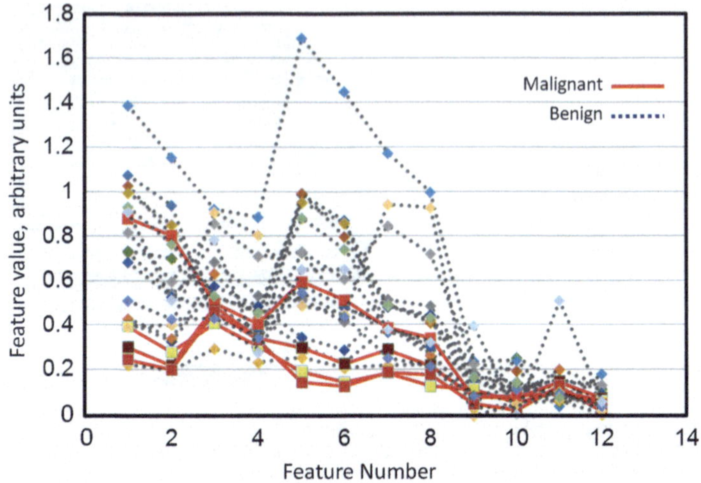

Fig. 9.47 Feature plot of breast tumours. Raw features were divided by age. Patient age range: 17–55 years. Feature numbers: impedance of unaffected side, 1–4, and of affected side, 5–8; slope of impedance between 200 kHz and 5 kHz of unaffected side, 9–11, and of affected side, 10–12. Red lines: confirmed malignancy from core biopsy. Dotted lines: non-malignant (some through biopsy, some through other clinical investigations, Al-Amin et al. 2014)

Table 9.1 Test for classification for malignancy detection using electrical impedance method (Al-Amin et al. 2014)

Parameter	Value	Parameter	Value
True positives (tp)	3	Positive predictive value	60%
True negatives (tn)	13	Negative predictive value	93%
False positives (fp)	2	Sensitivity [tp/(tp + fn)]	75%
False negative (fn)	1	Specificity [tn/(tn + fp)]	87%

tumour underneath could be used as a control. With this aim, two modalities were conceived, and simulation studies were carried out using COMSOL as described briefly below.

The first modality (Huda et al. 2016) uses two concentric FIM with different separation of electrodes, placed concentrically on the tumour as shown in Fig. 9.48 (left). The one with the smaller separation has more contribution from the tumour and less contribution of the surrounding normal tissue, while the one with the larger separation has less contribution from the tumour and more contribution of the surrounding normal tissue. Therefore, the latter measurement essentially acts as a control to the former. A comparison of the two measurements is expected to give an improved ability for distinguishing malignant tumours from benign ones. The second modality (Ahmed et al. 2016) used a series of FIM measurements such that the focused zones fall into adjacent elements of a matrix as shown in Fig. 9.48 (right). The placement should be chosen such that the tumour is at the centre of one of the matrix elements while the others have normal tissue underneath. Such a

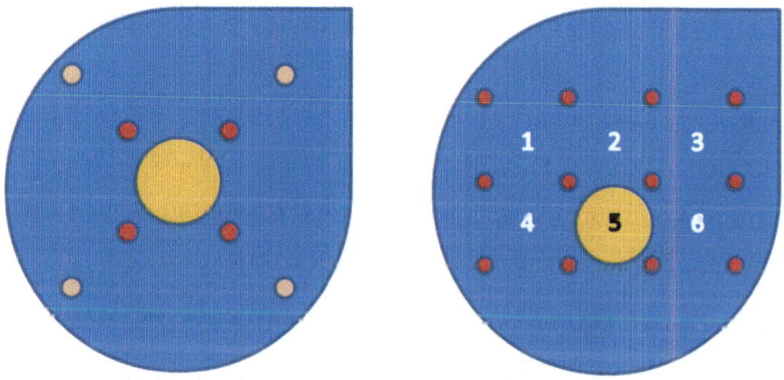

Fig. 9.48 Schematics of two modalities of FIM measurements for improved distinction of malignant and benign tumours. Dual separation four-electrode FIM (left) and multiple four-electrode FIM measurements in a matrix, known as Pigeon Hole Imaging (PHI) (right). Numbered matrix elements are relevant to Fig. 9.49

configuration has been termed Pigeon Hole Imaging (PHI) by the Dhaka University group earlier (Afroj et al. 2004; Rabbani 2010, 2012). This arrangement has the advantage that the adjacent zones that act as the control have no tumours underneath. The ratio of the impedance value measured at one of the adjacent normal matrix positions (1 or 2 in Fig. 9.48) to that in the position having the tumour (5 in Fig. 9.48) was chosen as the parameter for the test for malignancy. This ratio would be unity for a benign tumour if the volume was infinite, assuming a benign tumour to have the same impedivity as the normal breast tissue. However, for a finite volume of the size of a human breast the ratios will be slightly different because of boundary effects. On the other hand this ratio would be greater than unity if the tumour is malignant since the impedivity of malignant tumour tissue is expected to be much lower than that of normal breast tissue. Therefore, a threshold for this ratio may be chosen based on actual studies on a large number of human subjects, for different types of tumours, which will give possible distinctions between malignant and benign tumours.

To have a preliminary idea whether the percentage variations expected in the above ratios between malignant and benign tumours can be distinguished, COMSOL simulation was performed for a cubic volume of size 15 cm filled with a material having the electrical impedivity of normal breast tissue for a frequency of 500 kHz (Ahmed et al. 2016). A spherical tumour of 3 cm diameter was placed centrally near one of the sides on which the electrodes are attached. The separation of electrodes was 3 cm between adjacent electrodes, and the tumour was placed at different depths from the electrode plane. The impedivity of malignant tissue was assumed to be about half that of benign tumour tissue while the impedivity of the latter was assumed the same as that of normal breast tissue at this frequency. The percentage deviation of the ratios estimated between a malignant tumour and a benign tumour are plotted in Fig. 9.49 for two positions of normal tissue, position

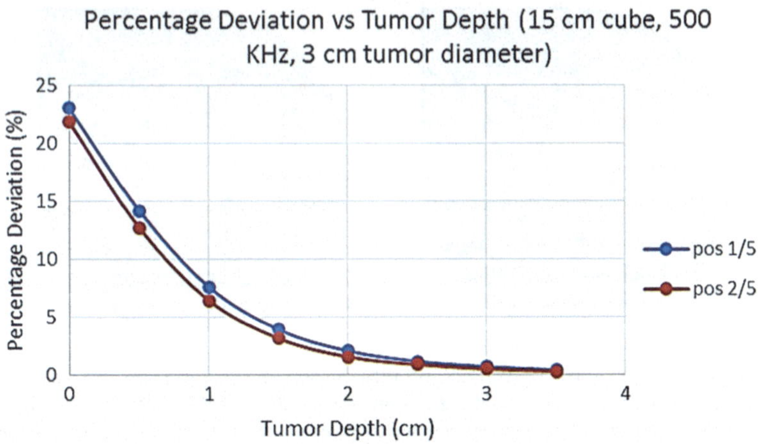

Fig. 9.49 Plots to show ability to distinguish malignant tumour through multiple FIM measurements in a matrix (PHI) at different depths (electrode separation, 3 cm; tumour diameter, 3 cm; frequency, 500 kHz) obtained using COMSOL FEM simulation. Assumed impedivity of malignant tissue is almost half that of normal and benign tissues. Matrix positions relate to those in Fig. 9.48. Values of measurement at positions 1 and 2 are divided by that at position 5, the element with the tumour. The slight difference is because of the boundary geometry and sensitivity of FIM

1 and 2, with the tumour in position 5. This essentially gives the distinguishing capability of this method. It can be seen that at zero depth (the tumour touching the surface, its centre at 1.5 cm depth), the percentage deviation is more than 20% which should be easily measurable. The small deviation between positions 1 and 2 is due to differences in sensitivity of FIM in adjacent matrix elements and due to boundary geometry. Assuming 3–5% deviation to be the minimum measurable in a practical instrumentation with clarity, the malignant tumour may be identified down to a depth of about 3 cm (position of its centre). Since breast tissue is usually soft, one may press down on the electrodes to bring the tumour up with respect to the electrodes, so a higher deviation value may be expected. Of course for smaller tumours the percentage deviations would be less, but using a smaller electrode separation, the figure may be increased. Thus, it looks very promising that such a method may be able to distinguish between malignant and benign tumours non-invasively, once it is palpated. It now remains to be seen how measurements turn out in practice on human subjects.

It however needs to be noted that actually the breast has several tissue types and a large person-to-person variation, so it is very difficult to estimate exact values of impedivity for normal and cancerous (malignant) breast tissues. Morimoto et al. (1993) reported a higher impedivity value for malignant tumours compared to that of normal breast tissue through in vivo measurements, combining the measurements on many subjects. Jossinet (1996) reported slightly higher impedivity of malignant tumours compared to that of normal mammary glands, again, combining values obtained from many subjects using excised tissues. On the other hand, Surowiec et al. (1988), making measurements on tissue samples collected from different regions of the same breast, reported malignant tumours to have lower impedivity compared to that of normal tissues at a distance of about 25 mm on the same breast, a totally opposite trend. Therefore, a haziness still remains in the impedivity values of malignant, benign and normal tissues of the breast. However, samples taken from adjacent regions of the same breast by Surowiec et al. (1988) provide a better comparison and make this trend more acceptable.

9.4 Measurement Challenges

9.4.1 Repeatability and Reproducibility

Any useful measurements should have good repeatability and reproducibility. The former demands that if the same observer performs a measurement repeatedly on the same subject, in the same sitting or at different sessions, the measurements should give the same output. The latter demands that if different observers perform the same measurement on the same subject, the outputs should be the same. This is a problem for all bioimpedance measurements since exact positioning of the electrodes may vary between measurements. It has been indicated that a fixed distance electrode probe gives a better performance in this regard (Moon et al. 2010). Keeping geometric or photographic records of the electrode placement may help in minimising the geometric error. Most of the FIM applications mentioned in this chapter use electrode probes with fixed geometries, which is expected to give better results based on the above-mentioned observation.

Another parameter of importance is the pressure applied on the electrodes. Since the human body surface is mostly soft and pliable, change of pressure on electrodes, even at the same position, gives a slightly different value of impedance since the whole geometry of the volume conductor and the electrode system is changed a little. Again, too little pressure may not make a good electrode-skin contact. Therefore, pressure on electrodes for all the measurements should be kept to a minimum, and it should be the same as much as possible. Even then, person-to-person variation for identical locations on the body would be difficult to eliminate. For measurements on the cervix, it has been shown that a firm pressure gives more reproducible measurements than a soft pressure on the electrode probe (Jokhi et al. 2009). This may be because of the natural firmness of the cervix which limits the deformation of the tissue at a certain maximum pressure. However, this may not be applicable at other measurement sites, such as over the stomach or on the breast, where the body is more pliable. Besides, a breast tumour may be soft or rigid which will affect the measurement even with a constant electrode pressure. Still, an attempt to achieve a constant electrode pressure would be better from an experimental point of view. A mechanical spring system with an indicator on the probe mount may be used for this purpose. Use of electronic sensors giving a record of the pressure could be useful. Alternatively, adhesive electrodes may be used to follow the natural contour of the breast, contributing to almost a "zero" pressure but good electrical contact.

9.4.2 Effect of Body Movements

The inner organs of the body move during respiration which complicates the measurements on the thorax. For studies of breast tumours and stomach emptying, the subject may be asked to breathe in (full inspiration) and to hold it for a few seconds when the measurement is performed. This phase is better than a breathe-out (full expiration) phase since the lungs offer higher impedance during full inspiration diverting more current to target organs outside the lungs. As a result of the higher current density, a greater sensitivity is achieved at such target organs. Again, for stomach emptying study, the electrode placement may be judiciously made to keep the lung region outside the focused zone as much as possible. However, one needs to consider that the size and position of the stomach also changes with the amount of food (solid or liquid) intake. Therefore, the electrode separation should be large enough to cover the whole of the stomach region throughout the movement. In Fig. 9.38 FIM-6 was used to advantage, since a large separation of the current electrodes helps in having a substantial depth sensitivity which is also needed for the stomach. For breast tumour studies, the contribution of the lung can be minimised through a smaller electrode separation, which reduces the depth sensitivity. On the other hand, for measurement on the lung itself, particularly to assess lung ventilation, one cannot escape some uncertainty. Electrodes should be judiciously placed such that the sensitive focused zone always covers some part of the lungs. Edges of the lung or organs outside the lung should not come within the focused zone as this would add contributions that would be difficult to isolate and interpret.

9.4.3 Noise Due to Skin-Electrode Movements

When the thorax moves during the respiratory cycle, the skin to body fluid interface under the electrodes goes through changes resulting in noise. However, the noise is predominantly of low frequency, around a few Hz, while most of the electrical impedance measurements are carried out at frequencies greater than 5 kHz. Therefore, such low-frequency noise due to skin-electrode movement can be filtered out relatively easily using a high-pass filter. Since measurements are carried out at a specific frequency at any instant, random noise may be eliminated using appropriate narrowband-pass filters or using phase-sensitive detection methods.

9.4.4 *Person-to-Person Variation*

The body size varies depending on age, sex and individual body frames. Therefore, it may be necessary to suit electrode separations according to body size and volume, and this will need specific studies for each application. Through such studies, compensation formula may be derived to get comparable values for different individuals, thus making interpretation of results easier. Therefore, FIM systems should have arrangements to change electrode separations, either continuously or discretely, or there should be a set of electrodes available with different electrode separations. Appropriate electrode sizes should be prescribed for specific measurement sites for specified body frames. Of course, when adhesive electrodes are used, it offers a greater freedom but at the expense of more time and cost.

9.5 Future Prospects

Being simple in methodology and instrumentation and being able to localise a region that was the major challenge in earlier simple techniques that excludes EIT, FIM appears to have a great variety of applications not only in biomedical areas but in industries, in geology and in oceanography where some electrical impedance techniques are already being used. As stated earlier, FIM may be useful in most applications where EIT has been suggested to target reasonably large organs.

Instrumentation to produce FIM values automatically using microcontroller data acquisition will allow portable equipment to be produced for widespread use (Aktharuzzaman et al. 2011). Multifrequency FIM systems (Kadir et al. 2012) will enhance the capabilities of the FIM systems.

FIM may be used to determine the volume of objects in 3D if the relevant impedivities and the depth of an object are known and some work is already in progress at University of Dhaka (Kadir et al. 2010; Ahmed et al. 2014). FIM may also be useful in the monitoring of artificial respiration in hospitals. ECG gated FIM may be used to perform localised lung perfusion studies which in combination with localised lung ventilation, again measured using the same FIM electrodes, may offer improved diagnosis of lung disorders showing effective lung regions. Bladder emptying for uroflow studies may be another potential area.

Attempts have been made by groups to use electrical impedance techniques to measure temperatures of tissue during hyperthermia treatment of tumours at deep tissue sites. FIM with a localised target may be suitable for such measurements. Brown et al. (2000b) has successfully employed TPIM measurements at multiple frequencies to detect abnormal precancerous conditions in the cervix and in the oesophagus. FIM with its ability to obtain better localisation may improve the detection ability.

Any measurement on the thorax is complicated by the fact that all the internal organs move with breathing. Similarly the heart moves during its beating cycle.

Therefore, Pigeon Hole Imaging (PHI), an extension of FIM, may be used to follow such movements and correct the data to get improved information. This may allow volume determination of different organs in the thorax, including that of the blood in the cardiac chambers.

Recently irreversible electroporation techniques have been successfully used to destroy cancer cells which use two or four long insulated needles with exposed tips to drive current pulses to the region of the tumour to be ablated (Davalos et al. 2005; Onik et al. 2007). Using the same four needles, FIM may be used to monitor the degree of ablation, while the procedure is in progress (Rabbani 2008).

Furthermore, for in vitro studies of blood and other body tissues, FIM performed at different frequencies may provide useful information. Using microelectrodes FIM may be extended to the study of bacteria and other microorganisms in water and other media.

9.6 Conclusion

It appears that electrical impedance techniques have a large potential in different fields of measurements, particularly in biomedical applications. However, these could not make the expected impact so far. This is possibly due to the wide region of sensitivity of traditional four-electrode measurement (TPIM) systems on one hand and the inaccuracy of pixel-level images in 2D electrical impedance tomography (EIT) on the other. 2D EIT is affected by objects in 3D, making the pixel values inaccurate. Besides, the complexity and cost possibly go against its benefits. 3D EIT, which is still under development, is far more complex and, therefore, is expected to find application in very limited and special areas when ready. FIM forming a bridge between TPIM and EIT appears to have the promise of bringing about the impact of electrical impedance techniques sooner if pursued earnestly. Having less number of electrodes to deal with and simple instrumentation and offering a reasonable localisation of the zone of measurement that can be manipulated through electrode spacing, FIM appears to be a technique with great potential.

Besides, FIM can be made into small portable devices making diagnosis available widely, even in remote areas of the world. The simplicity of technology brings a great saving in cost, which makes it all the more suitable for the low-resource countries of the world, inhabited by almost 80% of the global population. This vast majority of people have long been deprived of the modern technologies that can enhance the quality of life of human beings. Because of the huge economic disparity created in the modern world, which may be linked to a technology disparity, such a simple technique as FIM has the promise of taking modern technologies to this deprived majority quickly, reducing the technology gap, even if by a little. Scientists and engineers in these countries can themselves make and distribute such devices for their own population, which will also allow these devices to be made to adapt to local conditions. Besides, this will ensure long usable life of the instruments since expertise and spares for maintenance and repair would be available locally. If such a scenario can be attained within a foreseeable future, all the efforts behind the development of FIM would seem worthwhile.

References

Abir, R., Pettersen, F. J., Martinsen, O. G., & Rabbani, K. S. (2013). Effect of a spherical object in 4 electrode focused impedance method (FIM): Measurement and simulation. *Journal of Physics: Conference Series, 434*, 012009. https://doi.org/10.1088/1742-6596/434/1/012009

Abir, A. R., & Rabbani, K. S. (2014). Sensitivity study for a 4-electrode focused impedance method (FIM) using finite element method analysis. *Bangladesh Journal of Medical Physics, 7*, 1–7.

Afroj, K., Alam, N., Rahman, M., & Rabbani, K. S. (2004). Pigeon hole imaging (PHI) – An electrical admittance backprojection technique. *Bangladesh Journal of Medical Physics, 3*, 7–13.

Ahmed, S. F. B., Alam, S. K., & Rabbani, K. S. (2016, November). Breaset tumour characterisation using Pigeon hole imaging (PHI) – A simulation study. *Paper presented at 11th SARC International Cancer Conference & 3rd Bangladesh Cancer Congress.*

Ahmed, S. P., Kadir, M. A., Rahman, R., Al-Quaderi, G. D., & Rabbani, K. S. (2014). Determination of organ volume using focused impedance method (FIM): A simulation approach. *Bangladesh Journal of Medical Physics, 7*, 24–33.

Aktharuzzaman, Baig, T. N., & Rabbani, K. S. (2011). Design of a microcontroller based system to implement 4-electrode focused impedance method (FIM). *Bangladesh Journal of Medical Physics, 4*(2), 75–80.

Al-Amin, A., Parvin, S., Kadir, M. A., Tahmid, T., Alam, S. K., & Rabbani, K. S. (2014). Classification of breast tumour using electrical impedance and machine learning techniques. *Physiological Measurement, 35*, 965–974. https://doi.org/10.1088/0967-3334/35/6/965

Al-Quaderi, G. D., Ahmed, S. P., & Rabbani, K. S. (2014). Determination of the thickness of a resistive material layer in a finite volume conductor using focused impedance method (FIM) – A simulation study. *Bangladesh Journal of Medical Physics, 7*, 8–23.

Barber, D. C., Brown, B. H., & Freeston, I. L. (1983). Imaging spatial distributions of resistivity using applied potential tomography. *Electronics Letters, 19*, 933–935.

Brown, B. H., Barber, D. C., & Seagar, A. D. (1985). Applied potential tomography. Possible clinical applications. *Clinical Physics and Physiological Measurement, 6*, 109–121.

Brown, B. H., Wilson, A. J., & Bertemes-Filho, P. (2000a). Bipolar and tetrapolar transfer impedance measurements from a volume conductor. *Electronics Letters, 36*, 2060–2062.

Brown, H., Tidy, J. A., Boston, K., Blackett, A. D., Smallwood, R. H., & Sharp, F. (2000b). Relation between tissue structure and imposed electrical current flow in cervical neoplasia. *The Lancet, 355*, 892–895.

Davalos, R. V., Mir, M., & Rubinsky, B. (2005). Tissue ablation with irreversible electroporation. *Annals of Biomedical Engineering, 33*(2), 223–231. https://doi.org/10.1007/s10439-005-8981-8

Ferdous, H., Baig, T. N., & Rabbani, K. S. (2013). Thorax mapping for localized lung impedance change using focused impedance measurement (FIM): A pilot study. *Journal of Electrical Bioimpedance, 4*, 57–61.

Geselowitz, D. B. (1971). An application of electrocardiographic lead theory to impedance plethysmography. *IEEE Transactions on Biomedical Engineering, 18*, 38–41.

Haowlader, S., Baig, T. N., & Rabbani, K. S. (2010). Abdominal fat thickness measurement using focused impedance method (FIM) - Phantom study. *Journal of Physics: Conference Series, 224*, 012061. (IOP Publishing). https://doi.org/10.1088/1742-6596/224/1/012061

Haykin, S. S. (2009). *Neural networks and learning machines*. New York: Prentice Hall.

Huda, K., Rabbani, M., Kadir, M. A., & Rabbani, K. S. (2016, November). Characterisation of breast tumours using focused impedance method (FIM) with dual electrode separation, a simulation study. *Paper presented at 11th SARC International Cancer Conference & 3rd Bangladesh Cancer Congress.*

Iquebal, A. H. M., & Rabbani, K. S. (2010). 3D sensitivity of 6-electrode focused impedance method (FIM). *Journal of Physics: Conference Series, 224*, 012156. (IOP Publishing). https://doi.org/10.1088/1742-6596/224/1/012156

Iquebal, M., & Rabbani, K. S. (2013). Correlation of liquid volume in stomach to electrical transfer impedance measurements using FIM. *Bangladesh Journal of Medical Physics, 6*, 66–74.

Islam, N., Rabbani, K. S., & Wilson, A. (2010). The sensitivity of focused electrical impedance measurements. *Physiological Measurement, 31*, S97–S109. https://doi.org/10.1088/0967-3334/31/8/S08

Jokhi, R. P., Ghule, V. V., Brown, B. H., & Anumba, D. O. C. (2009). Reproducibility and repeatability of measuring the electrical impedance of the pregnant human cervix – The effect of probe size and applied pressure. *Biomedical Engineering OnLine, 8*, 10. https://doi.org/10.1186/1475-925X-8-10

Jossinet, J. (1996). Variability of impedivity in normal and pathological breast tissue. *Medical and Biological Engineering and Computing, 34*, 346–350. https://doi.org/10.1007/BF02520002

Jossinet, J. (1998). The impedivity of freshly excised human breast tissue. *Physiological Measurement, 19*(1), 61–76. https://doi.org/10.1088/0967-3334/19/1/006

Kadir, M. A., Ahmed, S. P., Al-Quaderi, G. D., Rahman, R., & Rabbani, K. S. (2013). Application of focused impedance method (FIM) to determine the volume of an object within a volume conductor. *Proceedings of the 2013 COMSOL Conference*, Bangalore, India. http://www.comsol.com/paper/download/182751/kadir_paper.pdf.

Kadir, M. A., Baig, T. N., & Rabbani, K. S. (2009). Application of 6-electrode focused impedance method (FIM) to study lungs ventilation. *Proceedings of EIT2009, Conference on Electrical Impedance Tomography*, Manchester, UK.

Kadir, M. A., Baig, T. N., & Rabbani, K. S. (2015). Focused impedance method to detect localized lung ventilation disorders in combination with conventional spirometry. *Biomedical Engineering: Applications, Basis and Communications, 27*, 1550029.

Kadir, M. A., Ferdous, H., Baig, T. N., & Rabbani, K. S. (2010). Ventilation mapping of chest using focused impedance method (FIM). *Journal of Physics: Conference Series, 224*, 012031. (IOP Publishing). https://doi.org/10.1088/1742-6596/224/1/012031

Kadir, M. A., Rabbani, K. S., & Wilson, A. (2012). Development of a multi-frequency system for medical applications of focused impedance method (FIM) appropriate for developing countries. *Proceedings of the 7th International Conference on Appropriate Healthcare Technologies for Developing Countries (AHT-2012)*, London, UK. https://doi.org/10.1049/cp.2012.1487.

Karal, M. A. S., & Rabbani, K. S. (2009). Sensitivity of four-electrode focused impedance method (FIM) for objects with different conductivity. *Proceedings of EIT2009 Conference on Electrical Impedance Tomography*, Manchester, UK.

Moon, J. R., Stout, J. R., Smith, A. E., Tobkin, S. E., Lockwood, C. M., Kendall, K. L., et al. (2010). Reproducibility and validity of bioimpedance spectroscopy for tracking changes in total body water: Implications for repeated measurements. *British Journal of Nutrition, 104*(9), 1384–1394. https://doi.org/10.1017/S0007114510002254

Morimoto, T., Kimura, S., Konishi, Y., Komaki, K., Uyama, T., Monden, Y., et al. (1993). A study of the electrical bio-impedance of tumors. *Journal of Investigative Surgery, 6*, 25–32.

Onik, G., Rubinsky, B., & Mikus, P. (2007). Irreversible electroporation: Implications for prostate ablation. *Technology in Cancer Research and Treatment, 6*(4), 295–300. https://doi.org/10.1177/153303460700600405

Pettersen, F. J., Ferdous, H., Kalvøy, H., Martinsen, Ø. G., & Høgetveit, J. O. (2014). Comparison of four different FIM configurations - A simulation study. *Physiological Measurement, 35*, 1067–1082. https://doi.org/10.1088/0967-3334/35/6/1067

Rabbani, K. S. (2010). Focused impedance method (FIM) and Pigeon hole imaging (PHI) for localized measurements – A review. *Journal of Physics: Conference Series, 224*, 012003. (IOP Publishing). https://doi.org/10.1088/1742-6596/224/1/012003

Rabbani, K. S. (2012). Focused impedance method (FIM) and Pigeon hole imaging (PHI) as two potentially low cost and simple modalities for different diagnostic applications. *Proceedings of the 7th International Conference on Appropriate Healthcare Technologies for Developing Countries (AHT-2012)*, September 2012, London, UK. https://doi.org/10.1049/cp.2012.1454.

Rabbani, K. S. (2008). Potentials of electrical techniques in imaging and therapy for Palliative care of cancer patients. *Proceedings of HTTTG Palliative Therapy Workshop*, Vietnam, pp. 94–101. http://iupesm.ifmbe.org/wp-content/uploads/2014/08/PalliativeRadioTherapy.pdf

Rabbani, K. S., & Kabir, H. (1991). Studies on the effect of the third dimension on a two dimensional electrical impedance tomography system. *Clinical Physics and Physiological Measurement, 12*, 393–402.

Rabbani, K. S., & Kadir, M. A. (2011). Possible applications of focused impedance method (FIM) in biomedical and other areas of study. *Bangladesh Journal of Medical Physics, 4*(2), 67–74.

Rabbani, K. S., & Karal, M. A. S. (2008). Variation in sensitivity within the focused zone of the new four-electrode focused impedance measurement (FIM) system. *Dhaka University Journal of Science, 56*(2), 221–224.

Rabbani, K. S., Sarker, M., Akond, M. H. R., & Akter, T. (1998). Focused impedance measurement (FIM) - A new technique with improved zone localization. *Proceedings, X International Conference on Electrical Bioimpedance*, Barcelona, Spain, 5–9 April, 1998 (pp. 31–34).

Rabbani, K. S., Sarker, M., Akond, M. H. R., & Akter, T. (1999). Focused impedance measurement (FIM) - A new technique with improved zone localization. In *Electrical bioimpedance methods, Annals of the New York Academy of Sciences* (Vol. 873, pp. 408–420). https://doi.org/10.1111/j.1749-6632.1999.tb09490.x

Saha, S. K., Al-Quaderi, G. D., & Rabbani, K. S. (2013). 3D sensitivity of 8-electrode FIM through experimental study in a phantom. *Bangladesh Journal of Medical Physics, 6*, 55–65.

Sarker, S. A., Mahalanabis, D., Bardhan, P. K., Alam, N. H., Rabbani, K. S., Kiber, A., et al. (1997). Noninvasive assessment of gastric acid secretion in man (application of electrical impedance tomography (EIT)). *Digestive Diseases and Sciences, 42*(8), 1804–1809.

Scharfetter, H., Schlager, T., Stollberger, R., Felsberger, R., Hutten, H., & Hinghofer-Szalkay, H. (2001). Assessing abdominal fatness with local bioimpedance analysis: Basics and experimental findings. *International Journal of Obesity, 25*, 502–511.

Surovy, N. J., Billah, M., Haowlader, S., Al-Quaderi, G. D., & Rabbani, K. S. (2012). Determination of abdominal fat thickness using dual electrode separation in focused impedance method (FIM). *Physiological Measurement, 33*, 707–718.

Surowiec, A. J., Stuchly, S. S., Barr, J. R., & Swarup, A. (1988). Dielectric properties of breast carcinoma and the surrounding tissues. *IEEE Transactions on Biomedical Engineering, 35*, 257–263.

Wagenaar, J., & Adler, A. (2016). Electrical impedance tomography in 3D using two electrode planes: Characterization and evaluation. *Physiological Measurement, 37*, 922–937.

Chapter 10
Clinical Applications of Electrical Impedance Spectroscopy

Carlos-Augusto González-Correa

The aim of this chapter is to give a general view of different applications where EBI and EBIS can play a role. We offer a global glimpse to the readers and some bibliographic references on them, so that, if somebody has a particular interest, he/she can have an initial approach into it. Applications of electrical impedance tomography (EIT) are not considered (see chapter in this book by R Bayford and also Bayford and Tizzard 2012 and Holder 2005), but we do have included some EBIS applications not mentioned in some of the reviews previously published in the literature (Bera 2014, Coffman and Cohen 2013, and the series published in 1996 in the volume 24 of the journal Critical Reviews in Biomedical Engineering by: Valentinuzzi (1996), Rigaud et al. (1996), Valentinuzzi et al. (1996), Rigaud and Morucci (1996), Morucci and Marsili (1996), and Morucci and Rigaud (1996)).

10.1 Overview

Electricity is a universal phenomenon, and it is considered as the backbone of modern industrial society (Jones 1991). It is present everywhere, and the human organism is not an exception. In this particular case, we have to consider it from two different perspectives: (a) the spontaneous, physiological occurrence of electrical phenomena inside our bodies and (b) the either passive or active response of our organism, as a whole or parts of it, to external or exogenic electric stimuli (Grimmnes and Martinsen 2000) or to exogenic stimuli producing electrical changes

---

C.-A. González-Correa (✉)
Universidad de Caldas, Grupo de Bio-Impedancia Eléctrica (GruBIE), Manizales, Caldas, Colombia
e-mail: c.gonzalez@ucaldas.edu.co

F. Simini, P. Bertemes-Filho (eds.), *Bioimpedance in Biomedical Applications and Research*, https://doi.org/10.1007/978-3-319-74388-2_10

Table 10.1 Different perspectives for the study of human bioelectricity

	Structure	Phenomena	Examples of techniques
Endogenic electric activity	Cell membrane	Voltage and current (transmembrane, transepithelial, along tissues and organs).	*Electrophysiology*: "patch clamp" (at cell level), electrocardiography (ECG), electroencephalography (EEG), electrooculography (EOG), electrogastrography (EGG), electromyography (EMG), galvanic skin potential (GSP).
	Epithelia		
	Tissues		
	Organs		
Response to exogenic electrical stimuli or exogenic stimuli evoking an electrical response	Passive	Resistance.	*Electrical impedance (EI) or electrical bioimpedance (EBI)*: BI analysis (BIA, for body composition, either whole body or segmental), EBI spectrometry (EBIS), EI tomography (EIT), electroacupuncture according to Voll (EAV) as a diagnostic tool, polygraphs, galvanic skin response/resistance (GSR).
	Active	Nerve and muscle, stimulation, wound healing.	*Electrotherapy*: for wound and fracture healing, transcutaneous electrical nerve stimulation (TENS or TNS, for pain relief), electroacupuncture according to VOLL (EAV) as a therapeutic device, electrostimulation ("passive gymnastics"), electroconvulsive therapy (ECT, formerly known as electroshock therapy or shock treatment), evoked potentials (EP).

or changes that affect the electrical properties of the objet under study. This concept is expressed in Table 10.1.

From all those situations that involve the study of electrical phenomena in our organisms (either active or passive), this chapter is focused on human medical applications based on the study of the passive electrical response of tissues, organs, and the organism as a whole that can be grouped under the umbrella of electrical bioimpedance (EBI) and electrical bioimpedance spectroscopy (EBIS). Electrical impedance tomography (EIT) is not considered. In the next section (10.2 "Basic principles"), the basic physical principles are reviewed after some considerations about a few biological aspects of the human body, necessary for a better understanding of how the technique operates. Figure 10.1 gives a general view of the different branches of the application of EBI in human clinical situations.

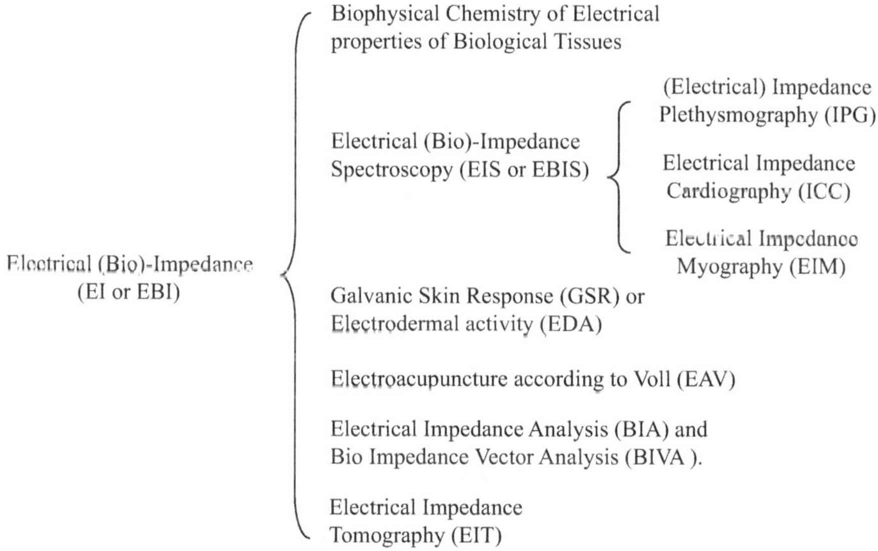

Fig. 10.1 Different branches of EI (EBI) in human clinical studies

10.2 Basic Principles

10.2.1 Biological Aspects

According to the biological cell theory, cells are the basic structural unit of any living being. In humans, as a complex multicellular organism, cells are organized in tissues, which we can define as a collection of specialized cells, with the same embryological origin, designed to carry out some specific physiological functions (there are four types of human tissues: nervous, muscular, epithelial, and connective). The next level of organization of the human body are organs (i.e., a bone, a muscle, the heart, the brain, the stomach, the colon, etc.), a structure made up of different tissues in order to carry out more complex functions. Different organs work together, forming 12 systems, which are in charge of still more complex functions around four gross ones (see Table 10.2): nutrition, defense, movement, and reproduction.

It is worth mentioning that, in the present century, the human microbiota (mainly bacteria), especially that inhabiting the intestinal tube, is being considered as an extra important organ or "meta"-organ or virtual organ (O'Hara and Shanahan 2006), without which we could not live (actually all living creatures and ecosystems on earth are dependent upon bacteria). Human beings (as practically every living plant or animal) would then be "meta"-organisms (Brown and Hazen 2015), also referred to as "super-organisms" (Sleator 2010). The major part of it is found in the colon, occupying the most external and looser layer of the mucus bilayer produced

Table 10.2 Human body systems and their main functions

General function	System	Specific main function(s)
Nutrition	Digestive	Breakdown of food and absorption of nutrients (including water)
	Respiratory	Oxygen intake (as O_2) and its elimination (as CO_2)
	Circulatory	Transport of nutrients and other biomolecules, cells, and metabolic waste
	Urinary	Elimination of metabolic waste and water homeostasis
Defense	Immune	Defense of the organism
	Lymphatic	Defense of the organism and cell transport
	Tegumentary	Protection against the outside world
Movement	Skeletal	Body support
	Muscular	Contraction to maintain position and move bones around joints (skeletal muscle)
	Nervous	Integration of information and control of movement
Reproduction	Endocrine	Growth and regulation of sexual function and metabolism
	Reproductive	Reproduction

by the intestinal epithelium. The inner and more compact layer, firmly adhered to the mucosa, acts as a physicochemical barrier that protects us from microorganisms and harmful xenobiotics. In the colon, the thickness of the mucus bilayer has an average of 600 μm (Johansson et al. 2013).

All human cells are very complex structures, but we are accustomed to represent them in the very simplistic form of small circles (representing the outer cell or plasma membranes), each with a still smaller circle embedded in it (representing the nucleus membrane). In reality, the intracellular space is full of different structures delimited by membranes (mitochondria, lysosomes, ribosomes, Golgi apparatus), and all these structures have been so far neglected in the analysis of EBIS data. The problem is that its contribution to the passive response of tissues is not clear, although there are some authors that have tried to take them into account (Shiffman 2013; Shiffman and Rutkove 2013).

Another important point to take in consideration is the fact that plasma cell membranes act as capacitors, as their constitution (made up of phospholipids) gives them the characteristics of a dielectric, and they are surrounded by conductive liquids (electrolytic solutions) on both sides: intracellular and extracellular. However, they do not behave as perfect capacitors, as they are leaky, allowing a permanent flow of sodium ions (Na^+) inward and of potassium ions (K^+) outward. A transmembrane (integral) protein known as Na^+/K^+ATP-asa (or Na^+/K^+ATP pump) works permanently in order to keep ionic intracellular chemical concentrations in resting conditions. Each molecule of this protein extracts three Na^+ ions per two K^+ ions taken inward. This is done against their respective chemical gradients, and, therefore, this process requires energy, taken from the hydrolysis of an ATP

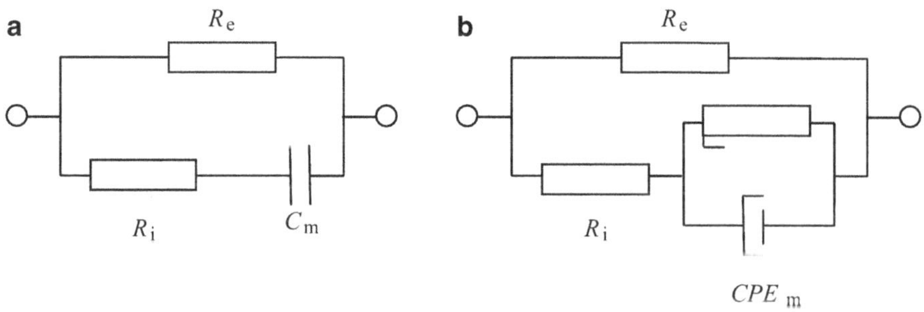

Fig. 10.2 Two electric models of biological tissues: (**a**) Debye and (**b**) Cole-Cole (see, for instance, Grimmes and Martinsen 2005)

molecule, of those mainly released by mitochondria (under aerobic conditions). All this is important because, as we will see later, a deficit in the production of ATP would mean that, somehow, cell membranes become "more leaky," as the Na^+/K^+ATP pump would not be able to maintain the physiological concentrations.

Although EBIS deals with tissues, organs, and the whole body, we have to mention some aspects of single cells in order to better understand some of the points that will be treated later.

10.2.2 Physical Aspects

In the electrical modeling of a biological tissue, we have the following main components: an extracellular (or intercellular) space, a cell membrane, and an intracellular space where, as already said, organelles and a lot of different sub-compartments are found.

The most simple and, initially, the more used electrical model of biological tissues represents them as two electric branches, one of them representing the resistance of the extracellular space (say R_e, although different authors use different nomenclatures[1]) and another one in parallel with a perfect capacitor representing the cell membrane (say C_m) and a resistance representing the resistance of the intracellular space (R_i). This is commonly known as the Debye model (Grimmes and Martinsen 2005) as shown in Fig. 10.2a.

A more sophisticated model is one similar to the Debye model, but replacing the perfect capacitor for an imperfect one, also known as constant phase element (CPE), to account for the cell membrane as an imperfect dielectric (CPE_m). This model is known as the Cole-Cole or simply the Cole model (see Fig. 10.2b).

[1] Brown et al. (2000), for instance, use R for R_e, S for R_i, and C for C_m, while Grimmes and Martinsen (2000) use the symbols G, R, and C, respectively.

The mathematical formulations of the Debye and the Cole model, in terms of electrical impedance and as a function of angular frequency (ω), are, respectively,

$$Z(\omega) = R_\infty + \frac{R_0 - R_\infty}{1 + j\omega\tau}$$

and

$$Z(\omega) = R_\infty + \frac{R_0 - R_\infty}{1 + (j\omega\tau)^{(1-\alpha)}}$$

It can be seen that there are four parameters in the equation expressing the Cole model (R_∞, R_0, τ, and α), whereas in the equation for the Debye model α is not present. The parameter α can assume a range of values between 0 and 1, and the Debye model is a specific case of the Cole model, when α equals 0.[2]

The relation between the formula for the Cole model with the four parameters and its electrical model can be explained as follows.

a. R_0 is equal to R_e, as it is considered that, at zero Hz, the current only flows through the extracellular space.
b. R_∞ is equal to the resulting resistance of R_e and R_i in parallel as, at high frequencies, either C in the Debye model or CPE_m in the Cole model is short-circuited (i.e., its capacitive effect disappears), and, therefore, $R_\infty = (R_e \times R_i)/R_e + R_i$.
c. When ω equals zero, $Z(\omega) = R_0$, the formula becomes $Z(\omega) = R_\infty + ((R_0 - R_\infty)/(1 + 0)) = R_\infty + ((R_0 - R_\infty)/1) = R_\infty + (R_0 - R_\infty) = R_\infty + R_0 - R_\infty = R_0$.

10.3 Changes in Tissues that Can Alter Their Passive Electrical Characteristics

10.3.1 Extracellular Space (ES)

The two main characteristics that can affect this component are its biochemical composition and its dimensions. In general, the first one can be considered stable, as potent systemic homeostatic mechanisms keep it in generally very narrow ranges (considered as "normal"). In contrast, the dimensions can vary very easily in different ways. The number/density, size, shape, separation, and orientation of cells are all changes that alter the cross-sectional area of the ES, directly affecting its resistance to current flow. As examples, it is worth mentioning two cases that we

[2]Some authors use α instead of $1 - \alpha$ (Brown et al. 1999). Attention has to be paid to the expression being used, as, according to which one is selected, both alphas are complements to 1 (or 10^0). For example, using the first expression, $\alpha = 1$ would mean $\alpha = 0$ if the second expression is to be used, being both equivalent to the Debye model, i.e., the latter is a specific case of the Cole-Cole model.

considered of paramount importance in the role that EBIS can play in clinical diagnosis: inflammation and increased intestinal epithelia permeability. One of the hallmarks of inflammation is an increase in vascular permeability, allowing plasma components and inflammatory cells to exit the bloodstream (Arroyo and Iruela-Arispe 2010). This increase in water content will be represented by a decrease in the electrical resistance of the inflamed tissue (see, for instance, Tornuev et al. 2014 and Gonzalez-Correa et al. 2003).

The increase in intestinal epithelial permeability due to disruption of the tight junctions is being now considered as a possible hallmark of a pathological condition of the intestinal wall heading toward many chronic diseases, not only of the intestinal tract itself but also to systemic (or extraintestinal) diseases like obesity, type 2 diabetes, autism, arthritis, lupus, allergies, depression, and many others (Gonzalez-Correa et al. 2017; Lerner and Matthias 2015, Viggiano et al. 2015; Bischoff et al. 2014; Soler et al. 1999).

10.3.2 Cells

As we have already mentioned, there are a number of cell variables that can affect the passive electrical response of tissues. Their number/density, size, separation, shape, and orientation, as well as their membrane's properties, play a role, but they are all intertwined. For instance, given a certain size of cells (red blood cells, for instance), their concentration would affect the electrical impedance of blood in two ways: the size of the extracellular space and the total capacitance, the latter being a function of total cell membrane surface and cell arrangement. Each cell can be considered as a tiny capacitor, in which the thickness can be assumed as constant ($\approx 7.5 \times 10^{-9}$ m; Yamamoto 1963) with a capacitance of $0.5\text{–}1.0 \times 10^{-2}$ Fm^{-2} and a resistivity of 15.5–22.0 Ωm^2 (Golowasch et al. 2009). Cell size also acts in a similar way, in that the larger the cells, the lesser extracellular space will be left in the tissue, and, at the same time, capacitance will vary for each cell (the larger the cell, the larger its individual capacitance, but total capacitance of the tissue may decrease, as there will be fewer cells). Separation between cells can be well reflected when measuring impedance, as it is the case, for instance, when studying epithelia permeability, something that eventually could become a very interesting biomarker, as we will show later. Muscle is a good example of how shape and orientation act on the passive electrical tissue properties, as it is well known that muscle impedance values are different if they are measured in the same direction of the myocytes or transversally (Brown et al. 1999, chap. 8 "Non-ionizing electromagnetic radiation: tissue absorption and safety issues, p. 235). Finally, membrane functional characteristics also can play a role, as, depending on how "leaky" they are or the availability of ATP, the α parameter may be affected, an issue not well addressed yet, but that will be commented on in Sect. 4.6.2 ("Muscle"), when we mention EBIS studies in fibromyalgia (Colina-Gallo et al. 2016b).

10.3.3 *Intracellular Space*

This compartment is still more complex than the extracellular one, due to the very different and abundant intracellular structures (organelles) present in there. Its contribution to the electric passive response is a field that has not been sufficiently explored. Most of the research has focused mainly in differences found at low frequencies (R_0), or the relation between R_∞ and R_0, used, for instance, in BIA (bioelectrical impedance analysis) to calculate total intra- and extracellular water. Nevertheless, some authors have tried to incorporate these substructures in the electrical model, as already mentioned (Shiffman 2013; Shiffman and Rutkove 2013).

10.4 EBIS Applications Different from Cancer and by Body Systems

EBIS has been used to explore early detection of cancer in different organs (skin, cervix, esophagus, urinary bladder, and colon). Breast cancer has been investigated by EIT, and there is a device in the market, approved by the FDA, which operates as adjuvant to mammography (TranScan TS 2000 by Siemens, Germany; see, for instance, Barter and Hicks 2000). For cervical cancer, a commercial device has been produced in England (ZedScan I by Zilico UK; see, for instance, Tidy et al. 2013). In this section, we will mention some clinical applications of EBIS by body systems and organs, especially those different from cancer, and, in Sect. 5 ("Cancer and EBIS"), we will consider some general aspects of its use in early detection of cancer.

10.4.1 *Digestive*

10.4.1.1 Mouth (Oral) Mucosa

From 1995 to 2003, Nicander and Ollmer with some others published the results of different studies on the mouth (oral) mucosa, especially in relation to irritation (Ollmar et al. 1995; Nicander and Ollmer 2004; Rantanen et al. 2003; Nicander and Ollmar 1999). More recently, the use of a new device for determining mouth mucosa dryness has been reported (oral moisture meter Mucus III; Fukushima et al. 2013).

10.4.1.2 Teeth

There are different aspects of dental health or situations for which EI has been used: enamel and dentine status, the presence of caries, the quality of fillings adjustment, length of the dental canal, and pulp status.

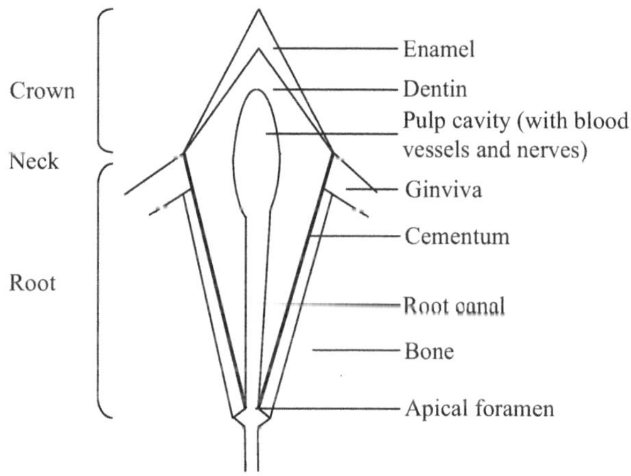

Fig. 10.3 Schematic representation of a tooth

Figure 10.3 shows a simplified structure of a tooth with its three major parts (crown, neck, and root), as well as its components: enamel, dentin, pulp cavity, tooth canal, and the two main surrounding structures of the root—gingiva (gum) and bone. Of these six components, three are considered as hard tissues with very high resistivities that decrease in the following order—enamel, dentin, and bone—while the other three are made up of soft tissue and, therefore, have a low resistivity: gingiva and the content of both the pulp cavity and the root canal. The high resistivity of hard tissue (despite being highly mineralized) is explained by a firm intercellular structure with low cellularity and low water content, while soft tissue has opposite characteristics.

10.4.1.3 Tooth Caries

Dental caries are defined as "localized destruction of the tooth surface initiated by decalcification of the enamel followed by enzymatic lysis of organic structures and leading to cavity formation. If left unchecked, the cavity may penetrate the enamel and dentin and reach the pulp" (MeSH[3]). There is a pandemic of this condition, with about 2.4 billion people having untreated caries in permanent teeth and 621 million children worldwide presenting untreated caries in deciduous teeth (Kassebaum et al. 2015). EBI has been used to detect and monitor the condition (Pretty and Ellwood 2013; Eldarrat et al. 2010; Liao et al. 2007), and different devices can be found in the market for this purpose (Somma et al. 2012), all of them operating under the same principles.

[3]MeSH is the thesaurus of Medline and PubMed by the National Center for Biotechnology Information (NCBI) and the National Library of Medicine (NLM) of the United States. Webpage: https://www.ncbi.nlm.nih.gov/mesh.

The principles of this EBI application are explained by Pretty and Ellwood (2013) as follows: Basically, the continuum of caries development implies an increase in the porosity of both the enamel, firstly, and, subsequently, of dentin. This means a greater contents of water and, therefore, a reduction in the electrical resistance, as there are more electrolytic fluids filling the pores. The electrical current being applied passes from one electrode put in contact with the tooth surface to a receiving electrode in the patient's hand. As the resistivity of body soft tissue is much lower than the resistivity of the dental hard tissue, the resistance measured by the devices mainly reflect the resistance of the dental surface being explored. Sensitivity and specificity of the methods have been explored by different authors and are around 74.8 and 87.6, respectively, and reproducibility is also considered as excellent. The same authors (Pretty and Ellwood 2013 but also Liao et al. 2007) consider that electrical caries monitor (ECM, as it is commonly known) "... represents a practical, objective and sensitive method of monitoring dental caries and has been proven to have significant utility in caries clinical trials."

In relation to dentine, Eldarrat et al. (2010) reported age-related changes in electrical impedance of dentin, due to occlusion of the tubules by peritubular dentine.

10.4.1.4 Length of Dental Canal

For many years, electronic devices based on EBI principles have been used by general dentists and endodontists in order to determine the length of the dental root canal and localize the apical foramen of teeth, when they have to treat the presence of bacteria and their by-products within it (Nekoofar et al. 2006). According to these authors, based on the main electrical properties being used by the different commercial devices available in the market, they can be classified as resistance-based, low-frequency oscillation, high-frequency devices (capacitive-based), capacitance and resistance, two frequencies, and multifrequency. Resistance-based devices were the first ones that were developed, and they measure the electric resistance between a conductor file introduced in the canal and an electrode attached to the oral mucosa. It is assumed that there is a fixed resistance of 6.5 kΩ between the instrument and the oral mucous membrane, which serves to determine the canal length, i.e., the location of the apical foramen. This measurement is of clinical relevance, due to the fact that treatment procedures should be confined within the root canal system (Nekoofar et al. 2006). More recently, different authors have compared the in vivo and ex vivo accuracy of different types of devices used for this purpose and available in the market (see, for instance, da Silva and Alves 2014; Somma et al. 2012). It seems that there are no statistically significant differences in the results obtained by the instruments that were tested.

10.4.1.5 Tongue

In 2016, Pacheck et al. published a study on the use of EBI as a possible biomarker for the drug efficacy in patients with amyotrophic lateral sclerosis (ALS), a disease characterized by degeneration of motor neurons and consequent muscle atrophy. McIlduff et al. (2016a, b) discussed the conditions to obtain good recordings of the tongue electrical impedance. Physical principles of this application are the same as those used in what is known as electrical impedance myography (EIM), which will be discussed later, under the subheading "Muscle Studies" (see Sect. 10.4.6.2). Studies about buccal cancer detection have also been carried out (Sun et al. 2010) as well as about the menstrual cycle (Rezác 2008).

10.4.1.6 Esophagus

One of the first aspects of the esophagus that has been investigated with EBIS is esophageal motility, with studies as early as 1994 (Fass et al.) and 1995 (Rao et al.), and, more recently, by authors such as Tambucci et al. (2015). The former had concluded in their time that "… impedance procedures may give additional significant information about bolus transport and esophageal wall movements."

Gastroesophageal reflux disease (GERD) is a medical condition in which gastric content moves backwardly. Its prevalence is increasing worldwide with estimates ranging between 18.1 and 27.8% (El-Serag et al. 2014). The most prominent symptom associated with this condition is heartburn, caused by the inflammatory irritation produced to the esophageal wall by the reflux content. When the reflux is very severe, it can even have other manifestations such as laryngeal symptoms, cough, asthma, recurrent otitis media, idiopathic pulmonary fibrosis, pharyngitis, and sinusitis (Bardhan et al. 2012). But, by far, a still worse consequence is the appearing of Barrett's esophagus, a metaplastic and precancerous condition, and, later on, of gastroesophageal adenocarcinoma (Nelsen et al. 2012).

The road from normal esophageal epithelium (of the squamous type) to cancer is reflux → inflammation → metaplasia (transformation of squamous to intestinal epithelium) → dysplasia → cancer. EBIS has been used to assess both the existence and scoring of the reflux itself, by a mixed technique called combined esophageal pH and multichannel intraluminal impedance monitoring (pH-MII), as well as the possible identification of the conversion of Barrett epithelium into cancerous tissue. For the first purpose, a set of concentric electrodes is added to conventional pH catheters, using between six and eight pairs of electrodes, separated 2 cm from each other, each pair going to a channel connected to an impedance transducer (Sifrim et al. 1999). The combined pH-MMI technique allows a more comprehensive physicochemical characterization of reflux episodes in terms of consistency of the refluxate (i.e., liquid, gas, or mixed), height reached, and presence/clearance of bolus and acid (Tutuian and Castell 2006). Other advantages of the method, as pointed out by Mousa et al. (2011), are reflux detection regardless of its pH value, direction of the flow (i.e., antegrade as in swallows or retrograde as in authentic reflux), and ability of detecting GERD-associated symptoms even if the patient is taking acid-suppression medications.

The Sheffield group working in EBIS carried out some research in the field of early detection of cancer in patients with Barrett's esophagus. They first showed, ex vivo, that it was possible to differentiate squamous from columnar epithelium in terms of their passive electrical characteristics (González-Correa et al. 1999). Subsequently, they studied the conditions for in vivo transendoscopic measurements in humans, using a 1.95-m-long, 3.2-mm-diameter, four-electrode probe specifically designed for that purpose (González-Correa et al. 2000). Finally, using resected tissue from 32 patients, they found (Gonzalez-Correa et al. 2003) small but statistically significant differences between non-inflamed non-dysplastic columnar tissue (with $R_0 = 4.9\ \Omega\text{m}$), columnar tissue showing inflammation ($R_0 = 4.2\ \Omega\text{m}$, $p = 0.016$), and columnar tissue showing dysplasia ($R_0 = 3.4\ \Omega\text{m}$, $p = 0.040$). Regrettably, this line of research was not further pursued but could be retaken under the theory of field cancerization (Fernández et al. 2016), where measurements either on the tongue or the esophagus itself could show early changes conducting to later cancer transformation.

There is also a technique known as impedance planimetry used to measure cross-sectional areas in the alimentary tract, which has been applied for studying motility disorders and the functioning of sphincters (Lawenko and Lee 2016). A device for this purpose is in the market under the commercial name of EndoFlip® by Crospon in the United States and Ireland. The principle of this device is shown in Fig. 10.4. Measurements are made along a catheter provided with a set of 10–12 electrodes, used in tetrapolar arrangements (more information about the principles of this device can be found in the webpage http://www.crospon.com/TechnologyBehindEndoFLIP.htm).

10.4.1.7 Stomach

Although electrogastrography (EGG) has been used for many decades with early recordings dating back to 1922 (Murakami et al. 2013), the application of EBIS for the study of the stomach has been scarce, despite very early initial studies, as those by Rehm (1953) and Rehm et al. (1955). The Sheffield group also made some research, with planar arrays placed on the abdominal wall for the study of gastric emptying (Smallwood et al. 1994). More recently, EBI has been used to study some stomach properties, both from outside (abdominal wall; Giouvanoudi et al. 2003) and inside it, with measurements taken directly on the gastric mucosa (Beltran et al. 2013). In 2015, Beltran and Sacristan presented a device developed by their research group in Mexico for the study of ischemia and damage of mucosa in the gut of critically ill patients, claiming that they consider it to be safe for this purpose and showing a good performance. It works with a current of 1 mA p-p in a bandwidth between 100 Hz and 1 MHz.

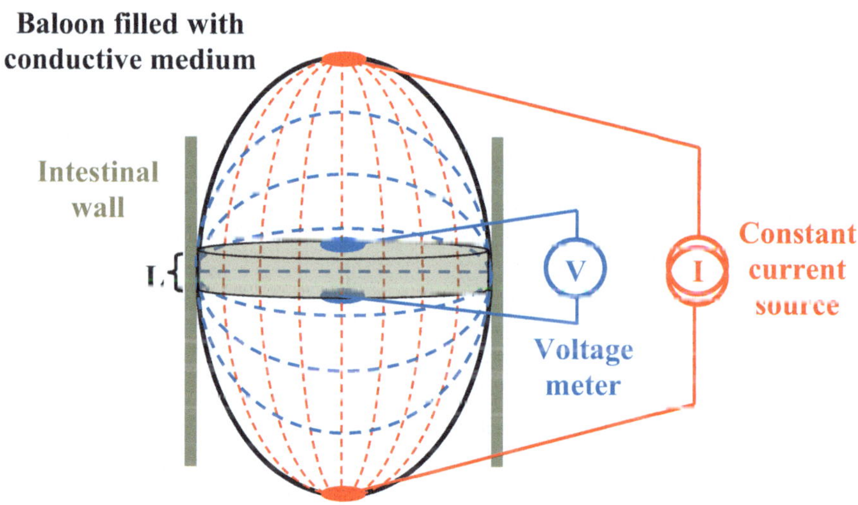

1. According to Ohm´s Law we obtain the resistance by $R = V/I$;
2. From $R = \rho * L / A$ and $A = \pi * (D / 2)^2$, we can replace A and get $R = (\rho * L) / (\pi * D^2 /4)$;
3. Therefore $D^2 = (4 * \rho * L) / (\pi * R)$ and $D = 2*[(\rho * L) / (\pi * R)]^{0.5}$.

Where: R is resistance,
V is the measured voltage,
I is the applied (known) current,
ρ is the resistivity of the conducting medium (known),
L is the distance between the voltage measuring electrodes,
A is the area of the cylindrical section, and
D is the diameter of the cylinder.

Fig. 10.4 Principles for the measurement of cross sectional areas of hollow organs (like the esophagus) by the technique known as impedance planimetry in devices such as EndoFlip®

Gastric motility has also been studied with a combination of simultaneous reading with electrogastrography (EGG) and bioimpedance measurements (Li et al. 2011). They made noninvasive measurements with a tetrapolar array over the gastric area, with an arrangement as shown in Fig. 10.5, comparing results between healthy subjects and subjects with functional dyspepsia or gastritis, concluding that it is a convenient and effective method.

10.4.1.8 Small Intestine

Due to the difficulty of reaching this organ, there are not many studies on the application of EBIS to explore it. Nevertheless, some approaches have been implemented, such as transendoscopic readings (Leontev et al. 2015; Nguyen et al.

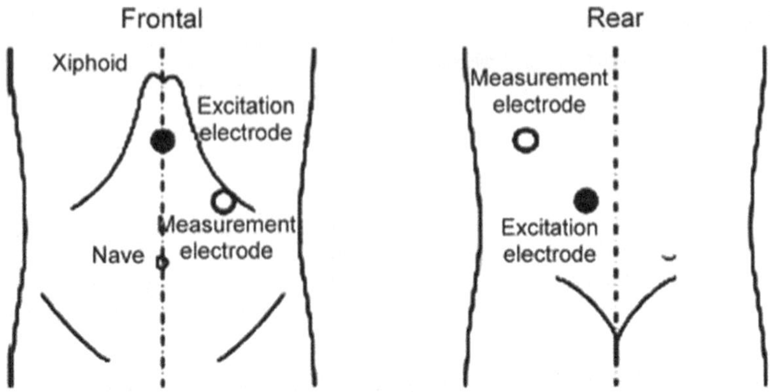

Fig. 10.5 Tetrapolar electrode arrangement for noninvasive study of gastric motility (Source: Li et al. 2011)

2010), intrasurgical measurements from the serosal side of the wall, ex vivo studies of biopsy specimens (Man et al. 2015; Vanheel et al. 2014), and cell cultures.

Much of the work has been performed in animals, and some of the aspects that have been subjected to exploration are mucosal permeability and cancerous changes. Ussing chambers (named after its inventor, Hans Henriksen Ussing) have been widely used, measuring transepithelial electrical resistance (TEER), paracellular permeability, and inflammation (Casselbrant et al. 2015; Vanheel et al. 2014). In 2016, Strand-Amundsen et al. published results on the use of EBIS for in vivo characterization of ischemic small intestine in pigs, using a Solartron 1260/1294 impedance gain-phase analyzer and concluding that the method could be used for "…real-time monitoring and assessment of the presence and duration of small intestinal ischemia."

10.4.1.9 Large Intestine (Colon and Rectum)

Although colon and rectum are more accessible than small intestine, research on the application of EBIS to its study is also scarce. In 1999, Soler et al. published an article where they concluded that increased tight junction permeability of the colon epithelium produces a decrease in epithelial barrier function, preceding the development of colon tumors. More recently, Chandler et al. (2015) propose a time-dependent element for the characterization of biological tissues by measuring their passive electrical resistance in order to distinguish healthy from diseased tissue and proved it on an ex vivo human colon tissue. In Sect. 5 ("Cancer and EBIS"), we will mention the possibilities for cancer detection in the colon and rectum with EBIS.

It is also worthwhile to highlight the increasing importance that the study of the colon is gaining among the scientific community in the sight of the advances in the twenty-first-century science (Sellers and Morton 2014). Among them, we have to mention the role of intestinal microbiota in health and disease (Kåhrström et al.

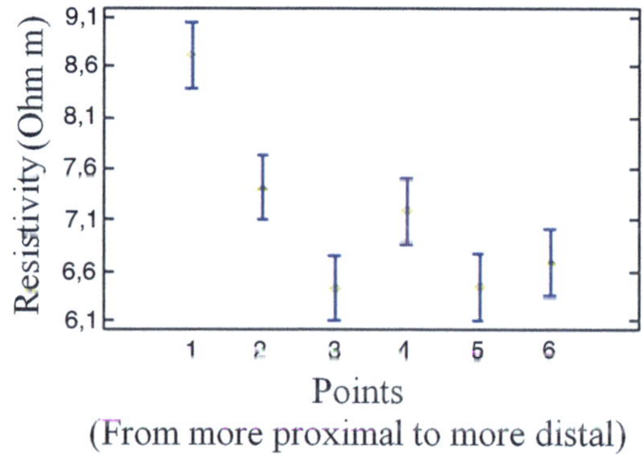

Fig. 10.6 EBIS reading on the left and right hemicolons in rabbits (Modified from Mulett-Vásquez et al. 2016a, b)

2016), as well as that of the mucous layer (Rodríguez-Piñeiro and Johansson 2015), with its double function of being its habitat and a protective barrier against host invasion by it. In the last section of this chapter (6 "The Future: Chronic Diseases and EBIS"), we will talk about the possible role of EBIS not only for the study of cancer but also of many other chronic diseases whose pathogenesis begins in the colon. Changes and the status of microbiota, mucus bilayer, and epithelium are all factors that can be reflected in EBIS measurements.

Now is the time for more studies of the colorectum using EBIS: firstly, characterizing it, and, secondly, looking for differences between the electric bioimpedance characteristics of the organ in healthy people versus those of people with diseases, especially with chronic diseases. In this respect, for instance, our research groups on Electrical Bioimpedance (GruBIE) at Universidad de Caldas (Colombia) and on Biological and Semiconductor Material Science (CIMBIOS) in Bucaramanga (Colombia) have begun a jointly work where, initially, differences between the right and the left hemicolons in rabbits were found (Mulett-Vásquez et al. 2016b, see Fig. 10.6). Subsequently, statistically significant differences in rectal EBIS measurements (REBIS) between people with cancer against people without cancer were also found (Mulett-Vásquez et al. 2016a).

10.4.2 Respiratory System

In this area, more research has been carried out using EIT (a review of this technique can be found in the chapter by in this book and also in Gong et al. 2015), rather than EBIS. Nevertheless, there have been some attempts to use the latter in situations such as the study of laryngeal function (Noll et al. 2011; Szczesniak et al. 2009) and pulmonary edema/congestion (Charles et al. 2015; Donadio et al. 2015; Weyer et al. 2014). For the lanrynx, a similar technique to that used for bolus transit in the digestive wall has been used, while, for the lungs, a BIA device was employed, measuring the relation between intracellular and extracellular water. BIA has also been used for determining malnutrition in people suffering of lung cancer (Daghfous et al. 2014; Gupta et al. 2009). Piccoli et al. (2012) also proposed what they call BIVA (bioelectrical impedance vector analysis) to differentiate between cardiac and non-cardiac apnea.

10.4.3 Cardiovascular

Plethysmography (from the Greek plethynein, to increase, and graphein, to record) is defined as the study of changes in volume of an organ, part of the body, or the whole body. When used in the thorax with electrical bioimpedance techniques, it is known as impedance electrocardiography (ICG) and is mainly used for determination of hemodynamic parameters (Siedlecka et al. 2015; Miller and Horvath 1978), whereas, when used for the study of blood vessels, it is known as impedance plethysmography (IPG). Early studies were initiated in the 1940s, and further development in this field was carried out by the US National Aeronautics and Space Administration (NASA) in the 1960s for its application in chest studies (Miller and Horvath 1978). These two authors are considered as pioneers in the application of ICG in psychophysiology.

According to Siedlecka et al. (2015), there are 12 hemodynamic parameters that can be assessed by ICG: cardiac output (CO), cardiac index (CI), stroke volume (SV), stroke index (SI), thoracic fluid content (TFC), systemic vascular resistance (SVR), systemic vascular resistance index (SVRI), velocity index (VI), acceleration index (ACI), left ventricular ejection time (LVET), pre-ejection period (PEP), and systolic times ratio (STR). Of all them, stroke volume and cardiac output have been the most investigated (Bayram and Yancy 2009), in conditions such as blood hypertension, cardiac insufficiency, dyspnea in intensive care wards, and cardiac rehabilitation (Siedlecka et al. 2015) but also in postural stress, pacemaker optimization, sleep studies, pregnancy monitoring, and Holter-type monitoring, with one device being approved for this last purpose by the Food and Drug Administration (FDA) in the United States in 2010 (Cybulski et al. 2012). However, despite all this, the limitations of the technique have hindered its wider use. The electrode arrangement for this technique is shown in Fig. 10.7.

IPG has been used mainly for detection of deep venous thrombosis in the legs but also for monitoring blood pressure and atherosclerosis (Mašanauskienė et al. 2014). When studying venous thrombosis, the method is based on the measurement of blood flow after it has been cut off by inflating a pneumatic cuff around the thigh. In the presence of, for instance, venous thrombosis, the emptying will show a delay in a characteristic way, a variable that can be picked up by electric impedance. IPG is considered as the gold standard method in this case (Harel et al. 2005). IPG has been also proposed as an alternative for the study of peripheral arterial disease (Mašanauskienė et al. 2014), and even an implantable device has been developed for that purpose and proved in vivo with pigs (Theodor et al. 2014). Bernstein et al. (2015) have experimented with what they call "transbrachial

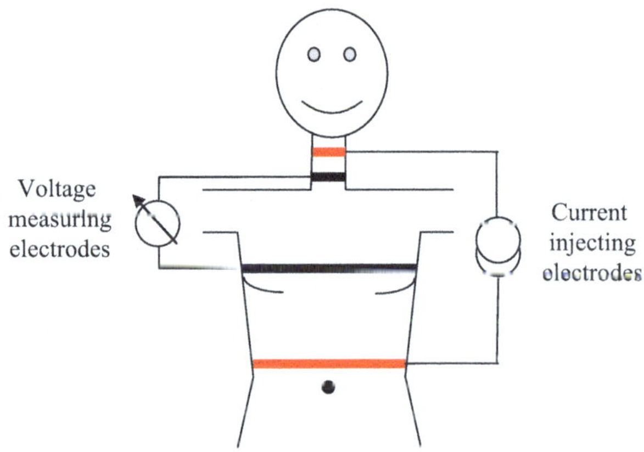

Fig. 10.7 Schematic representation of the tetrapolar arrangement used for impedance measurements in the thorax or impedance electrocardiography (ICG)

electrical bioimpedance velocimetry (TBEV)," a combination of ICG and IPG, where they study cardiac hemodynamics with operational signals acquired in the arm.

10.4.4 Genitourinary System

In relation to this system, we have to mention the following areas of research using EBIS: estimation of fluid volumes in hemodialysis patients (Piccoli 2014), determination of urinary bladder volume (Zariffa et al. 2016), and cancer (Keshtkar et al. 2006; Khan et al. 2016). The work done by the group led by Prof. Brian Brown derived the design of an impedance device that has begun to be widely used, called Zedscan® and produced by the English firm Zilico (Manchester UK). The principle is that the different status of the exocervical epithelium produces different impedance spectra or profiles (see Fig. 10.8). Research on detection of breast cancer by EBIS has been done mainly using EIT (Khan et al. 2016; Halter et al. 2009a, b; Barrio et al. 2015).

10.4.5 Skin

The recording and study of electrical phenomena associated to the skin has a long history dating back to the mid-nineteenth century, when Du Bois-Reymond carried out some experiments (Boucsein 2012, Chap. 1 "Principles of Electrodermal Phenomena," p. 1). One special field is known as galvanic skin response (GSR) but also as electrodermal activity (EDA). GRS appears as an entry in the Medical Subject Headings (MeSH, the thesaurus of the public service PubMed), while EDA

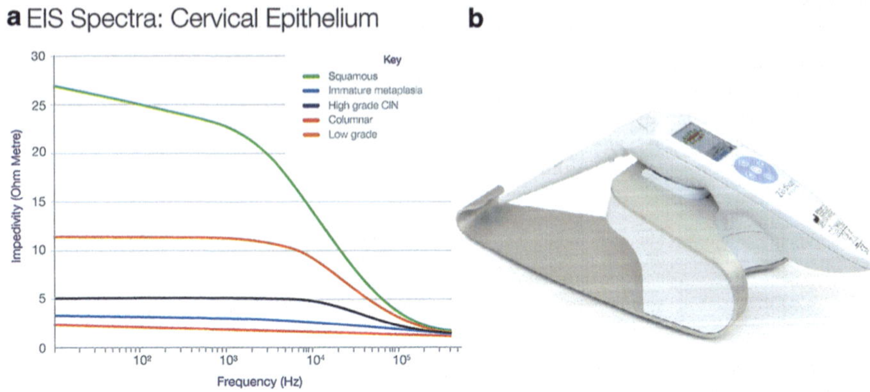

Fig. 10.8 (**a**) Different EIS spectra for cervical tissue. (**b**) The ZedScan device by Zilico (Source: images taken from https://zilico.co.uk)

is the name adopted by the Society for Psychophysiological Research. There are two variants for recordings of EDA: electrodermal potentials (EDP), also known as endosomatic (when potential differences are originated in the skin itself), and electrodermal response (EDR), also known as exosomatic response, when either direct current (DC) or alternating current is applied to it (Boucsein 2012, Chap. 1 "Principles of Electrodermal Phenomena," p. 2). The main principle of EDR is that there is a link between emotional and sympathetic responses of an individual and his/her sweating activity, which alters the resistance/conductance of the skin. This is one of the bases behind the devices known as polygraphs or "lie detectors." So far, research has been done mostly using DC current, but, attending the call by Boucesein et al. to use alternating current, as a way to overcome some of the errors produced when using DC, Pabst et al. (2016) made a publication showing an "... excellent agreement between a 20 Hz AC method and standard DC method ...." They also highlight that, with AC, more information can be gathered, as susceptance can be analyzed, adding capacitance to the traditional resistance/conductance analysis.

Wound healing is another field in which some authors have explored the utility of EBI/EBIS for monitoring purposes. Spence and Pomeranz (1996) used transcutaneous electrical resistance measurements for repeated noninvasive monitoring of epidermal repair, but not wound strength. In 2010, Weber et al. proposed the use of an electrode array for wound mapping based on EBIS, a device that could be incorporated into standard commercial occlusive dressings, allowing, for instance, telemonitoring of chronic ulcer healing. More recently, by EBI measurements, Solmaz et al. (2016) studied the effect of laser irradiation on wound healing, and they concluded that it may be considered as a noninvasive supplementary method for following the healing process.

Much of recent research done with EBI on the skin has been carried out by the research group led by Stig Ollmer in association with Ingrid Nicander from Sweden. He even founded the Swedish medical technology company called

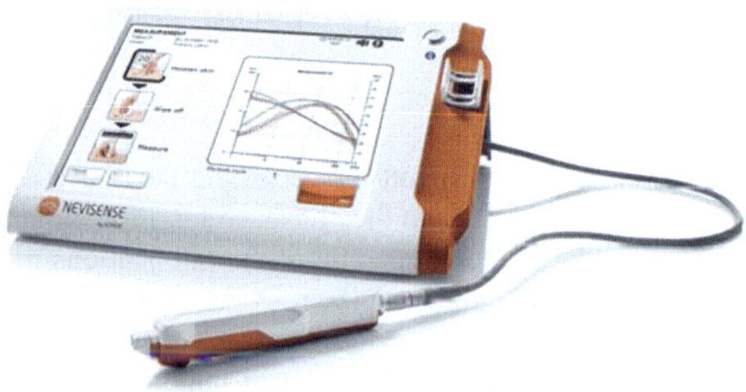

Fig. 10.9 Nevisense by Scibase for detection of skin cancer (Source: www. http://scibase.se/en/the-nevisense-product/)

SciBase in 1988, which has developed an EBIS-based device called Nevisense (see Fig. 10.9) for detection of malignant melanoma (see http://scibase.se/en/). Most of their work has focused on skin cancer (see, for instance, Aberg et al. 2004), but they have also explored other skin conditions such as inflammation (dermatitis) and atopy (Nicander et al. 2003; Nicander and Ollmer 2004). Recently, a group, with the participation of Stig Ollmer, has presented a battery-less implantable sensor for bioimpedance applications (Rodriguez et al. 2016). Some other authors have explored other aspects related to the skin such as skin hydration (Tagami 2014, Rim et al. 2005).

Another field that has been explored, although not in the mainstream occidental conceptions of physiology, pathology, health, and disease, is the relation of dermal points (i.e., acupuncture points; Tseng et al. 2014; Shima et al. 2012; Ahn and Martinsen 2007) or skin zones (dermatomes; Boytsov and Belousova 2015; Itsekson et al. 2010; Weitzen et al. 2007; Zimlichman et al. 2005) with diseases affecting different organs. Both techniques are based on the principles established by the German physician Reinhold Voll (1909–1989), initially known as electroacupuncture according to Voll (EAV), and, at present, referred to as electrodermal screening or electrodermal screening test (EDT). There are plenty of commercial branches worldwide offering electronic devices designed for this purpose, initially known as Dermatron. Basically, they pass a small electrical direct current from one point in one hand to an electrode hold in the other hand and measure the associated resistance and capacitance, transforming the reading to an arbitrary scale from 0 to 100. More recently, EBIS has been also used in this field (Ahn et al. 2010). This could also be linked to the theory of cancerogenic and etiologic field effect and a new emerging paradigm of a common course for most chronic diseases, two subjects that we will consider in the final part of this chapter.

10.4.6 Musculoskeletal System

10.4.6.1 Bone

Because bones are difficult to access and they have a very high resistivity, studies with EBIS are scarce, and they have, so far, little clinical application. Nevertheless, we would like to mention some of the publications made on the subject. In 1994, Collier and Ntui published their work, where they claim that healthy bone has small phase delays compared to fractured bone. More recently, Yoshida et al. (2009) proposed the measurement of bone electrical impedance, as a method for determining the appropriate time for the removal of external fixation when it is used, i.e., to determine the time when the bone has healed, something that is usually made by radiographic and clinical examination.

10.4.6.2 Muscle

As with the skin, research on the passive electric characteristics of skeletal muscle dates back to the late nineteenth and early twentieth centuries (Rutkove 2009). This author, in collaboration with Carl Shiffman and Roland Aaron, is the researcher who, in the present century and since 2005, has most contributed to the advance of the application of EBIS to the study of neuromuscular diseases (see, for instance, Rutkove et al. 2005, one of their earliest publications). They have developed what is now known as electrical impedance myography (EIM; Sanchez and Rutkove 2016), for the noninvasive study of neuromuscular diseases. A search in PubMed (December 2016), with the strategy "Rutkove SB[Author] and "electrical impedance myography," gives a result of 62 items, the first of them being Rutkove et al. (2005). Initial measurements were carried out at a single frequency (EBI at 50 kHz), but, at present, they are working in different directions, both with animals and humans, using multifrequency (EBIS) and trying to develop a robust and portable device for this specific purpose.

Among different diseases that they are studying with EIM are dystonic muscle hypertrophy, myositis, amyotrophic lateral sclerosis, disuse atrophy, unilateral radiculopathy, muscular dystrophy, Duchenne muscular dystrophy, and sarcopenia. Some of the aspects that they have considered so far are the possibility of distinguishing between myopathic diseases and neurogenic conditions, severity of some diseases, and response to treatments. Disease-associated changes in the muscle that can produce changes in its passive electrical response are fiber atrophy, endomysial fat and connective tissue content, edema, inflammatory infiltration, blood flow, and, as a particular characteristic of this tissue, fiber anisotropy due to its micro- and macroscopic structure.

Finally, it is also worth mentioning that our research groups (GruBIE and CIMBIOS) have begun to investigate the possibility of using EIS for the diagnosis of fibromyalgia. So far, it seems that there are no statistical significant differences

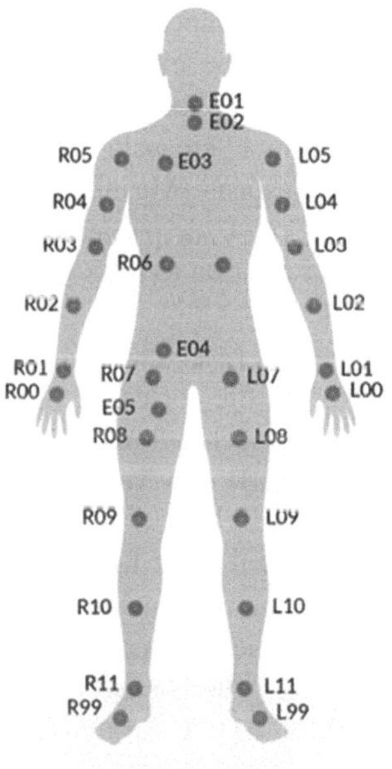

R00/L00	......	Posterior surface of hand at distal end of third metacarpal.
R01/L01	......	Posterior surface of wrist, between radius and ulna.
R02/L02	......	Midpoint between radial process and olecranon.
R03/L03	......	Just over lateral epicondyle.
R04/L04	......	Midpoint between olecranon and acromial process.
R05/L05	......	Acromial process.
R06/L06	......	Lower costal margin on medial axillar line.
R07/L07	......	Anterior surface of thigh on the midline, at gluteal crease.
R08/L08	......	Midpoint between R07/L07 and upper edge of patella.
R09/L09	......	Just over external joint line of knee.
R10/L10	......	Midpoint between R09/L09 and external melleoli.
R11/L11	......	Anterior surface of the ankle, between internal and external malleoli.
R99/L99	......	Anterior surface of foot at distal end of third metatarsal.
E01	......	5,0 cm proximal to sternal notch, in the midline on anterior surface of the neck.
E02	......	At sternal notch.
E03	......	5,0 cm medial ro R05.
E04	......	5,0 cm proximal to R07.
E05	......	5,0 cm distal to R07.

Fig. 10.10 Proposed standardized nomenclature for the anatomical positions of electrode placements (Source: Colina-Gallo et al. 2016a)

between normal subjects and subjects with fibromyalgia in the parameters R_0 R_∞ and τ (Colina-Gallo et al. 2016b), but the α parameter could probably be useful for such purpose (data not yet published). Our hypothesis is that fibromyalgia could be a mitochondriopathy, with insufficient production of ATP (adenosine triphosphate) having, as a consequence, a malfunctioning of the Na^+-K^+-ATPase. This would mean a decreasing in the resting membrane potential and, therefore, a cell triggering of action potential with, under normal circumstances, subthreshold stimuli. The need of anaerobic production of ATP and consequent increase of lactate production could also explain the generalized pain characteristic of this condition. Previous to this work, a publication was also made (Colina-Gallo et al. 2016a) where the validity of multiple segmental measurements is proposed with the displacement of just one voltage lead, as well as a nomenclature for the naming of different points for the placement of the electrodes (Fig. 10.10).

10.5 Cancer and EBIS

EBIS has been used for the diagnosis of different types of cancer such as breast (Malich et al. 2001), cervix (Walker et al. 2003; Brown et al. 2000), prostate (Halter et al. 2009a, b), skin (Blad and Baldetorp 1996), tongue (Ching et al. 2010), and colon (Soler et al. 1999). With the exception of prostate, most of the results have shown a decrease in the real part of the electrical impedance in cancerous tissue when compared to normal tissue. This finding could probably be explained by changes in the extracellular space, such as a lowering of its pH (Goodwin et al. 2014) as well as disruption of cell adhesion and gap junctions (Yamasaki 1991), which are recognized as universal in carcinogenesis and metastasis (Mesnil et al. 2005). So, emphasis has been placed in only one of the four Cole-Cole parameters, namely, R_0, but there is still much to be learned, especially in relation to the role of the cell membrane and the intracellular space and organelles.

Another aspect to take into account is the fact that most of the research has been carried out to detect tissue that is already precancerous (dysplastic) or cancerous. We hypothesize that, under the theoretical umbrella of "field cancerization" theory (Lochhead et al. 2015; Slaughter et al. 1953), the role of EBIS in very early detection could be much more prominent. In 2016, for instance, Mulett-Vásquez et al. published some initial readings made on the rectum, showing the feasibility of using EIS for taking in vivo rectal readings in humans (17 patients undergoing colonoscopy). For this purpose, they used the same technique and equipment reported by Brown et al. (2000) in their study on cervical cancer, where a four-gold-electrode probe is used for passing a small electrical current of 40 μA through one pair, and the associated voltage is measured between the other pair of electrodes. These preliminary results seem to show that it could be possible to detect the presence of cancer in the colon, with EBIS measurements taken in the rectum. The important issue here is the possibility of studying the rectum (a very easy accessible organ) for detection of colon diseases, including colorectal cancer, in a similar way as what the group led by Backman and Roy (2013) is doing with optic methods. One point to highlight is that the follow-up of the passive electrical characteristics of the rectum in an individual could signal early changes associated, for instance, with inflammation and augmented mucosal permeability, both of them associated with colon cancer development but also with many other chronic diseases, as we will mention in the next and final section of this chapter.

The concept of "field cancerization" (Slaughter et al. 1953) or its more advanced formulation by Lochhead et al. (2015) as "etiologic field effect" establishes that cancers anywhere arise from tissues that, due to environmental stimuli, have undergone a transformation that makes them a fertile soil for cancerous transformation. This effect is mediated by epigenetic changes that could be detected very early, before cancer appears, i.e., it would make possible to detect people at risk of developing cancer, rather than the diagnosis of cancer even if it is done at an early stage. For instance, the pioneering work carried out by the group of Backman and Roy (2013) could be mirrored by studies using EBIS instead of optic techniques. The fact that these early changes extend to large parts of specific tissues, organs, or even the whole organism makes it possible to detect changes in organs difficult to reach (colon, intestine, stomach, pancreas, liver, lungs, etc.) by examining more accessible organs, like the mouth or the rectum. In other words, changes in the latter would reflect what is happening in the former.

10.6 The Future: Chronic Diseases and EBIS

In the previous section, we have already mentioned the possibilities for EBIS in early detection of cancer, but there are far more chronic diseases that could be detected at early stages and followed up by EBIS. This technique has many advantages: it is easy to use, non- or minimally invasive, not expensive, cost saving, and portable; also, electricity can be applied practically to any tissue or organ under appropriate conditions. It is also feasible to obtain information not only of the extracellular space but also from the intracellular compartment and even from the cell membrane status. For this, more advanced measurement, models, and analysis techniques have to be developed, but the future is very promising.

One fact that is worth mentioning is that most chronic diseases seem to have a common pathway that has emerged recently, thanks to advances in techniques associated with molecular biology, especially in the field of the so-called omics (metagenomics, transcriptomics, proteomics, and many others). As shown in Fig. 10.11, in this pathway, one key factor is intestinal permeability (Bischoff et al. 2014). This means that, monitoring the status of the intestinal wall via EBIS, we could detect when a person is beginning to present changes in the intestinal wall that suggest that he/she has begun the pathway of a chronic disease, including cancer. Factors of this picture that can be reflected in the EBIS measurements of either the rectal or the colonic wall are amount of bacteria, thickness and composition of the mucus layer, mucosal permeability, and low-grade chronic inflammation (also named as "meta-inflammation"; see Minihane et al. 2015; León-Pedroza et al. 2015).

Finally, as already mentioned in Sect. 4.5 ("Skin"), it seems that more people are now taking interest in studying EBI changes of the skin in relation to different disease conditions. In 2004, Gonzalez-Correa, upon the principles of Chinese acupuncture, proposed the idea of a "binary medicine" and, based on it, a device that he called "Binatron." The main idea behind this proposal is that a diagnosis of the different meridian status could be done in terms of −1 (hypotonic), 0 (normo- or eutonic), and +1 (hypertonic). For a hypotonic meridian, treatment would be +1 (electric stimulation), and for a hypertonic meridian, treatment would be −1 (electric sedation), which will produce, in either case, a 0. Stimulation would be achieved by injecting current in the direction of the *chí* or *ki* flow (vital energy, according to the traditional Chinese medicine, TCM), i.e., orthodromically, while sedation would be achieved injecting current in the opposite direction, i.e., antidromically. This is illustrated in Fig. 10.12, and its bases are not only the concepts of TCM but also the findings such as those by Friedman et al. (1989).

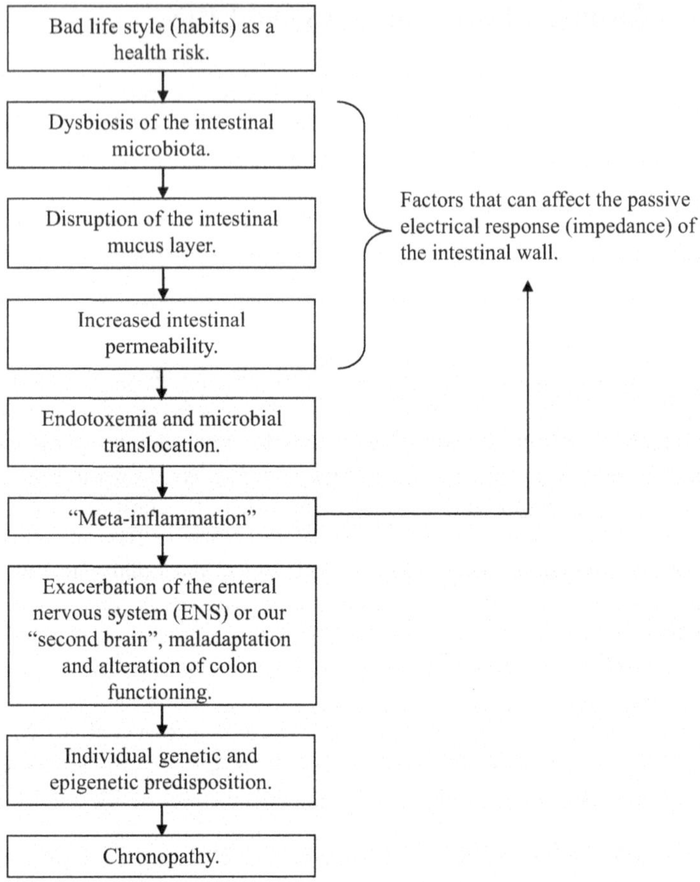

Fig. 10.11 Common pathway or physiopathology of chronic diseases (including different cancer types), according to the emerging paradigm of the twenty-first century (Source: based upon Gonzalez-Correa et al. 2017)

	Meridians on both body sides					
	Left			Right		
Number of the meridian	1	2	3......	1	2	3......
Electrodiagnosis (by EBIS)	0	-1	+1	+1	0	0
Electrotreatment	0	+1	-1	-1	0	0
Result	0	0	0	0	0	0

Fig. 10.12 Principle of acupunctural diagnosis using EBIS and treatment by electrostimulation, according to the theory of "Binary Medicine" (Gonzalez-Correa 2004)

References

Aberg, P., Nicander, I., Hansson, J., Geladi, P., Holmgren, U., & Ollmar, S. (2004). Skin cancer identification using multifrequency electrical impedance—A potential screening tool. *IEEE Transactions on Biomedical Engineering, 51*(12), 2097–2102.

Ahn, A. C., & Martinsen, O. G. (2007). Electrical characterization of acupuncture points: Technical issues and challenges. *Journal of Alternative and Complementary Medicine, 13*(8), 817–824.

Ahn, A. C., Park, M., Shaw, J. R., McManus, C. A., Kaptchuk, T. J., & Langevin, H. M. (2010). Electrical impedance of acupuncture meridians: The relevance of subcutaneous collagenous bands. *PLoS One, 5*(7), e11907.

Arroyo, A. G., & Iruela-Arispe, M. L. (2010). Extracellular matrix, inflammation, and the angiogenic response. *Cardiovascular Research, 86*(2), 226–235.

Backman, V., & Roy, H. K. (2013). Advances in biophotonics detection of field carcinogenesis for Colon cancer risk stratification. *Journal of Cancer, 4*(3), 251–261.

Bardhan, K. D., Strugala, V., & Dettmar, P. W. (2012). Reflux revisited: Advancing the role of pepsin. *International Journal of Otolaryngology, 2012*, 646901.

Barrio, A. V., Eaton, A., & Frazier, T. G. (2015). A prospective validation study of bioimpedance with volume displacement in early-stage breast cancer patients at risk for lymphedema. *Annals of Surgical Oncology, 22*(Suppl 3), S370–S375.

Barter, S. J., & Hicks, I. P. (2000). Electrical impedance imaging of the breast (TranScan TS 2000): Initial UK experience. *Breast Cancer Research, 2*(Suppl 2), A11.

Bayford, R., & Tizzard, A. (2012). Bioimpedance imaging: An overview of potential clinical applications. *Analyst, 137*(20), 4635–4643.

Bayram, M., & Yancy, C. W. (2009). Transthoracic impedance cardiography: A noninvasive method of hemodynamic assessment. *Heart Failure Clinics, 5*(2), 161–168.

Beltran, N. E., Ceron, U., Sanchez-Miranda, G., Remolina, M., Godinez, M. M., Peralta, I. Y., et al. (2013). Incidence of gastric mucosal injury as measured by reactance in critically ill patients. *Journal of Intensive Care Medicine, 28*(4), 230–236.

Beltran, N. E., & Sacristan, E. (2015). Gastrointestinal ischemia monitoring through impedance spectroscopy as a tool for the management of the critically ill. *Experimental Biology and Medicine (Maywood, N.J.), 240*(7), 835–845.

Bera, T. K. (2014). Bioelectrical impedance methods for noninvasive health monitoring: A review. *Journal of Medical Engineering, 2014*, 381251.

Bernstein, D. P., Henry, I. C., Lemmens, H. J., Chaltas, J. L., DeMaria, A. N., Moon, J. B., et al. (2015). Validation of stroke volume and cardiac output by electrical interrogation of the brachial artery in normals: Assessment of strengths, limitations, and sources of error. *Journal of Clinical Monitoring and Computing, 29*(6), 789–800.

Bischoff, S. C., Barbara, G., Buurman, W., Ockhuizen, T., Schulzke, J. D., Serino, M., et al. (2014). Intestinal permeability—A new target for disease prevention and therapy. *BMC Gastroenterology, 18*(14), 1077–1085.

Blad, B., & Baldetorp, B. (1996). Impedance spectra of tumour tissue in comparison with normal tissue; a possible clinical application for electrical impedance tomography. *Physiological Measurement, 17*(Suppl 4A), A105–A115.

Boucsein, W. (2012). *Electrodermal activity* (2nd ed.). New York: Springer.

Boytsov, I. V., & Belousova, T. E. (2015). Interrelations between electrodermal activity and internal diseases. *Fiziologiia Cheloveka, 41*(6), 104–113.

Brown, B. H., Smallwood, R. H., Barber, D. C., Lawford, P. V., & Hose, D. R. (1999). *Medical physics and biomedical engineering. (Chap. 8 "Non-ionizing electromagnetic radiation: Tissue absorption and safety issues", p. 231)*. Bristol, UK: Institute of Physics (IOP).

Brown, B. H., Tidy, J. A., Boston, K., Blackett, A. D., Smallwood, R. H., & Sharp, F. (2000). Relation between tissue structure and imposed electrical current flow in cervical neoplasia. *Lancet, 355*(9207), 892–895.

Brown, J. M., & Hazen, S. L. (2015). The gut microbial endocrine organ: Bacterially derived signals driving cardiometabolic diseases. *Annual Review of Medicine, 66*, 343–359.

Casselbrant, A., Elias, E., Fändriks, L., & Wallenius, V. (2015). Expression of tight-junction proteins in human proximal small intestinal mucosa before and after Roux-en-Y gastric bypass surgery. *Surgery for Obesity and Related Diseases, 11*(1), 45–53.

Chandler, J. H., Culmer, P. R., Jayne, D. G., & Neville, A. (2015). A time-dependent model for improved biogalvanic tissue characterisation. *Medical Engineering & Physics, 37*(10), 956–960.

Charles, C. J., Rademaker, M. T., Melton, I. C., Gutfinger, D., Eigler, N. L., Qu, F., et al. (2015). Thoracic impedance measures tissue characteristics in the vicinity of the electrodes, not intervening lung water: Implications for heart failure monitoring. *Journal of Clinical Monitoring and Computing, 29*(1), 65–76.

Ching, C. T., Sun, T. P., Huang, S. H., Hsiao, C. S., Chang, C. H., Huang, S. Y., et al. (2010). A preliminary study of the use of bioimpedance in the screening of squamous tongue cancer. *International Journal of Nanomedicine, 5*, 213–220.

Coffman, F. D., & Cohen, S. (2013). Impedance measurements in the biomedical sciences. *Studies in Health Technology and Informatics, 185*, 185–205.

Colina-Gallo, E., González-Correa, C. A., Dussán-Lubert, C., & Miranda-Mercado, D. A. (2016a). Segmental electrical bioimpedance measurements with a single lead (Electrode) displacement. *IFMBE Proceedings, 54*, 20–23.

Colina-Gallo, E., González-Correa, C. A., & Miranda-Mercado, D. A. (2016b). Correlation between algometry and electrical bioimpedance in subjects with and without fibromyalgia. *IFMBE Proceedings, 54*, 72–75.

Collier, R. J., & Ntui, J. A. (1994). In vivo measurements of the phase constants of transverse mechanical waves in a human tibia from 100 to 1000 Hz. *Medical Engineering & Physics, 16*(5), 379–383.

Cybulski, G., Strasz, A., Niewiadomski, W., & Gąsiorowska, A. (2012). Impedance cardiography: Recent advancements. *Cardiology Journal, 19*(5), 550–556.

Da Silva, T. M., & Alves, F. R. (2014). Ex vivo accuracy of Root ZX II, Root ZX Mini and RomiApex A-15 apex locators in extracted vital pulp teeth. *The Journal of Contemporary Dental Practice, 15*(3), 312–314.

Daghfous, H., El Ayeb, W., Alouane, L., & Tritar, F. (2014). Evaluation of nutritional status in lung cancer using bio electrical impedance analysis and mini nutritional assessment. *La Tunisie Médicale, 92*(12), 737–742.

Donadio, C., Bozzoli, L., Colombini, E., Pisanu, G., Ricchiuti, G., Picano, E., et al. (2015). Effective and timely evaluation of pulmonary congestion: Qualitative comparison between lung ultrasound and thoracic bioelectrical impedance in maintenance hemodialysis patients. *Medicine (Baltimore), 94*(6), e473.

Eldarrat, A. H., High, A. S., & Kale, G. M. (2010). Age-related changes in ac-impedance spectroscopy studies of normal human dentine: Further investigations. *Journal of Materials Science. Materials in Medicine, 21*(1), 45–51.

El-Serag, H. B., Sweet, S., Winchester, C. C., & Dent, J. (2014). Update on the epidemiology of gastro-oesophageal reflux disease: A systematic review. *Gut, 63*(6), 871–880.

Fass, J., Silny, J., Braun, J., Heindrichs, U., Dreuw, B., Schumpelick, V., et al. (1994). Measuring esophageal motility with a new intraluminal impedance device. First clinical results in reflux patients. *Scandinavian Journal of Gastroenterology, 29*(8), 693–702.

Fernández, P. J., Méndez-Sánchez, S. C., Gonzalez-Correa, C. A., & Miranda, D. A. (2016). Could field cancerization be interpreted as a biochemical anomaly amplification due to transformed cells? *Medical Hypotheses, 97*, 107–111.

Friedman, M. F., Birxh, S., & Tiller, W. A. (1989). Towards the development of a mathematical model for acupuncture meridians. *Acupuncture & Electro-Therapeutics Research, 14*, 217–226.

Fukushima, Y., Yoda, T., Kokabu, S., Araki, R., Murata, T., Kitagawa, Y., et al. (2013). Evaluation of an oral moisture-checking device for screening dry mouth. *Open Journal of Stomatology, 3*, 440–446.

Giouvanoudi, A., Amaee, W. B., Sutton, J. A., Horton, P., Morton, R., Hall, W., et al. (2003). Physiological interpretation of electrical impedance epigastrography measurements. *Physiological Measurement, 24*(1), 45–55.

Golowasch, J., Thomas, G., Taylor, A. L., Patel, A., Pineda, A., Khalil, C., et al. (2009). Membrane capacitance measurements revisited: Dependence of capacitance value on measurement method in nonisopotential neurons. *Journal of Neurophysiology, 102*(4), 2161–2175.

Gong, B., Krueger-Ziolek, S., Moeller, K., Schullcke, B., & Zhao, Z. (2015). Electrical impedance tomography: Functional lung imaging on its way to clinical practice? *Expert Review of Respiratory Medicine, 9*(6), 721–737.

González-Correa, C. A. (2004). Toward a binary interpretation of acupuncture theory: Principles and practical consequences. *Journal of Alternative and Complementary Medicine, 10*(3), 573–579.

González-Correa, C. A., Brown, B. H., Smallwood, R. H., Kalia, N., Stoddard, C. J., Stephenson, T. J., et al. (1999). Virtual biopsies in Barrett's esophagus using an impedance probe. *Annals of the New York Academy of Sciences, 873*, 313–321.

González-Correa, C. A., Brown, B. H., Smallwood, R. H., Kalia, N., Stoddard, C. J., Stephenson, T. J., et al. (2000). Assessing the conditions for in vivo electrical virtual biopsies in Barrett's oesophagus. *Medical & Biological Engineering & Computing, 38*(4), 373–376.

Gonzalez-Correa, C. A., Brown, B. H., Smallwood, R. H., Stephenson, T. J., Stoddard, C. J., & Bardhan, K. D. (2003). Low frequency electrical bioimpedance for the detection of inflammation and dysplasia in Barrett's oesophagus. *Physiological Measurement, 24*(2), 291–296.

Gonzalez-Correa, C. A., Mulett-Vásquez, E., Miranda, D. A., Gonzalez Correa, C. H., & Gómez Buitrago, P. A. (2017). The colon revisited or the key to wellness, health and disease. *Medical Hypotheses, 108*, 133–143.

Goodwin, J., Yachi, K., Nagane, M., Yasui, H., Miyake, Y., Inanami, O., et al. (2014). In vivo tumour extracellular pH monitoring using electron paramagnetic resonance: The effect of X-ray irradiation. *NMR in Biomedicine, 27*(4), 453–458.

Grimmes, S., & Martinsen, G. (2000). *Bioimpedance & bioelectricity basics. (Chapter 1, Introduction, p. 1, and chap. 7 "Data and models", p. 214)*. London, UK: Academic Press.

Grimmes, S., & Martinsen, G. (2005). Cole electrical impedance model—A critique and an alternative. *IEEE Transactions on Biomedical Engineering, 52*(1), 132–135.

Gupta, D., Lammersfeld, C. A., Vashi, P. G., King, J., Dahlk, S. L., Grutsch, J. F., et al. (2009). Bioelectrical impedance phase angle in clinical practice: Implications for prognosis in stage IIIB and IV non-small cell lung cancer. *BMC Cancer, 9*, 37.

Halter, R. J., Schned, A., Heaney, J., Hartov, A., & Paulsen, K. D. (2009a). Electrical properties of prostatic tissues: I. Single frequency admittivity properties. *The Journal of Urology, 182*(4), 1600–1607.

Halter, R. J., Schned, A., Heaney, J., Hartov, A., & Paulsen, K. D. (2009b). Electrical properties of prostatic tissues: II. Spectral admittivity properties. *The Journal of Urology, 182*(4), 1608–1613.

Harel, F., Dupuis, J., Benelfassi, A., Ruel, N., & Grégoire, J. (2005). Radionuclide plethysmography for noninvasive evaluation of peripheral arterial blood flow. *American Journal of Physiology. Heart and Circulatory Physiology, 289*(1), H258–H262.

Holder, D. S. (2005). *Electrical impedance tomography: Methods, history and applications*. Bristol, UK: Institute of Physics Publishing (IOP).

Itsekson, A., Shepshelovich, D., Kanevsky, A., & Seidman, D. S. (2010). Measurement of electrical resistance of dermal-visceral zones as a diagnostic tool for gynecologic disorders. *The Israel Medical Association Journal, 12*(6), 334–337.

Johansson, M. E., Gustafsson, J. K., Holmén-Larsson, J., Jabbar, K. S., Xia, L., Xu, H., et al. (2013). Bacteria penetrate the normally impenetrable inner colon mucus layer in both murine colitis models and patients with ulcerative colitis. *Gut, 63*(2), 281–291.

Jones, D. A. (1991). Electrical engineering: The backbone of society. *Proceedings of the IEE: Science, Measurement and Technology, 138*(1), 1–10.

Kåhrström, C. T., Pariente, N., & Weiss, U. (2016). Intestinal microbiota in health and disease. *Nature, 535*(7610), 47.

Kassebaum, N. J., Bernabé, E., Dahiya, M., Bhandari, B., Murray, C. J., & Marcenes, W. (2015). Global burden of untreated caries: A systematic review and metaregression. *Journal of Dental Research, 94*(5), 650–658.

Keshtkar, A., Keshtkar, A., & Smallwood, R. H. (2006). Electrical impedance spectroscopy and the diagnosis of bladder pathology. *Physiological Measurement, 27*(7), 585–596.

Khan, S., Mahara, A., Hyams, E. S., Schned, A. R., & Halter, R. J. (2016). Prostate cancer detection using composite impedance metric. *IEEE Transactions on Medical Imaging, 35*(12), 2513–2523.

Lawenko, R. M., & Lee, Y. Y. (2016). Emerging roles of the Endolumenal Functional Lumen Imaging Probe (EndoFLIP) in gastrointestinal motility disorders. *Journal of Neurogastroenterology and Motility, 23*. [Epub ahead of print].

León-Pedroza, J. I., González-Tapia, L. A., del Olmo-Gil, E., Castellanos-Rodríguez, D., Escobedo, G., & González-Chávez, A. (2015). Low-grade systemic inflammation and the development of metabolic diseases: From the molecular evidence to the clinical practice. *Cirugia y Cirujanos, 83*(6), 543–551.

Leontev, A. S., Korotkevich, A. G., Repnikova, R. V., Merzlyakov, M. V., & Shestak, I. S. (2015). Endoscopic diagnosis and parietal impedancometry in the assessment of changes in the mucosa of the duodenum in patients with suspicion on postcholecystectomy syndrome. *Eksp Klin Gastroenterol*, (4), 34–37.

Lerner, A., & Matthias, T. (2015). Changes in intestinal tight junction permeability associated with industrial food additives explain the rising incidence of autoimmune disease. *Autoimmunity Reviews, 14*(6), 479–489.

Li, Z. Y., Ren, C. S., Zhao, S., Sha, H., & Deng, J. (2011). Gastric motility functional study based on electrical bioimpedance measurements and simultaneous electrogastrography. *Journal of Zhejiang University. Science. B, 12*(12), 983–989.

Liao, Y. M., Feng, Z. D., & Chen, Z. L. (2007). In situ tracing the process of human enamel demineralization by electrochemical impedance spectroscopy (EIS). *Journal of Dentistry, 35*(5), 425–430.

Lochhead, P., Chan, A. T., Nishihara, R., Fuchs, C. S., Beck, A. H., Giovannucci, E., et al. (2015). Etiologic field effect: Reappraisal of the field effect concept in cancer predisposition and progression. *Modern Pathology, 28*(1), 14–29.

Malich, A., Boehm, T., Facius, M., Freesmeyer, M. G., Fleck, M., Anderson, R., et al. (2001). Differentiation of mammographically suspicious lesions: Evaluation of breast ultrasound, MRI mammography and electrical impedance scanning as adjunctive technologies in breast cancer detection. *Clinical Radiology, 56*(4), 278–283.

Man, A. L., Bertelli, E., Rentini, S., Regoli, M., Briars, G., Marini, M., et al. (2015). Age-associated modifications of intestinal permeability and innate immunity in human small intestine. *Clinical Science (London, England), 129*(7), 515–527.

Mašanauskienė, E., Sadauskas, S., Naudžiūnas, A., Unikauskas, A., & Stankevičius, E. (2014). Impedance plethysmography as an alternative method for the diagnosis of peripheral arterial disease. *Medicina Lithuania, 50*(6), 334–339.

McIlduff, C., Yim, S., Pacheck, A., Geisbush, T., Mijailovic, A., & Rutkove, S. B. (2016a). An improved electrical impedance myography (EIM) tongue array for use in clinical trials. *Clinical Neurophysiology, 127*(1), 932–935.

McIlduff, C. E., Yim, S. J., Pacheck, A. K., & Rutkove, S. B. (2016b). Optimizing electrical impedance myography of the tongue in ALS. *Muscle & Nerve*. https://doi.org/10.1002/mus.25375. [Epub ahead of print].

Mesnil, M., Crespin, S., Avanzo, J. L., & Zaidan-Dagli, M. L. (2005). Defective gap junctional intercellular communication in the carcinogenic process. *Biochimica et Biophysica Acta, 1719*(1-2), 125–145.

Miller, J. C., & Horvath, S. M. (1978). Impedance cardiography. *Psychophysiology, 15*(1), 80–91.

Minihane, A. M., Vinoy, S., Russell, W. R., Baka, A., Roche, H. M., Tuohy, K. M., et al. (2015). Low-grade inflammation, diet composition and health: Current research evidence and its translation. *The British Journal of Nutrition, 114*(7), 999–1012.

Morucci, J. P., & Marsili, P. M. (1996). Bioelectrical impedance techniques in medicine. Part III: Impedance imaging. Second section: Reconstruction algorithms. *Critical Reviews in Biomedical Engineering, 24*(4-6), 599–654.

Morucci, J. P., & Rigaud, B. (1996). Bioelectrical impedance techniques in medicine. Part III: Impedance imaging. Third section: Medical applications. *Critical Reviews in Biomedical Engineering, 24*(4-6), 655–677.

Mousa, H. M., Rosen, R., Woodley, F. W., Orsi, M., Armas, D., Faure, C., et al. (2011). Esophageal impedance monitoring for gastroesophageal reflux. *Journal of Pediatric Gastroenterology and Nutrition, 52*(2), 129–139.

Mulett-Vásquez, E., Correa-Florez, A., Dussán-Lubert, C., Miranda-Mercado, D. A., & González-Correa, C. A. (2016a). In vitro luminal measurements of colon electrical impedance in rabbits. *IFMBE Proceedings, 54*, 28–31.

Mulett-Vásquez, E., Gonzalez-Correa, C. A., Miranda-Mercado, D. A., Osorio-Chica, M., & Dussan-Lubert, C. (2016b). *In vivo* Electrical-Impedance Spectroscopy (EIS) readings in the human rectum. *IFMBE Proceedings, 54*, 68–71.

Murakami, H., Matsumoto, H., Ueno, D., Kawai, A., Ensako, T., Kaida, Y., et al. (2013). Current status of multichannel electrogastrography and examples of its use. *Journal of Smooth Muscle Research, 49*, 78–88.

Nekoofar, M. H., Ghandi, M. M., Hayes, S. J., & Dummer, P. M. (2006). The fundamental operating principles of electronic root canal length measurement devices. *International Endodontic Journal, 39*(8), 595–609.

Nelsen, E. M., Hawes, R. H., & Iyer, P. G. (2012). Diagnosis and management of Barrett's esophagus. *The Surgical Clinics of North America, 92*(5), 1135–1154.

Nguyen, N. Q., Bryant, L. K., Burgstad, C. M., Fraser, R. J., Sifrim, D., & Holloway, R. H. (2010). Impact of bolus volume on small intestinal intra-luminal impedance in healthy subjects. *World Journal of Gastroenterology, 16*(17), 2151–2157.

Nicander, I., & Ollmar, S. (1999). Electrical bioimpedance related to structural differences and reactions in skin and oral mucosa. *Annals of the New York Academy of Sciences, 873*, 221–226.

Nicander, I., & Ollmar, S. (2004). Clinically normal atopic skin vs. non-atopic skin as seen through electrical impedance. *Skin Research and Technology, 10*(3), 178–183.

Nicander, I., Rantanen, I., Rozell, B. L., Söderling, E., & Ollmar, S. (2003). The ability of betaine to reduce the irritating effects of detergents assessed visually, histologically and by bioengineering methods. *Skin Research and Technology, 9*(1), 50–58.

Noll, L., Rommel, N., Davidson, G. P., & Omari, T. I. (2011). Pharyngeal flow interval: A novel impedance-based parameter correlating with aspiration. *Neurogastroenterology and Motility, 23*(6), 551–556.

O'Hara, A. M., & Shanahan, F. (2006). The gut flora as a forgotten organ. *EMBO Reports, 7*(7), 688–693.

Ollmar, S., Eek, A., Sundström, F., & Emtestam, L. (1995). Electrical impedance for estimation of irritation in oral mucosa and skin. *Medical Progress through Technology, 21*(1), 29–37.

Pabst, O., Tronstad, C., Grimnes, S., Fowles, D., Martinsen, Ø. G. (2016). Comparison between the AC and DC measurement of electrodermal activity. *Psychophysiology*. [Epub ahead of print].

Pacheck, A., Mijailovic, A., Yim, S., Li, J., Green, J. R., McIlduff, C. E., et al. (2016). Tongue electrical impedance in amyotrophic lateral sclerosis modeled using the finite element method. *Clinical Neurophysiology, 127*(3), 1886–1890.

Piccoli, A. (2014). Estimation of fluid volumes in hemodialysis patients: Comparing bioimpedance with isotopic and dilution methods. *Kidney International, 85*(4), 738–741.

Piccoli, A., Codognotto, M., Cianci, V., Vettore, G., Zaninotto, M., Plebani, M., et al. (2012). Differentiation of cardiac and noncardiac dyspnea using bioelectrical impedance vector analysis (BIVA). *Cardiac Failure, 18*(3), 226–232.

Pretty, I. A., & Ellwood, R. P. (2013). The caries continuum: Opportunities to detect, treat and monitor the re-mineralization of early caries lesions. *Journal of Dentistry, 41*(Suppl 2), S12–S21.

Rantanen, I., Jutila, K., Nicander, I., Tenovuo, J., & Söderling, E. (2003). The effects of two sodium lauryl sulphate-containing toothpastes with and without betaine on human oral mucosa in vivo. *Swedish Dental Journal, 27*(1), 31–34.

Rao, S. S., Hayek, B., & Summers, R. W. (1995). Impedance planimetry: An integrated approach for assessing sensory, active, and passive biomechanical properties of the human esophagus. *The American Journal of Gastroenterology, 90*(3), 431–438.

Rehm, W. S. (1953). Electrical resistance of resting and secreting stomach. *The American Journal of Physiology, 172*(3), 689–699.

Rehm, W. S., Dennis, W. H., & Schlesinger, H. (1955). Electrical resistance of the mammalian stomach. *The American Journal of Physiology, 181*(2), 451–470.

Rezác, P. (2008). Potential applications of electrical impedance techniques in female mammalian reproduction. *Theriogenology, 70*(1), 1–14.

Rigaud, B., & Morucci, J. P. (1996). Bioelectrical impedance techniques in medicine. Part III: Impedance imaging. First section: General concepts and hardware. *Critical Reviews in Biomedical Engineering, 24*(4-6), 467–597.

Rigaud, B., Morucci, J. P., & Chauveau, N. (1996). Bioelectrical impedance techniques in medicine. Part I: Bioimpedance measurement. Second section: Impedance spectrometry. *Critical Reviews in Biomedical Engineering, 24*(4–6), 257–351.

Rim, J. H., Jo, S. J., Park, J. Y., Park, B. D., & Youn, J. I. (2005). Electrical measurement of moisturizing effect on skin hydration and barrier function in psoriasis patients. *Clinical and Experimental Dermatology, 30*(4), 409–413.

Rodriguez, S., Ollmar, S., Waqar, M., & Rusu, A. (2016). A batteryless sensor ASIC for implantable bio-impedance applications. *IEEE Transactions on Biomedical Circuits and Systems, 10*(3), 533–544.

Rodríguez-Piñeiro, A. M., & Johansson, M. E. (2015). The colonic mucus protection depends on the microbiota. *Gut Microbes, 6*(5), 326–330.

Rutkove, S. B. (2009). Electrical impedance myography: Background, current state, and future directions. *Muscle & Nerve, 40*(6), 936–946.

Rutkove, S. B., Esper, G. J., Lee, K. S., Aaron, R., & Shiffman, C. A. (2005). Electrical impedance myography in the detection of radiculopathy. *Muscle & Nerve, 32*(3), 335–341.

Sanchez, B., & Rutkove, S. B. (2016). Electrical impedance myography and its applications in neuromuscular disorders. *Neurotherapeutics*. [Epub ahead of print].

Sellers, R. S., & Morton, D. (2014). The colon: From banal to brilliant. *Toxicologic Pathology, 42*(1), 67–81.

Shiffman, C. A. (2013). Circuit modeling of the electrical impedance: Part III. Disuse following bone fracture. *Physiological Measurement, 34*(5), 487–502.

Shiffman, C. A., & Rutkove, S. B. (2013). Circuit modeling of the electrical impedance: I. Neuromuscular disease. *Physiological Measurement, 34*(2), 203–221.

Shima, R., Jiang, Z., Fen, S. Y., Monnavar, A. A., & Ali, K. (2012). Development and evaluation of a novel four-electrode device system for monitoring skin impedance. *African Journal of Traditional, Complementary, and Alternative Medicines, 9*(4), 599–606.

Siedlecka, J., Siedlecki, P., & Bortkiewicz, A. (2015). Impedance cardiography - Old method, new opportunities. Part I. Clinical applications. *International Journal of Occupational Medicine and Environmental Health, 28*(1), 27–33.

Sifrim, D., Silny, J., Holloway, R. H., & Janssens, J. J. (1999). Patterns of gas and liquid reflux during transient lower oesophageal sphincter relaxation: A study using intraluminal electrical impedance. *Gut, 44*(1), 47–54.

Slaughter, D. P., Southwick, H. W., & Smejkal, W. (1953). Field cancerization in oral stratified squamous epithelium; clinical implications of multicentric origin. *Cancer, 6*(5), 963–968.

Sleator, R. D. (2010). The human superorganism - of microbes and men. *Medical Hypotheses, 74*(2), 214–215.

Smallwood, R. H., Mangnall, Y. F., & Leathard, A. D. (1994). Transport of gastric contents. *Physiological Measurement, 15*(Suppl 2a), A175–A188.

Soler, A. P., Miller, R. D., Kathleen, V., Laughlin, K. V., Carp, N. Z., Klurfeld, D. M., et al. (1999). Increased tight junctional permeability is associated with the development of colon cáncer. *Carcinogenesis, 20*(8), 1425–1431.

Solmaz, H., Dervisoglu, S., Gulsoy, M., & Ulgen, Y. (2016). Laser biostimulation of wound healing: Bioimpedance measurements support histology. *Lasers in Medical Science, 31*(8), 1547–1554.

Somma, F., Castagnola, R., Lajolo, C., Paternò Holtzman, L., & Marigo, L. (2012). In vivo accuracy of three electronic root canal length measurement devices: Dentaport ZX, Raypex 5 and ProPex II. *International Endodontic Journal, 45*(6), 552–556.

Spence, D. W., & Pomeranz, B. (1996). Surgical wound healing monitored repeatedly in vivo using electrical resistance of the epidermis. *Physiological Measurement, 17*(2), 57–69.

Strand-Amundsen, R. J., Tronstad, C., Kalvøy, H., Gundersen, Y., Krohn, C. D., Aasen, A. O., et al. (2016). In vivo characterization of ischemic small intestine using bioimpedance measurements. *Physiological Measurement, 37*(2), 257–275.

Sun, T. P., Ching, C. T., Cheng, C. S., Huang, S. H., Chen, Y. J., Hsiao, C. S., et al. (2010). The use of bioimpedance in the detection/screening of tongue cancer. *Cancer Epidemiology, 34*(2), 207–211.

Szczesniak, M. M., Rommel, N., Dinning, P. G., Fuentealba, S. E., Cook, I. J., & Omari, T. I. (2009). Intraluminal impedance detects failure of pharyngeal bolus clearance during swallowing: A validation study in adults with dysphagia. *Neurogastroenterology and Motility, 21*(3), 244–252.

Tagami, H. (2014). Electrical measurement of the hydration state of the skin surface in vivo. *The British Journal of Dermatology, 171*(Suppl 3), 29–33.

Tambucci, R., Thapar, N., Saliakellis, E., Pescarin, M., Quitadamo, P., Cristofori, F., et al. (2015). Clinical relevance of esophageal baseline impedance measurement: Just an innocent bystander. *Journal of Pediatric Gastroenterology and Nutrition, 60*(6), 776–782.

Theodor, M., Ruh, D., Ocker, M., Spether, D., Förster, K., Heilmann, C., et al. (2014). Implantable impedance plethysmography. *Sensors (Basel), 14*(8), 14858–14872.

Tidy, J. A., Brown, B. H., Healey, T. J., Daayana, S., Martin, M., Prendiville, W., et al. (2013). Accuracy of detection of high-grade cervical intraepithelial neoplasia using electrical impedance spectroscopy with colposcopy. *British Journal of Obstetrics and Gynaecology, 120*, 400–411.

Tornuev, Y. V., Koldysheva, E. V., Lapiy, G. A., Molodykh, O. P., Balakhnin, S. M., Bushmanova, G. M., et al. (2014). Bioimpedancemetry in the diagnostics of inflammatory process in the mammary gland. *Bulletin of Experimental Biology and Medicine, 156*(3), 381–383.

Tseng, Y. J., Hu, W. L., Hung, I. L., Hsieh, C. J., & Hung, Y. C. (2014). Electrodermal screening of biologically active points for upper gastrointestinal bleeding. *The American Journal of Chinese Medicine, 42*(5), 1111–1121.

Tutuian, R., & Castell, D. O. (2006). Review article: Oesophageal spasm - diagnosis and management. *Alimentary Pharmacology & Therapeutics, 23*(10), 1393–1402.

Valentinuzzi, M. E. (1996). Bioelectrical impedance techniques in medicine. Part I: Bioimpedance measurement. First section: General concepts. *Critical Reviews in Biomedical Engineering, 24*(4-6), 223–255.

Valentinuzzi, M. E., Morucci, J. P., & Felice, C. J. (1996). Bioelectrical impedance techniques in medicine. Part II: Monitoring of physiological events by impedance. *Critical Reviews in Biomedical Engineering, 24*(4-6), 353–466.

Vanheel, H., Vicario, M., Vanuytsel, T., Van Oudenhove, L., Martinez, C., Keita, Å. V., et al. (2014). Impaired duodenal mucosal integrity and low-grade inflammation in functional dyspepsia. *Gut, 63*(2), 262–271.

Viggiano, D., Ianiro, G., Vanella, G., Bibbò, S., Bruno, G., Simeone, G., et al. (2015). Gut barrier in health and disease: Focus on childhood. *European Review for Medical and Pharmacological Sciences, 19*(6), 1077–1085.

Walker, D. C., Brown, B. H., Blackett, A. D., Tidy, J., & Smallwood, R. H. (2003). A study of the morphological parameters of cervical squamous epithelium. *Physiological Measurement, 24*(1), 121–135.

Weber, S. A., Watermann, N., Jossinet, J., Byrne, J. A., Chantrey, J., Alam, S., et al. (2010). Remote wound monitoring of chronic ulcers. *IEEE Transactions on Information Technology in Biomedicine, 14*(2), 371–377.

Weitzen, R., Epstein, N., Shoenfeld, Y., & Zimlichman, E. (2007). Diagnosing diseases by measurement of electrical skin impedance: A novel technique. *Annals of the New York Academy of Sciences, 1109*, 185–192.

Weyer, S., Zink, M. D., Wartzek, T., Leicht, L., Mischke, K., Vollmer, T., et al. (2014). Bioelectrical impedance spectroscopy as a fluid management system in heart failure. *Physiological Measurement, 35*(6), 917–930.

Yamamoto, T. (1963). On the thickness of the unit membrane. *The Journal of Cell Biology, 17*(2), 413–421.

Yamasaki, H. (1991). Aberrant expression and function of gap junctions during carcinogenesis. *Environmental Health Perspectives, 93*, 191–197.

Yoshida, T., Kim, W. C., Kawamoto, K., Hirashima, T., Oka, Y., & Kubo, T. (2009). Measurement of bone electrical impedance in fracture healing. *Journal of Orthopaedic Science, 14*(3), 320–329.

Zariffa, J., Grouza, V., Popovic, M. R., & Hassouna, M. M. (2016). A phase-based electrical plethysmography approach to bladder volume measurement. *Annals of Biomedical Engineering, 44*(4), 1299–1309.

Zimlichman, E., Lahad, A., Aron-Maor, A., Kanevsky, A., & Shoenfeld, Y. (2005). Measurement of electrical skin impedance of dermal-visceral zones as a diagnostic tool for inner organ pathologies: A blinded preliminary evaluation of a new technique. *The Israel Medical Association Journal, 7*(10), 631–634.

Chapter 11
Body Composition by Bioelectrical Impedance Analysis

Clara Helena González-Correa

11.1 Introduction

This module reviews the main concepts of bioelectrical impedance analysis (BIA) for the evaluation of body composition and its use for the diagnosis and follow-up of individuals with different health conditions. Before making a description of the classical and alternative approaches of BIA, body composition and BIA main concepts are outlined. Then, some types of measurements through the body are mentioned and BIA parameters to estimate body components are listed. Afterwards, factors influencing BIA results are discussed, before mentioning clinical applications of bioimpedance analysis.

11.2 Body Composition

Body composition is related to the division of total body mass into components. Wang et al. (1992) developed a model of five levels of increasing complexity: I, atomic; II, molecular; III, cellular; IV, tissue system; and V, whole body. However, the two-compartment model introduced by Siri (1961) and Brozek et al. (1963) that comprises fat mass (FM) and fat-free mass (FFM) continues to be the most practical model and will be used in this module. These components are the main targets of studies and techniques related to the body composition assessment. FM consists of essential fat for normal physiological function and storage fat that comprises adipose tissue. FFM includes muscle, water, bone, connective tissue, and

C. H. González-Correa (✉)
Research Group on Bioelectrical Impedance, Department of Basic Sciences for Health, Universidad de Caldas, Manizales, Colombia
e-mail: clara.gonzalez@ucaldas.edu.co

F. Simini, P. Bertemes-Filho (eds.), *Bioimpedance in Biomedical Applications and Research*, https://doi.org/10.1007/978-3-319-74388-2_11

viscera. The proportion of each component is an important health-related indicator. Thus, while low levels of fat (<14% for women and <8% for men) are associated to exercise addition, eating disorders (Heyward and Stolarczyk 1996), decreased fecundity in women of reproductive age (Ziomkiewicz et al. 2008), and increased mortality of older women (Zaslavsky et al. 2016), high levels of body fat (>32% for adult women and >25% for adult men) are associated with increased risk of cardiometabolic disease, stroke, osteoarthritis, and diabetes (Jahangir et al. 2014; Kim et al. 2013; Bouchi et al. 2015; Zheng and Chen 2015; García-Jiménez et al. 2016). It is known that these risk factors are present throughout life, so it is important to evaluate body composition as an integral part of health care.

11.2.1 Practical Uses of Body Composition Assessment

- Monitor children's growth and development in body composition.
- Identify health risk associated with total body fat and educate patients about these topics.
- Monitor changes associated with specific diseases that alter body composition.
- Evaluate how exercise and nutrition intervention programs are working.
- Estimate individual adequate body weight and prescribe diet and physical exercise.
- Investigate the relationship between body composition and increased morbidity and mortality and between body composition and decreased function in the elderly.
- Personalize exercise program to optimize athletic performance and evaluate their effectiveness (Heyward and Stolarczyk 1996; Smith 2004).

11.2.2 Methods for Body Composition Assessment

After cadaver analysis, which is considered the gold standard for body composition assessment, only four component models (measurement of body mass, total body volume, total body water, and bone mineral), estimated by different techniques, are appropriately accurate as the reference method for assessing body composition and for evaluating new methods. In vivo methods, only measure body composition indirectly. However, they all produce errors that must be taken into account when analyzing the results they produce (Wells and Fewtrell 2006; Lee and Gallagher 2008).

Common methods for assessing body composition are divided into laboratory tests and field tests, according to their complexity. Laboratory tests, mainly for research because of their low availability, high cost, and technical complexity, include isotope dilution for measuring total body water (TBW), hydrodensitometry, total body potassium (TBK^{40}), air displacement plethysmography (ADP), ultrasound, computed tomography (CT), dual-energy X-ray absorptiometry (DEXA), specific regional bone density, three-dimensional photonic scanning, quantitative magnetic resonance, and magnetic resonance imaging or spectroscopy (MRI/MRS) (Wagner and Heyward 1999). Field techniques, more affordable and simpler, include anthropometric measurements, skinfolds, near-infrared interactance (NIR), and bioelectrical impedance analysis (BIA) (Wagner and Heyward 1999; Lee and Gallagher 2008).

11.3 Electrical Bioimpedance Concept

11.3.1 Definition

Bioelectrical impedance analysis (BIA) has been defined as the measurement of "the opposition of body cells and tissues to the flow of a radiofrequency alternating electrical current" (Ellis and Wong 1998). Alternating current is needed because a continuous current produces an effect of polarization at the level of the electrodes which prevents the continuous current from going through (Foster and Schwan 1989). Another reason is the need to avoid stimulation of tissues and organs.

11.3.2 Principles and Assumptions

BIA is based in the fact that the voltage associated to electrical current changes as the current flows through different materials. These changes are related to the following properties of these materials (Mikes et al. 1999):

(a) Substances contained in those materials: biological materials, for example, containing fluid and electrolytes offer less resistance to the current flow than those containing fat (Baumgartner 1996).
(b) Geometry: if an isotropic cylinder is considered, the longer it is, the greater the resistance to the current flow, and the wider it is, the smaller the resistance.
(c) Temperature: in metals, the resistance increases with temperature but in electrolytes decreases (EE Trade Group 1971).

The application of BIA to studies of body composition started by comparing the human body to a geometrical isotropic conductor. In the conductor showed in Fig. 11.1, the impedance is related to its length and configuration, its cross-sectional area, and the applied signal frequency.

The volume of the conductor is estimated by multiplying the right side of this equation by *L*/*L*

$$Z = \rho L^{*} L/\mathrm{LA}$$

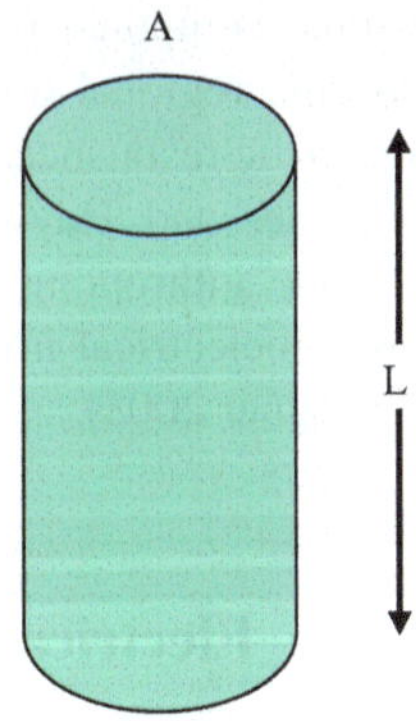

Fig. 11.1 The volume of an isotropic conductor can be estimated by measuring its length (L), cross-sectional area (A), and resistance and knowing its specific resistivity

As LA = volume, then

$$\text{Volume} = \rho L^2/Z$$

If ρ is constant, then, $V = L^2/Z$ (1) (Kushner and Schoeller 1986).

$$Z^2 = R^2 + X^2$$

If X is assumed negligible, then,

$$\sqrt{Z^2} = \sqrt{R^2} \quad \text{and} \quad Z = R$$

where Z = Impedance (ohm); ρ = Specific resistance depending on the material (ohm/m); L = Length of the conductor (m); A = Cross sectional area (m^2)

$$V = L^2/Z$$

Applying the above theory, it has been assumed that:

1. The geometry of the human body can be assimilated to a cylindrical shape with its length proportional to the subject's height (Ht).
2. The reactance (X) term contributing to the body's impedance (Z) is small, such that the resistance component (R) can be considered equivalent to body impedance.
3. The human body has a constant specific resistance.
4. The current can be injected and distributed in a constant form.

When these assumptions are combined, it can be shown that the conducting volume is proportional to the term Ht^2/Z, called the impedance index.

At low frequencies (1–5 kHz), cells behave electrically as nonconductors, and current flows through extracellular spaces characterized in the model by R resistor. However, at higher frequencies (50–above 100 kHz), current passes through cell membranes because the reactance decreases (represented by *C*) and the current crosses the intracellular spaces (*S* resistor) (Baumgartner 1996).

Therefore, impedance measurements made over a range of low to high frequencies allow the development of prediction equations relating impedance measures at low frequencies to extracellular fluid volume and at high frequencies to total body fluid volume.

11.3.3 Validation of BIA

The resistance values obtained with BIA are considered indirect body parameters and therefore must be calibrated with a more direct method of body composition, such as TBW by dilution techniques, body potassium, hydrodensitometry, dual-energy X-ray absorptiometry (DXA), magnetic resonance imaging (RMI), or a combination of some of them.

Prediction equations include height, as a surrogate measure for the interelectrode distance, but may also include weight, gender, or age as variables. Many studies have been undertaken not only to validate particular prediction equations but also to identify other variables, which may improve the quality of the predictor. The overall precision of any given prediction equation is the sum of the accuracy associated with each independent variable. Typically, standard errors of the estimate (SEE) for TBW are of the order of 3–10%, slightly higher for FFM. It may be clear that prediction equations are population-specific. The researcher must choose a prediction equation that is appropriate for his/her group of subjects (Gonzalez-Correa et al. 2007).

11.3.4 Precision and Reproducibility of BIA Measurements

Currently available impedance machines require calibration using resistor or resistor-capacitor (RC) circuits of precisely known component values. Electronic accuracy is generally high, although differences between the analyzers of different manufacturers or indeed between different machines from the same manufacturer have been noted (up to 10%). The causes of such differences are unknown but may result from inadequate calibration.

When attention is given to subject preparation, coefficients of variation within a measurement session range from 0.3 to 3%. Daily or weekly variations are greater, ranging from 1 to 5% for single frequency analyzers. Reliability varies with frequency for MF-BIA analyzers: It shows coefficients of reliability greater than 95% for inter- and intra-observer differences. In the case of resistance, changes occur with frequencies <500 kHz and for reactance with frequencies <100 kHz. It is likely that the single largest source of error in reproducibility lies in inaccurate or variable electrode replacement (Graves et al. 1989). Kyle et al. (2004a, b) reported, from the literature, variations within-day *R* measurements are $\approx$1–2%, daily or weekly intraindividual variability from ≈ 2 to 3.5%, and reproducibility/precision from 2.7 to 4.0%. Prediction errors are 3–8% for TBW and 3.5–6% for FFM.

11.3.5 Advantages and Limitations

1. Since BIA is portable, it becomes a very useful bedside method for both critically ill patients at hospital and field epidemiological studies. In addition, BIA is low time consuming, noninvasive, and safe. In terms of costs, the method is cheap in comparison with laboratory tests such as dilution, TBK, or DXA techniques and can be used for repeated measurements in short time (Lingwood et al. 1999). However, BIA has limitations with regard to assumptions and other types of limitations and restrictions. The first of them has to do with the geometry of the human body: human body is not a cylinder; instead, five cylinders connected in series can represent it better (arms, legs, and trunk, excluding the head). Since R is inversely proportional to the cross-sectional area, the upper and lower extremities with the smallest cross-sectional area will have the most influence on the whole-body R measurements. In contrast, the trunk, which contains approximately 50% of the body mass, will contribute <5–12% of total body resistance (Patterson 1989). We must also keep in mind that the electrophysical models vary (Lukaski 1996) and that in biological conductors ρ is not constant and will change according to tissue microstructure, hydration status, and concentration of electrolytic ions (Kushner 1992).
2. Given that the human body has insulators such as the lungs or intramuscular fat, the current density will not be uniformly distributed through different body segments (Kushner 1992). The reactance component of the body's impedance is not zero and several researchers have used the phase angle to assess body composition in various clinical conditions.
3. The results can change depending on the manufacturer of the machine from <1% to about 20% (Graves et al. 1989; Smye et al. 1993; Oldham 1996), and the paired cables of the impedance meter can produce additional capacitance (Xitron Technologies, Operating manual).
4. Changes in electrode position can cause changes when injecting the current (Deurenberg et al. 1991).
5. The size of the electrodes: small electrodes affect the uniformity in which the current is injected (Baker 1989).
6. BIA is age, race, gender, ethnicity, nutritional, and clinical status dependent (Chumlea et al. 1994), requiring an equation for each situation.
7. Disturbed hydration conditions can affect the results.
8. The method requires a good standardized protocol (Gonzalez-Correa and Caicedo-Eraso 2012).

11.4 BIA Approaches

11.4.1 Single Frequency Bioimpedance Analysis (SF-BIA)

Since the 1980s, BIA work has been done using the impedance index (Ht^2/Z) and measurements at a fixed frequency of 50 kHz for prediction of total body water (TBW) (Ellis and Wong 1998). This is based on the high correlation with whole-body impedance (Hoffer et al. 1969). TBW is the sum of extra- and intracellular fluids, respectively (about 25% and 75%) (Kyle et al. 2004a, b part I). However, SF-BIA does not discriminate between ECF and ICF. Instead, it estimates ECF and TBW; thus, ICF is calculated by subtracting the former from the latter. In subjects with normal hydration (Lukaski et al. 1986), body components are estimated using population-specific equations based on regression analysis (Lukaski et al. 1985) and using variables such as weight, height, gender, age, BMI, and reactance, among others. As many manufacturers of BIA devices do not report the equation included in the software, or the devices themselves do not show the raw data to recalculate the compartments of the body, many errors in the assessment of body composition can occur if the appropriate equation in not used.

After TBW estimation, FFM can be calculated under the assumption that FFM is hydrated at 73.2%. Then, FM can be derived by subtracting FFM from body weight (Mulasi et al. 2015). However, it is has been shown that in certain conditions, as in obese subjects, renal patients, or children, FFM hydration can be higher than this figure, in which cases the SF-BIA tends to overestimate the FFM (Mulasi et al. 2015).

Moreover, some authors suggest that SF-BIA at 50 kHz is not high enough to get over the barrier of membrane capacitance in order to differentiate ICW and ECW (Mulasi et al. 2015). For that reason, researchers continue looking for new ways to use BIA data without the need of equations or body compartments estimations. One of these attempts is bioelectrical impedance vector analysis (BIVA), a method that will be described later.

11.4.2 Multiple Frequency Bioimpedance Analysis (MF-BIA)

Initially, MF-BIA used two frequencies: low frequencies (1–20 kHz) representing ECF and high frequencies current (50 to $\geq$100 kHz) TBW. This method also needs multiple regression equations to estimate FFM, TBW, ICW, and ECW. The use of this approach, initiated in 1962 by Thomasset (1962), increased in the 1990s when researchers also began to use more than two frequencies (0, 1, 5, 50, 100, 200, and 500 kHz) to estimate the same parameters (Mialich et al. 2014; Ellis et al. 1999). However, it has been reported that at frequencies below 5 kHz and above 200–500 kHz, the BIA reproducibility decreases (Hannan et al. 1994). The MF-BIA method supposes advantages over the SF-BIA for estimation of body fluids as the former distinguishes clearly ECF from ICF [2] (Jaffrin and Morel 2008).

11.4.3 Bioelectrical Spectroscopy (BIS)

BIS uses a frequency range between zero to infinite. The analysis of its results to predict body volumes (ECF and TBW) is performed by nonlinear mathematical modeling instead of population-specific equations (Cornish and Ward 1998). The results are analyzed with reference to the measurement of body volumes by the technique of isotope dilution. The measured impedance data are fitted to mathematical models where the resistance at zero and at infinite frequencies is extrapolated and related to the extracellular resistances and the total resistances, respectively. From them, the intracellular resistance is calculated (Zink et al. 2015).

However, significant differences between predicted and measured values have been commonly found in healthy adults and groups with various pathologies. In addition, similar to SF-BIA and MF-BIA, BIS overestimates TBW when used with patients who have fluid overload, so their use is under discussion for clinical applications as in very obese patients (Lukaski 2013). This method will be explained in the next module. BIS has been used as a system for fluid management in heart failure patients in which fluid overload is frequently associated to pulmonary and peripheral edema (Weyer et al. 2014). The authors measured 25 patients by foot-to-foot segmental BIA that could be a guide to diagnosis cardiac decompensation. This allows physicians to provide a timely treatment.

11.4.4 Bioimpedance Vector Analysis (BIVA)

Bioimpedance Vector Analysis (BIVA) is another method for the use of SF-BIA at 50 kHz without the need to use regression equations (Fig. 11.2). It is useful for the assessment of hydration and FFM status. The method consists on drawing (R-Xc graph) resistance and reactance standardized for height to create a vector comparable to reference values obtained from gender- and race-specific healthy subjects. This method avoids some assumptions of BIA that are not met for the estimation of body composition (Piccoli et al. 1994).

The graph represents three ellipses named "tolerance ellipses" at 50, 75, and 95% tolerance and two vectors. When the minor vector falls in the left half of the ellipse, an increase in body mass is determined. The opposite happens when the minor vector falls in the right side. When the major vector falls in the lower and upper halves of the 50% ellipse, it means increased or decreased hydration ratios. This method has subsequently been successfully used by other authors (Cox-Reijven and Soeters 2000) in studies with patients who had different pathologies such as women with respiratory diseases and anorexia. Moreover, it has been useful

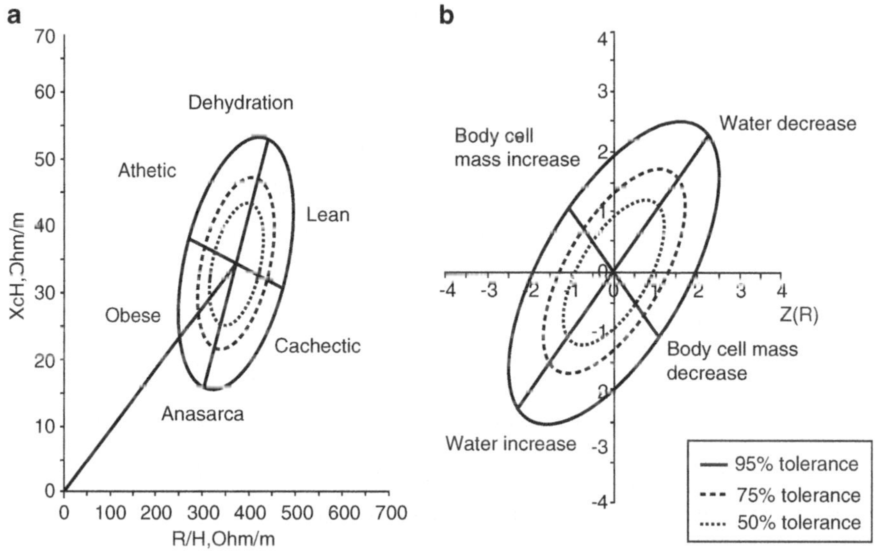

Fig. 11.2 An example of BIVA ellipses. From Bronhara et al. (2012)

for hemodialysis patients to calculate dry weight. Espinosa-Cuevas et al. (2007) reported that the BIVA tolerance ellipses in 439 Mexican subjects aged 18–82 years were quite different from Italians and Americans. This is how, as with other BIA approaches, population-specific tolerance ellipses are required for body composition assessment due to the effects of race, sex, and body size, among others. To overcome some of these difficulties, Marini et al. (2013) designed a specific BIVA method to correct errors due to body size in elders. Bronhara et al. (2012) show an illustrative graph:

Guldrís (2011) informs that BIVA has a specificity of 86%, a positive predictive value of 98%, and a negative predictive value of 20%. The error of the technique for obese people is 9%. As predicted by this author, additional applications of BIVA in critically ill, renal syndrome, cardiac syndrome, and high-risk pregnancy are being developed (Jones Sarah et al. 2015; Piuri et al. 2016).

11.4.5 Differential Impedance Analysis (DIA)

Differential Impedance Analysis (DIA) or real-time processing of bioimpedance was proposed by Sanchez et al. (2013). This new approach for bioimpedance data analysis, based on Fricke–Morse model of living beings, considers the tissue as a suspension of many of cells surrounded by an extracellular liquid. This novel approach has been used to gather information about lung perfusion and ventilation in real time. Since it is a relatively new proposal, it still requires more studies.

11.5 Types of Measurement in the Body Composition Assessment by Electrode Configuration

11.5.1 Hand to Foot Bioimpedance

Whole-Body Electrical Impedance is the most popular method used for estimating whole-body composition. It is measured through a hand to foot electrode configuration. The subject lies on a nonconductive bed, with arms separated from the trunk and legs between them. Four surface electrodes are placed either ipsilaterally or contralaterally on the dorsal surfaces of the right hand and foot at the distal metacarpals and metatarsals, respectively, and the distal prominences of the radius and ulna and between the medial and lateral malleoli at the ankle. The impedance analyzer delivers currents that vary from 100 up to 800 μA in a range of frequencies from 1 to 1000 kHz and is passed between the outer electrodes (Lukaski et al. 1985). The voltage drop is measured between the two inner electrodes. The use of this four-electrode arrangement allows to bypass the high skin impedance (Foster and Lukaski 1996). Other conditions apply when measuring body impedance: the subject should be with limbs abducted slightly, and measurements must be taken 10 min after adopting this posture; the subject should be fasted but not dehydrated and have not undertaken, within the past 12 h, strenuous exercise and ingested alcohol or diuretics. The skin surface at the sites of electrode placement should be cleaned with alcohol and electrodes should be accurately placed with reference to anatomical markers (Kushner et al. 1996).

However, there are two more approaches: foot-to-foot or leg-to-leg method and hand-to-hand method.

11.5.2 Foot-to-Foot or Leg-to-Leg Bioimpedance

Foot-to-Foot or Leg-to-Leg Bioimpedance measurements are carried out through an electronic pathway of the lower extremities. Introduced by Nuñez et al. (1997), this approach has many advantages because of the replacement of traditional disposable gel electrodes by convenient reusable stainless steel contact electrodes that require less care than disposable electrodes. Furthermore, the subject is standing instead of being in a supine position. Thus, the current flows through the lower extremities. This technique is now extensively used to appraise the whole-body composition. Wu et al. (2015) found a good accurate FFM prediction using foot-to-foot BIA bioelectrical impedance in healthy Asian individuals. However, Ziai et al. (2014) did not find good agreement between this BIA approach and DXA to measure FFM, FM, and % FM in adults with cystic fibrosis. Nevertheless, this approach continues to be used in other applications because of its advantages (Kasvis et al. 2015).

11.5.3 Hand-to-Hand Bioimpedance (HH-BIA)

Hand-to-Hand Bioimpedance (HH-BIA) measurements were proposed by Ghosh et al. (1997) in subjects with malnutrition. Deurenberg and Deurenberg Yap (2002) validated the hand-to-hand method on 298 Singaporean subjects and reported that readings were acceptable for them (Khalil 2014). Lu et al. (2016), in an epidemiological study with 704 healthy Chinese, compared the prediction of FFM by standing position to hand to foot and to (DXA). There was a good correlation for these compared measurements, although HH-BIA performed better for men than for women. Although this technique is convenient, it is less used in research and more in gyms and clinical applications that does require the same scientific rigor.

11.5.4 Segmental Bioimpedance Analysis (SBIA)

Segmental Bioimpedance Analysis (SBIA) measurement test approximates body geometry to five cylinders, including four limbs and the trunk. Although the trunk represents 50% of the body mass changes in lean mass, main changes occurring in fat-free mass estimations of limbs dominate the results of the BIA technique. Therefore, accumulations of abdominal fluid are not adequately reflected in total body measurements. This technique attempts to solve the disagreement between the trunk-to-limb ratio resistances (Kyle et al. 2004a). This is how its initial use intended to calibrate the technique for the prediction of total body composition by adding the results of the five segments. However, the improvement was very small (Ward 2012). This was how researchers began to look at the use for individual limbs and particular health conditions. Tanaka et al. (2007) stated that this technique achieves better estimation of skeletal muscle mass (SMM) than whole-body bioimpedance analysis (SEE of 6.1%) compared to MRI in 30 male subjects. Ward (2012) has made an update of this approach. Besides body composition assessment, SBIA has been useful to determine the hydration status of patients undergoing hemodialysis (Zhu et al. 2007) or lymphedema in patients with breast cancer (Ward et al. 2011).

There are more than four SBIA approximations according to electrode position; however, the most common are those proposed by (Organ et al. 1994; Scheltinga et al. 1992; Zhu et al. 1999; Jaffrin and Morel 2008). With small differences, SBIA underestimates fat-free mass and overestimates fat mass when compared to reference techniques (Ward 2012). As in the other BIA applications, the way to improve SBIA results is by using the specific equation for each population.

11.5.5 Segment Localized Electrical Bioimpedance Analysis: SLBIA

This method, instead of body composition assessment, is dedicated to appraise individual or muscle groups and smaller sections of the body. As well as several of the BIA applications, SLBIA requires the measurement of the length of the body segments to be evaluated (Aaron and Shiffman 2000). For these applications, geometry is simpler and, hence, enhances calculations. Thus, the technique is useful to evaluate limb segments such as forearm, thigh, or tongue muscles and abdominal fat. Rutkove et al. (2002), has been able to use this application to the diagnosis and monitoring of neuromuscular diseases (Colina et al. 2016).

11.6 Factors Affecting Body Composition Estimations by BIA

The accuracy of BIA measurements is limited by different sources of error such as physical model, cross-sectional area, ethnicity, body hydration, age, and level of body fat, among other variables. Equation for each population is required as they can produce overestimation when manufacturer's equations are used. However, there is a lack for an accepted standard of BIA procedures. This is compounded when there is not a good report of the BIA study's methodology (Barrera-Ariza et al. 2009); hence, the comparability between the results is poor and reduces the reliability of the method. Perhaps, standardization of methods would be the first step for BIA studies to move forward and subsequently improve its accuracy. Many variables influencing BIA estimations of body components have been studied and quantified in some way. Gonzalez-Correa et al. (1999) reported that if skin temperature, hydration status, subject posture, and prandial condition were not controlled, errors of up to 2.2 L above and 1.4 L below could occur in the estimation of TBW by BIA. That is why several attempts have been made to standardize the procedures (Gonzalez et al. 2000; Barrera-Ariza et al. 2009) in such a way as to control each and every one of the reported and new variables that appear such as silicone breast prostheses (Yamaguchi et al. 2012; Caicedo-Eraso et al. 2012). These authors developed a scoring system to assess MF-BIA technique and to identify methodological components which are essential to obtain adequate results by this technique, plan prospective studies, and critically evaluate information in the topic (Gonzalez-Correa et al. 2002). Later, Gonzalez-Correa and Caicedo-Eraso (2012), published a thorough list of 37 variables to consider and a checklist to standardize BIA protocol that would produce comparable results of studies performed with the classic hand-foot configuration in adults. In addition, Khalil (2014), document that age, sex, anthropometry, and ethnic group influence FM and FFM prediction since resistance and capacitance measurements contribute only by 0–20% to these estimations, but the percentage increases up to 11–53% when the height is included.

11.7 BIA Parameters

11.7.1 Evolution from Formulas to Raw Data: Resistance, Reactance, Impedance, and Phase Angle

After many years of use and dozens of predictive equations of body components through BIA, more importance is now being given to the use of raw data from bioimpedance meters such as the phase angle for the evaluation of physiological disorders and for the follow-up of treatments in subjects with different diseases. Lukaski (2013), stated that the applied multiple regression approaches and physical modules in bioimpedance analysis have limited utilization at individual level. Trying to overcome these difficulties, more research is being done to use the primary BIA parameters (Mulasi et al. 2015) to assess clinical outcomes or to evaluate the nutritional status without the need to use equations. It is how phase angle is now widely used as predictor of clinical status and disease prognosis, especially because it is indepent of geometry. Barbosa-Silva et al. (2005), have obtained phase angle reference by using data from 1967 adults subjects aged from 18 to 94. The figures vary between 6.97 and 4.77 degrees in men to 5.9 and 4.2 in women as the age increases. Selberg and Selberg (2002), found that PA identified hospitalized patients that would have shorter survival than normal subjects (PA: 6.6_ (0.6)_ vs. 4.9_ (1.2)_, $P < 0.001$). Bosy-Westphal et al. (2006) also identified phase angle reference values by gender and age from 15,605 German children and adolescents and 214,732 adults. When they were compared to the United States and Switzerland data, the results did not agree. They recommended the use of population-specific and impedance-analyzer-specific reference values for phase angle.

Nowadays it is possible to estimate body composition and evaluate tissues or body segments in different health conditions from the raw data obtained through electrical bioimpedance measurements (resistance, reactance, impedance, and phase angle). Although the technique basically detects changes in body fluids and cell membranes status, from these data, investigators have developed mathematical models and formulas that allow to evaluate fat mass (FM), fat-free mass (FFM), and lean body mass (LBM). From these fat-free mass and body fat mass indices (FFMI and BFMI) can be calculated. The indexes would compensate for differences in height and may identify individuals with normal BMI but who are at risk because of decreased FFM or elevated body fat.

11.8 Clinical Applications

If the limitations are taken into account, the BIA technique, used in a judicious and standardized way, can be very useful in clinical and epidemiological studies and practice. That is how US health authorities decided to include bioimpedance analysis in the NHANES program to assess sarcopenia, sarcopenic obesity in elderly, and nutritional status and health of subjects (Batsis et al. 2014).

Since changes in body fluids are key factors related to health, and body composition changes during growth and other physiological and pathological states, the estimation of body compartments is critical to health care. This is how the BIA has been used by most medical specialties. However, every day a new application appears and we must be attentive to recognize its potential. In Latin America, a region where the use of this method would be very convenient, the technique and the equipment to carry it out have not been sufficiently disseminated among the medical professions. This topic has been mainly confined to research in the physical and engineering aspects and less to its clinical applications. Recently, through the initiative of Brazilian researchers, the small Latin American community of researchers around BIA decided to exchange information and show their results in 2012 through the first Latin American Conference on Electrical Bioimpedance (CLABIO 2012). That is why it is necessary to show a small sample of what can be done with the technique and what are the limitations that still remain.

The following are the most common applications of BIA measurements:

The assessment of body composition is desirable in almost all normal or pathological conditions through the different stages of life. Although one of the main limitations for BIA use is for subjects with altered hydration status, BIVA method for monitoring body fluid variation seems useful for this purpose. Subjects with regional edema, lymphedema, wound healing, neuromuscular diseases, cancer, malnutrition, and prognosis for chronic diseases, in conditions of pulmonary, renal, cardiac, hepatic, endocrine, immunological, vascular, or musculoskeletal failures, among others, may be susceptible to be evaluated through one of the different approaches of BIA (Lukaski 2013; Mialich et al. 2014).

As the aging population increases, skeletal mass (SM) is now an important variable for sarcopenia diagnoses in elderly (loss of muscle mass and function). From this estimation, the prediction of skeletal muscle mass index is made. SM is considered a good marker of protein inside the body (Marini et al. 2012). Janssen et al. (2000) developed an equation to estimate SM. However, differences in anthropometry may change across populations and the results can vary (Mialich et al. 2014). In summary, SBIA should be a practical method for specific applications. However, as in many other approaches, it needs standardization and more specific devices since they are designed according to different settings: laboratory/research, field investigations, or clinical use. Thus, it is important to know the purpose for which the device was built, the proper equation to use, and the limitations of the results. Every day there is a new clinical or investigative utility for the use of the BIA. Table 11.1 shows some of the last applications of bioimpedance analysis in body composition assessment organized according to the interest of the various medical specialists to whom the technique can be useful.

Some clinical circumstances such as abnormal conditions of hydration, obesity, severe malnutrition, and neurological diseases are the main limitations for the use of BIA. In these cases, the recommendations are to use the segmental approach, to

Table 11.1 Applications of BIA in body composition assessment according to the interest of the several medical specialists to whom the technique can be useful

Medical clinical especialties	BIA parameters	Health conditions	Approach	Observed results	Authors
Neurology	FFM	22 patients with Huntington's disease		Body composition analysis by BIA was an useful marker of skeletal muscle mass in Huntington's disease	Cubo et al. (2015)
		14 healthy subjects			
Cardiology	ECF	130 acute decompensated heart	MF-BIA	BIA estimated ECW indicating return to nomovolemia	Sakaguchi et al. (2015)
		21 with pleural effusion and 24 healthy subjects		There was a significant positive correlation between	
Neumology	Ri, Re	21 with pleural effusion and 24 healthy subjects	BIS	Pleural effusion volume correlated with BIA increase thus, it was possible to detect fluid shift after thoracentesis	Zink et al. (2015)
Gastroenterology	% Body fat	235 patients with Barrett's oesophagus and 244 healthy subjects		Total body fat by BIA was associated with an increase in risk of Barrett's oesophagus	Kendall et al. (2014)
Nephrology	ECF, ICF, TBW	269 hemodialysis (HD) patients	BIS	Bioimpedance guided fluid management had benefits on blood pressure, arterial stiffness, and left ventricular hypertrophy management and thus, on the survival of the patients	Wizemann et al. (2009)
Endocrinology	% FM	71 women aged of 18 and 24, and 160 neonates		Authors detected energy deficiency in women having early motherhood	Aguilar-Moreno et al. (2015)
Orthopedics	ECF	30 subjects with limb amputation	MF-BIA	It was possible to compare residual limb fluid volume changes after doffing which is useful for socket design	Sanders et al. (2012)

(continued)

Table 11.1 (continued)

Medical clinical especialties	BIA parameters	Health conditions	Approach	Observed results	Authors
Physiatry	ECF-ICF	112 healthy children at different ages	MF-BIA	BIA was useful to evaluate muscle development and muscle cross-sectional areas in children of different ages	Uchiyama et al. (2017)
Dermatology	% FM	114 healthy adults	MF-BIA	% FM measured by BIA was inversely correlated to vitamin D levels	Bi et al. (2016)
Intensive care	ECF, ICF, TBW, BMC FFM/height^2, BCM/height^2	27 sepsis patients in ICU treated with continuous renal replacement therapy (CRRT)	MF-BIA	BIA could assess patient's hydration and body composition to allow clinical decisions	Wu et al. (2016)
Pediatrics	FFM	255 healthy Chinese children and adolescents	SF-BIA	High correlation $R^2 = 0.95$; SEE = 2.45 kg of FFM as measured by BIA and DXA	Wang et al. (2014)
Geriatrics	R, Xc	195 healthy adult Australians	SF-BIA	5 BIA equations were analyzed. Two formulas performed well for healthy Australians	Yu et al. (2016)
Infectology		120 patients with sepsis and acute kidney injury and patients pharyngolaryngeal or orthopedic surgery		AP strongly correlated with ECW and VE ECW and VE might be discriminative parameters of risk for increased intra-abdominal pressure	Dąbrowski et al. (2015)
Deportology	% FM	6048 young athletes	MF-BIA	BIA identified athletes at risk of undernutrition	Kalnina et al. (2015)
Nutriology	PA	25 patients, having cirrhosis by hepatitis C virus	SF-BIA	PA detect malnourished patients	Nunes et al. (2016)
Oncology	FFM, ECW and TBW	6 patients undergoing treatment for cancer	MF-BIA	Ketogenic diet might be helpful to preserve muscle mass as measured by BIA	Klement and Sweeney (2016)

have careful analysis when evaluating obese or very thin subjects, and to develop repeated measurements to follow up the patients (Mialich et al. 2014).

This work may have limitations because the magnitude of BIA literature. It should be admitted that in this overview important studies would be missing.

11.9 Conclusions

BIA is a fast, safe, portable, low-cost, and easy-to-use tool for evaluating body composition. It can be used in real time during dialysis or surgery. More and more clinical applications of the technique are being registered, and new uses are found for its different parameters. BIA method for body composition estimation seems to apply to many healthy and clinical conditions. Compartments as FM, FFM (skeletal muscle mass), and total body water, (ECW plus ICW) can be predicted by different approaches of BIA, using the proper method of analysis. Nowadays, the use of the technique is advancing from the evaluation of the static body components toward the estimation of their functional changes.

References

Aaron, R., & Shiffman, C. A. (2000). Using localized impedance measurements to study muscle changes in injury and disease. *Annals of the New York Academy of Sciences, 904*, 171–180.

Aguilar-Moreno, M., Galicia-Castillo, O. R., Aguilera-Reyes, U., Varea-Gonzalez, C., Bernis-Carro, C., & García-López, G. I. (2015). Hormonal state comparison (progesterone, estradiol, and leptin) of body fat and body mass indices in Mexican women as a risk factor for neonatal physiologic condition. *Journal of Pediatric and Adolescent Gynecology, 28*(3), 149–156. https://doi.org/10.1016/j.jpag.2014.06.007.

Baker, L. E. (1989). Principles of the impedance technique. *IEEE Engineering in Medicine and Biology Magazine, 8*, 11–15.

Barbosa-Silva, M. C., Barros, A. J., Wang, J., Heymsfield, S. B., & Pierson, R. N., Jr. (2005). Bioelectrical impedance analysis: Population reference values for phase angle by age and sex. *The American Journal of Clinical Nutrition, 82*(1), 49–52.

Barrera-Ariza, L., Gonzalez-Correa, C. H., & Gonzalez-Correa, C. A. (2009). Quality of reporting of bioelectrical impedance analysis (BIA) studies evaluating body fluid volumes: The need for standardization. Ifmbe. 2009. Proceedings ISSN: 1680-0737 ed: Springer v.25 fasc.VII (pp. 244–246).

Batsis, J. A., Mackenzie, T. A., Barre, L. K., Lopez-Jimenez, F., & Bartels, S. J. (2014). Sarcopenia, sarcopenic obesity and mortality in older adults: Results from the National Health and Nutrition Examination Survey III. *European Journal of Clinical Nutrition, 68*(9), 1001–1007. https://doi.org/10.1038/ejcn.2014.117. Epub 2014 Jun 25.

Baumgartner, R. N. (1996). Electrical impedance and total body electrical conductivity in human body composition. Human Kinetics United States. In F. R. Alex, B. H. Steven, & G. L. Timothy (Eds.), *Human body composition* (pp. 79–107). Champaign: Human Kinetics.

Bi, X., Tey, S. L., Leong, C., Quek, R., & Henry, C. J. (2016). Prevalence of vitamin D deficiency in Singapore: Its implications to cardiovascular risk factors. *PLoS One, 11*(1), e0147616. https://doi.org/10.1371/journal.pone.0147616. eCollection 2016.

Bosy-Westphal, A., Danielzik, S., Dörhöfer, R. P., Later, W., Wiese, S., & Müller, M. J. (2006). Phase angle from bioelectrical impedance analysis: Population reference values by age, sex, and body mass index. *JPEN Journal of Parenteral and Enteral Nutrition, 30*(4), 309–316.

Bouchi, R., Takeuchi, T., Akihisa, M., Ohara, N., Nakano, Y., Nishitani, R., et al. (2015). High visceral fat with low subcutaneous fat accumulation as a determinant of atherosclerosis in patients with Type 2 diabetes. *Cardiovascular Diabetology, 14*(1), 136.

Bronhara, B., Piccoli, A., & Pereira, J. C. (2012). Fuzzy linguistic model for bioelectrical impedance vector analysis. *Clinical Nutrition, 31*, 710–716.

Brozek, J., Grande, F., Anderson, J. T., & Keys, A. (1963). Densitometric analysis of body composition: Revision of some quantitative assumptions. *Annals of the New York Academy of Sciences, 110*, 113–140.

Caicedo-Eraso, J. C., Gonzalez-Correa, C. H., & Gonzalez-Correa, C. A. (2012). Use of electrocardiogram (ECG) electrodes for bioelectrical impedance analysis (BIA). In *Journal of Physics: Conference Series ISSN: 1742-6588* ed: Institute of Physics Publishing (IOP) v.407 fasc.012008 (pp. 1–8).

Chumlea, W. C., Guo, S. S., Bellisari, A., Baumgartner, R. N., & Siervogel, R. M. (1994). Reliability of multiple frequency bioelectric impedance. *American Journal of Human Biology, 6*, 195–202.

CLABIO. (2012). First Latin-American Conference on Bioimpedance. CLABIO 2012. Consulted on October 16, 2016. http://www.wikicfp.com/cfp/servlet/event.showcfp?eventid=20311&copyownerid=2.

Colina, G. E., González Correa, C. A., & Miranda, M. D. A. (2016). Miografía por impedancia eléctrica, electrical impedance myography. *Revista Colombiana de Medicina Física y Rehabilitación, 26*(1), 38–49.

Cornish, B. H., & Ward, L. C. (1998). Data analysis in multiple-frequency bioelectrical impedance analysis. *Physiological Measurement, 19*, 275–283.

Cox-Reijven, P., & Soeters, P. (2000). Validation of bio-impedance spectroscopy: Effects of degree of obesity and ways of calculating volumes from measured resistance values. *International Journal of Obesity, 24*, 271–280.

Cubo, E., Rivadeneyra, J., Gil-Polo, C., Armesto, D., Mateos, A., & Mariscal-Pérez, N. (2015). Body composition analysis as an indirect marker of skeletal muscle mass in Huntington's disease. *Journal of the Neurological Sciences, 358*(1–2), 335–338. https://doi.org/10.1016/j.jns.2015.09.351. Epub 2015 Sep 15.

Dąbrowski, W., Kotlinska-Hasiec, E., Jaroszynski, A., Zadora, P., Pilat, J., Rzecki, Z., et al. (2015). Intra-abdominal pressure correlates with extracellular water content. *PLoS One, 10*(4), e0122193. https://doi.org/10.1371/journal.pone.0122193. eCollection 2015.

Deurenberg, P., & Deurenberg Yap, M. (2002). Validation of skinfold thickness and hand-held impedance measurements for estimation of body fat percentage among Singaporean Chinese, Malay and Indian subjects. *Asia Pacific Journal of Clinical Nutrition, 11*, 1–7.

Deurenberg, P., Van der Kooy, K., Leenen, R., Weststrate, J., & Seidell, J. (1991). Sex and age specific prediction formulas for estimating body composition from bioelectrical impedance: A cross-validation study. *International Journal of Obesity, 15*, 17–25.

Electronic Engineering Trade Group. (1971). *Basic electricity* (2nd ed.). London: H.M. Stationary Off.

Ellis, K. J., & Wong, W. W. (1998). Human hydrometry: Comparison of multifrequency bioelectrical impedance with $2H_2O$ and bromine dilution. *Journal of Applied Physiology, 85*, 1056–1062.

Ellis, K. J., Bell, S. J., Chertow, G. M., Chumlea, W. C., Knox, T. A., Kotler, D. P., et al. (1999). Bioelectrical impedance methods in clinical research: A follow-up to the NIH Technology Assessment Conference. *Nutrition, 15*(11–12), 874–880.

Espinosa-Cuevas, M. A., Rivas-Rodríguez, L., González-Medina, E. C., Atilano-Carsi, X., Miranda-Alatriste, P., & Correa-Rotter, R. (2007). Vectores de impedancia en población mexicana. *Revista de Investigación Clínica, 59*(1), 15–24.

Foster, K. R., & Lukaski, H. C. (1996). Whole-body impedance. What does it measure? *American Journal of Clinical Nutrition, 64*, 388s–396s.

Foster, K. R., & Schwan, H. P. (1989). Dielectric properties of tissues and biological materials: A critical review. *Critical Reviews in Biomedical Engineering, 17*, 25–104.

García-Jiménez, C., Gutierrez-Salmeron, M., Chocarro-Calvo, A., Garcia-Martinez, J. M., Castaño, A., & De la Vieja, A. (2016). From obesity to diabetes and cancer: Epidemiological links and role of therapies. *British Journal of Cancer, 114*(7), 716–722.

Ghosh, S., Meister, D., Cowen, S., Hannan, J. W., & Ferguson, A. (1997). Body composition at the bedside. *European Journal of Gastroenterology & Hepatology, 9*, 783–788.

Gonzalez, C. H., Oldroyd, B., Evans, J. A., Smye, S. W., & Holland, P. (2000). Standardized limb abduction for bioimpedance measurements using position restraints. *Physiological Measurement, 21*(2), 263–270.

Gonzalez-Correa, C. A., Gonzalez Correa, C. H., & Ramos Rodriguez, A. (2007). Best published equation for the calculation of body fat in a sample of Colombian young males using bioelectrical impedance analysis. In H. Scharfetter & R. Merwa (Eds.), *Proceedings of the 13th International Conference on Electrical Bioimpedance and the 8th Conference on Electrical Impedance Tomography: ICEBI 2007*, August 29th–September 2nd 2007, Graz, Austria (pp. 803–806). Berlin, Heidelberg: Springer.

Gonzalez-Correa, C. H., & Caicedo-Eraso, J. C. (2012). Bioelectrical impedance analysis (BIA): A proposal for standardization of the classical method in adults. In *Journal of Physics: Conference Series ISSN: 1742-6588* ed: Institute of Physics Publishing (IOP) v.407 fasc.012018 (pp. 1–13).

Gonzalez-Correa, C. H., Evans, A., Smye, S., & Holland, P. (2002). Total body water measurement using bioelectrical impedance analysis, isotope dillution and total body potassium: A scoring system to facilitate intercomparison. In *Netherlands European Journal of Clinical Nutrition ISSN: 0954-3007* ed: John Libbey v.56 fasc.N/A (pp. 326–337).

Gonzalez-Correa, C. H., Evans, J. A., Smye, S., & Holland, P. (1999). Variables affecting BIA measurements of body water. *Medical & Biological Engineering & Computing, 37*(2), 106–107.

Graves, J. E.. Pollock, M. L. Colvinl, A. B. Van Loan M., And Lohman T. G. Comparison of different bioelectrical impedance analyzers in the prediction of body composition. American Journal Of Human Biology. 1989 1 :603–611.

Guldrís, S. C. (2011). Aplicaciones futuras de la bioimpedancia vectorial (BIVA) en nefrología. Future uses of vectorial bioimpedance (BIVA) in nephrology. *Nefrología (Madrid), 31*(6), 635–643.

Hannan, W. J., Cowen, S. J., Fearson, K. C. H., Plester, C. E., Falconer, J. S., & Richardson, R. A. (1994). Evaluation of multi-frequency bioimpedance analysis for the assessment of extracellular and total body water in surgical patients. *Clinical Science, 86*, 479–485.

Heyward, V. H., & Stolarczyk, L. M. (1996). *Applied body composition assessment*. Champaign: Human Kinetics.

Hoffer, E. C., Meador, C. K., & Simpson, D. C. (1969). Correlation of whole-body impedance with total body water volume. *Journal of Applied Physiology, 27*, 531–534.

Jaffrin, M. Y., & Morel, H. (2008). Body fluid volumes measurements by impedance: A review of bioimpedance spectroscopy (bis) and bioimpedance analysis (bia) methods. *Medical Engineering & Physics, 30*, 1257–1269.

Jahangir, E., De Schutter, A., & Lavie, C. J. (2014). The relationship between obesity and coronary artery disease. *Translational Research, 164*(4), 336–344. https://doi.org/10.1016/j.trsl.2014.03.010. Review.

Janssen, I., Heymsfield, S. B., Baumgartner, R. N., & Ross, R. (2000). Estimation of skeletal muscle mass by bioelectrical impedance analysis. *Journal of Applied Physiology, 89*, 465–471.

Jones Sarah, L., Tanaka, A., Eastwood, G. M., Young, H., Peck, L., Bellomo, R., et al. (2015). Bioelectrical impedance vector analysis in critically ill patients: A prospective, clinician blinded investigation. *Critical Care, 19*, 290. https://doi.org/10.1186/s13054-015-1009-3.

Kalnina, L., Sauka, M., Timpka, T., Dahlstrom, O., Nylander, E., Selga, G., et al. (2015). Body fat in children and adolescents participating in organized sports: Descriptive epidemiological study of 6048 Latvian athletes. *Scandinavian Journal of Public Health, 43*(6), 615–622. https://doi.org/10.1177/1403494815581696

Kasvis, P., Cohen, T. R., Loiselle, S. È., Kim, N., Hazell, T. J., Vanstone, C. A., et al. (2015). Foot-to-foot bioelectrical impedance accurately tracks direction of adiposity change in overweight and obese 7- to 13-year-old children. *Nutrition Research, 35*(3), 206–213. https://doi.org/10.1016/j.nutres.2014.12.012.

Kendall, B. J., Macdonald, G. A., Prins, J. B., O'Brien, S., & Whiteman, D. C. (2014). Study of digestive health. Total body fat and the risk of Barrett's oesophagus - A bioelectrical impedance study. *Cancer Epidemiology, 38*(3), 266–272. https://doi.org/10.1016/j.canep.2014.03.006. Epub 2014 Apr.

Khalil, S. F. (2014). The theory and fundamentals of bioimpedance analysis in clinical status monitoring and diagnosis of diseases. *Sensors, 14*(6), 10895–10928. https://doi.org/10.3390/s140610895.

Kim, J. Y., Kim, J. Y., Han, S. H., & Yang, B. M. (2013). Implication of high-body fat percentage on cardiometabolic risk in middle-aged, healthy, normal-weight adults. *Obesity (Silver Spring), 21*(8), 1571–1577.

Klement, R. J., & Sweeney, R. A. (2016). Impact of a ketogenic diet intervention during radiotherapy on body composition: I. Initial clinical experience with six prospectively studied patients. *BMC Research Notes, 9*, 143. https://doi.org/10.1186/s13104-016-1959-9.

Kushner, R. F. (1992). Bioelectrical impedance analysis: A review of principles and applications. *Journal of the American College of Nutrition, 11*, 199–209.

Kushner, R. F., Gudivaka, R., & Schoeller, D. A. (1996). Clinical characteristics influencing bioelectrical impedance analysis measurements. *The American Journal of Clinical Nutrition, 64*, 423S–427S.

Kushner, R. F., & Schoeller, D. A. (1986). Estimation of total wáter by bioelectrical impedance analysis. *The American Journal of Clinical Nutrition, 44*(3), 417–424.

Kyle, U. G., Bosaeus, I., De Lorenzo, A. D., Deurenberg, P., Elia, M., Gómez, J. M., et al. (2004a). Bioelectrical impedance analysis—Part I: Review of principles and methods. *Clinical Nutrition, 23*, 1226–1243.

Kyle, U. G., Bosaeus, I., De Lorenzo, A. D., Deurenberg, P., Elia, M., Manuel Gomez, J., et al. (2004b). Bioelectrical impedance analysis—Part II: Utilization in clinical practice. *Clinical Nutrition, 23*, 1430–1453.

Lee, S. Y., & Gallagher, D. (2008). Assessment methods in human body composition. *Current Opinion in Clinical Nutrition & Metabolic Care, 11*(5), 566–572.

Lingwood, B. E., Coghlan, J. P., Ward, L. C., Charles, B. G., & Colditz, P. B. (1999). Prediction of aminoglycoside distribution space in neonates by multiple frequency impedance analysis. *European Journal of Pharmacology, 55*, 671–676.

Lu, H.-K., Chiang, L.-M., Chen, Y.-Y., Chuang, C.-L., Chen, K.-T., Dwyer, G. B., et al. (2016). Hand-to-hand model for bioelectrical impedance analysis to estimate fat free mass in a healthy population. *Nutrients, 8*(10), 654. Published online 2016 Oct 21. https://doi.org/10.3390/nu8100654

Lukaski, H. (2013). Evolution of bioimpedance: A circuitous journey from estimation of physiological function to assessment of body composition and a return to clinical research. *European Journal of Clinical Nutrition, 67*, S2–S9.

Lukaski, H. C. (1996). Bioelectrical impedance analysis in body composition measurement: National Institutes of Health Technology Assessment Conference Statement. *The American Journal of Clinical Nutrition, 64*, 524S–532S.

Lukaski, H. C., Bolonchuk, W. W., Hall, C. B., & Siders, W. A. (1986). Validation of tetrapolar bioelectrical impedance method to assess human body composition. *Journal of Applied Physiology, 60*, 1327–1332.

Lukaski, H. C., Johnson, P. E., Bolonchuk, W., & Lykken, G. (1985). Assessment of fat-free mass using bioelectrical impedance measurements of the human body. *The American Journal of Clinical Nutrition, 41*, 810–817.

Marini, E., Sergi, G., Succa, V., Saragat, B., Sarti, S., Coin, A., et al. (2013). Efficacy of specific bioelectrical impedance vector analysis (biva) for assessing body composition in the elderly. *The Journal of Nutrition, Health & Aging, 17*, 515–521.

Marini, E., Buffa, R., Saragat, B., Coin, A., Toffanello, E. D., Berton, L., et al. (2012). The potential of classic and specific bioelectrical impedance vector analysis for the assessment of sarcopenia and sarcopenic obesity. *Clinical Interventions in Aging, 7*, 585–591. https://doi.org/10.2147/CIA.S38488.

Mialich, M. S., Sicchieri, J. M. F., & Junior, A. A. J. (2014). Analysis of body composition: A critical review of the use of bioelectrical impedance analysis. *International Journal of Clinical Nutrition, 2*, 1–10.

Mikes, D. M., Cha, B. A., Dym, C. L., Baumgartner, J., Hartzog, A. G., Tracey, A. D., et al. (1999). Bioelectrical impedance analysis revisited. *Limphology, 32*, 157–165.

Mulasi, U., Kuchnia, A. J., Cole, A. J., & Earthman, C. P. (2015). Bioimpedance at the bedside: Current applications, limitations, and opportunities. *Nutrition in Clinical Practice, 30*(2), 180 193. https://doi.org/10.1177/0884533614568155

Nunes, F. F., Bassani, L., Fernandes, S. A., Deutrich, M. E., Pivatto, B. C., & Marroni, C. A. (2016). Food consumption of cirrhotic patients, comparison with the nutritional status and disease staging. *Arquivos de Gastroenterologia, 53*(4), 250–256. https://doi.org/10.1590/S0004-28032016000400008.

Nuñez, C., Gallagher, D., Visser, M., Pi-Sunyer, F. X., Wang, Z., & Heymsfield, S. B. (1997). Bioimpedance analysis: Evaluation of leg-to-leg system based on pressure contact footpad electrodes. *Medicine and Science in Sports and Exercise, 29*, 524–531.

Oldham, N. M. (1996). Overview of bioelectrical impedance analyzers. *The American Journal of Clinical Nutrition, 64*(3 Suppl), 405S–412S.

Organ, L. W., Bradham, G. B., Gore, D. T., & Lozier, S. L. (1994). Segmental bioelectrical impedance analysis. Theory and application of a new technique. *Journal of Applied Physiology (1985), 77*(1), 98–112.

Patterson, R. (1989). Body fluid determinations using multiple impedance measurements. *IEEE Engineering in Medicine and Biology Magazine, 8*, 16–18.

Piccoli, A., Rossi, B., Pillon, L., & Bucciante, G. (1994). A new method for monitoring body fluid variation by bioimpedance analysis: The rxc graph. *Kidney International, 46*, 534–539.

Piuri, G., Ferrazzi, E., Bulfoni, C., Mastricci, L., Di Martino, D., & Speciani, A. F. (2016). Longitudinal changes and correlations of bioimpedance and anthropometric measurements in pregnancy: Simple possible bed-side tools to assess pregnancy evolution. *The Journal of Maternal-Fetal & Neonatal Medicine, 14*, 1–7. [Epub ahead of print].

Rutkove, S. B., Aaron, R., & Shiffman, C. A. (2002). Localized bioimpedance analysis in the evaluation of neuromuscular disease. *Muscle & Nerve, 25*, 390–397.

Sakaguchi, T., Yasumura, K., Nishida, H., Inoue, H., Furukawa, T., Shinouchi, K., et al. (2015). Quantitative assessment of fluid accumulation using bioelectrical impedance analysis in patients with acute decompensated heart failure. *Circulation Journal, 79*(12), 2616–2622. https://doi.org/10.1253/circj.CJ-15-0723. Epub 2015 Oct 16.

Sanchez, B., Bandarenka, A. S., Vandersteen, G., Schoukens, J., & Bragos, R. (2013). Novel approach of processing electrical bioimpedance data using differential impedance analysis. *Medical Engineering & Physics, 35*(9), 1349–1357. https://doi.org/10.1016/j.medengphy.2013.03.006. Epub 2013 Apr 17.

Sanders, J. E., Harrison, D. S., Cagle, J. C., Myers, T. R., Ciol, M. A., & Allyn, K. J. (2012). Post-doffing residual limb fluid volume change in people with trans-tibial amputation. *Prosthetics and Orthotics International, 36*(4), 443–449. https://doi.org/10.1177/0309364612444752. Epub 2012 May 15.

Scheltinga, M. R., Jacobs, D. O., Kimbrough, T. D., & Wilmore, D. W. (1992). Identifying body fluid distribution by measuring electrical impedance. *The Journal of Trauma, 33*, 665–670.

Selberg, O., & Selberg, D. (2002). Norms and correlates of bioimpedance phase angle in healthy human subjects, hospitalized patients, and patients with liver cirrhosis. *European Journal of Applied Physiology, 86*(6), 509–516.

Siri, W. E. (1961). Body composition from fluid space and density. In J. Brozek & A. Hanschel (Eds.), *Techniques for measuring body composition* (pp. 223–244). Washington, DC: National Academy of Science.

Smith, J. (2004). Body composition assessment and relationship to disease. http://www.ideafit.com/fitness-library/body-composition-and-relationship-to-diseaseassess-ment.

Smye, S. W., Sutcliffe, J., & Pitt, E. A. (1993). Comparison of four commercial systems used to measure whole-body electrical impedance. *Physiological Measurement, 14*(4), 473–478.

Tanaka, N. I., Miyatani, M., Masuo, Y., Fukunaga, T., & Kanehisa, H. (2007). Applicability of a segmental bioelectrical impedance analysis for predicting the whole body skeletal muscle volume. *Journal of Applied Physiology, 103*, 1688–1695.

Thomasset, A. (1962). Bio-electrical properties of tissue impedance measurements. *Lyon Médical, 207*, 107–118.

Uchiyama, T., Nakayama, T., & Kuru, S. (2017). Muscle development in healthy children evaluated by bioelectrical impedance analysis. Muscle development in healthy children evaluated by bioelectrical impedance analysis. *Brain Dev, 39*(2), 122–129. https://doi.org/10.1016/j.braindev.2016.08.013. Epub 2016 Sep 21.

Wagner, D. R., & Heyward, V. H. (1999). Techniques of body composition assessment: A review of laboratory and field methods. *Research Quarterly for Exercise and Sport, 70*(2), 135–149.

Wang, L., Hui, S. S., & Wong, S. H. (2014). Validity of bioelectrical impedance measurement in predicting fat-free mass of Chinese children and adolescents. *Medical Science Monitor, 20*, 2298–2310. https://doi.org/10.12659/MSM.890696.

Wang, Z. M., Pierson, R. N., Jr., & Heymsfield, S. B. (1992). The five-level model: A new approach to organizing body-composition research. *The American Journal of Clinical Nutrition, 56*, 19–28.

Ward, L., Winall, A., Isenring, E., et al. (2011). Assessment of bilateral limb lymphedema by bioelectrical impedance spectroscopy. *International Journal of Gynecological Cancer, 21*, 409–418.

Ward, L. C. (2012). Segmental bioelectrical impedance analysis: An update. *Current Opinion in Clinical Nutrition and Metabolic Care, 15*(5), 424–429.

Wells, J. C. K., & Fewtrell, M. S. (2006). Measuring body composition. *Archives of Disease in Childhood, 91*(7), 612–617. https://doi.org/10.1136/adc.2005.085522.

Weyer, S., Zink, M. D., Wartzek, T., Leicht, L., Mischke, K., Vollmer, T., et al. (2014). Bioelectrical impedance spectroscopy as a fluid management system in heart failure. *Physiological Measurement, 35*(6), 917.

Wizemann, V., Wabel, P., Chamney, P., Zaluska, W., Moissl, U., Rode, C., et al. (2009). The mortality risk of overhydration in haemodialysis patients. *Nephrology Dialysis Transplantation., 24*, 1574–1579.

Wu, C., Wang, X., Yu, W., Li, P., Liu, S., Li, J., et al. (2016). Short-term consequences of continuous renal replacement therapy on body composition and metabolic status in sepsis. *Asia Pacific Journal of Clinical Nutrition, 25*(2), 300–307. https://doi.org/10.6133/apjcn.2016.25.2.29.

Wu, C.-S., Chen, Y.-Y., Chuang, C.-L., Chiang, L.-M., Dwyer, G. B., Hsu, Y.-L., et al. (2015). Predicting body composition using foot-to-foot bioelectrical impedance analysis in healthy Asian individuals. *Nutrition Journal, 14*, 52. Published online 2015 May 19. https://doi.org/10.1186/s12937-015-0041-0

Yamaguchi, C. M., Faintuch, J., Silva, M. M., Modolin, M., Hayashi, S. Y., & Cecconello, I. (2012). Interference of silicone breast implants on bioimpedance measurement of body fat. *Clinical Nutrition, 31*(4), 574–576.

Yu, S. C., Powell, A., Khow, K. S., & Visvanathan, R. (2016). The performance of five bioelectrical impedance analysis prediction equations against dual X-ray absorptiometry in estimating appendicular skeletal muscle mass in an adult Australian population. *Nutrients, 8*(4), 189, https://doi.org/10.3390/nu8040189.

Zaslavsky, O., Rillamas-Sun, E., Li, W., Going, S., Datta, M., Snetselaar, L., et al. (2016). Association of dynamics in lean and fat mass measures with mortality in frail older women. *The Journal of Nutrition, Health & Aging, 21*(1), 112–119.

Zheng, H., & Chen, C. (2015). Body mass index and risk of knee osteoarthritis: Systematic review and meta-analysis af prospective studies. *BMJ Open, 5*(12), e007568.

Ziai, S., Coriati, A., Chabot, K., Mailhot, M., Richter, M. V., & Rabasa-Lhoret, R. (2014). Agreement of bioelectric impedance analysis and dual-energy X-ray absorptiometry for body composition evaluation in adults with cystic fibrosis. *Journal of Cystic Fibrosis, 13*, 585–588. https://doi.org/10.1016/j.jcf.2014.01.006

Zhu, F., Kuhlman, M., Kotanko, P., Handelman, G., Leonard, E., & Levin, N. (2007). A device for monitoring hydration state in hemodialysis patients using a calf bioimpedance technique. In *Proceedings of the 13th International Conference on Electrical Bioimpedance and the 8th Conference on Electrical Impedance Tomography, Graz, Austria*, 29 August–2 September 2007 (pp. 775–778).

Zhu, F., Schneditz, D., & Levin, N. W. (1999). Sum of segmental bioimpedance analysis during ultrafiltration and hemodialysis reduces sensitivity to changes in body position. *Kidney International, 56*, 692–699.

Zink, M. D., Weyer, S., Pauly, K., Napp, A., Dreher, M., Leonhardt, S., et al. (2015). Feasibility of bioelectrical impedance spectroscopy measurement before and after thoracentesis. *BioMed Research International, 2015*, 810797. https://doi.org/10.1155/2015/810797.

Ziomkiewicz, A., Ellison, P. T., Lipson, S. F., Thune, I., & Jasienska, G. (2008). Body fat, energy balance, and estradiol levels: A study based on hormonal profiles from complete menstrual cycles. *Human Reproduction, 23*(11), 2555–2563.

Chapter 12
Bioimpedance for Analysis of Body Composition in Sports

Lucia Malá, František Zahálka, and Tomas Maly

12.1 Introduction

Body composition (BC) is an important indicator of physical fitness and health in athletes. Generally body mass could be discriminate to active mass (lean body mass (LBM), body cell mass (BCM), muscle mass (MM), extra and intracellular mass ratio (ECM/BCM), intracellular water (ICW), phase angle (PhA)) and inactive mass (fat mass (FM). BIA is based on five level model of body composition. In sport will be the molecular level and in sport practice we can come across the terms lipid and fat. It is based on a multicomponent model and thus more accurately estimates indirectly measurable parameters assuming that constants for calculation are kept (Heymsfield et al. 2005).

Bioimpedance analysis (BIA) of BC in sports can be divided into the following fields:

- Conditions for the use of BIA in athletes
- Characteristics of the selected parameters of the BIA method in sports
- Effect of a given sport on active components: lean body mass (LBM), body cell mass (BCM), muscle mass (MM), extra and intracellular mass ratio (ECM/BCM), intracellular water (ICW) and phase angle (PhA)
- Proportion of fat mass (FM)
- Proportion of these components among athletes from various disciplines, differences between higher- and lower-level performance athletes, gender differences and differences during ontogenetic development and before and after sport performance

L. Malá · F. Zahálka (✉) · T. Maly
Faculty of Physical Education and Sport, Charles University, Prague, Czech Republic
e-mail: zahalka@ftvs.cuni.cz

F. Simini, P. Bertemes-Filho (eds.), *Bioimpedance in Biomedical Applications and Research*, https://doi.org/10.1007/978-3-319-74388-2_12

- Changes in BC during short-term and long-term interventions (fitness training, effect of training and nutrition, changes during sport training periodisation and short-term body weight reduction accompanied by dehydration in combat sports)
- Differences in BC parameters of athletes with respect to field position
- Morphological asymmetries detected using multifrequency bioimpedance (proportion of MM, FM and PhA in the limbs, e.g. comparison of the dominant and non-dominant limbs)

12.2 The Use of BIA in Sports

Bioelectrical impedance methods have many advantages, as they are inexpensive, noninvasive, portable, quick and safe, do not induce discomfort and do not require a high degree of technical skill to perform. Owing to these characteristics, BIA can be used not only in laboratories but also directly in sporting practice, in sport nutrition (Moon 2013). Standardised conditions of measurement are required to provide meaningful results (Kyle et al. 2004); however, this is not physiologically simple in athletes, as hydration status, concentration of electrolytes and carbohydrate loading can result in increased water retention (Warner et al. 2004) and, furthermore, these factors are affected by athletes' training load and nutrition. In athletes, error of the estimate is influenced by biological factors such as hydration status, protein to mineral ratio and differences in bone mineral density (Fornetti et al. 1999). Using equations for estimation of parameters that are not directly measurable (such as LBM, FM, BCM and extracellular mass ECM, MM) may also introduce error. Hortobagyi et al. (1992) reported that the LBM of African-American and Caucasian collegiate football players is systematically underestimated. Williams and Bale (1998) reported a similar result for male and female soccer players. The authors reported that prediction error for lean body mass ranged from 2.9 to 6.3 kg. Swartz et al. (2002) reported that in highly active and moderately active young men, BIA significantly overestimated the percent of FM. Specific equations for use with multifrequency BIA methods are available for the sporting population (Fornetti et al. 1999; Hortobagyi et al. 1992; Pichard et al. 1997; Yannakoulia et al. 2000). These equations were developed in studies comparing a reference method and multifrequency BIA in elite athletes (Esco et al. 2015; Mala et al. 2016).

Estimation of parameters that are not directly measurable is performed by inserting values for somatometric parameters (body height (BH) and body mass (BM)) and recorded resistances (reactance, resistance and impedance) or constant values into a prediction equation. In multifrequency analyses, equations have been developed for the general population along with those for active individuals. However, even though specific equations are available for athletes, they do not account for individual differences between athletes.

Mala et al. (2016) in Fig. 12.1 determined the accuracy of measurement based on intra-trial concurrent and convergent validity of BC measures obtained using BIA compared with those obtained using hydrodensitometry in elite (national

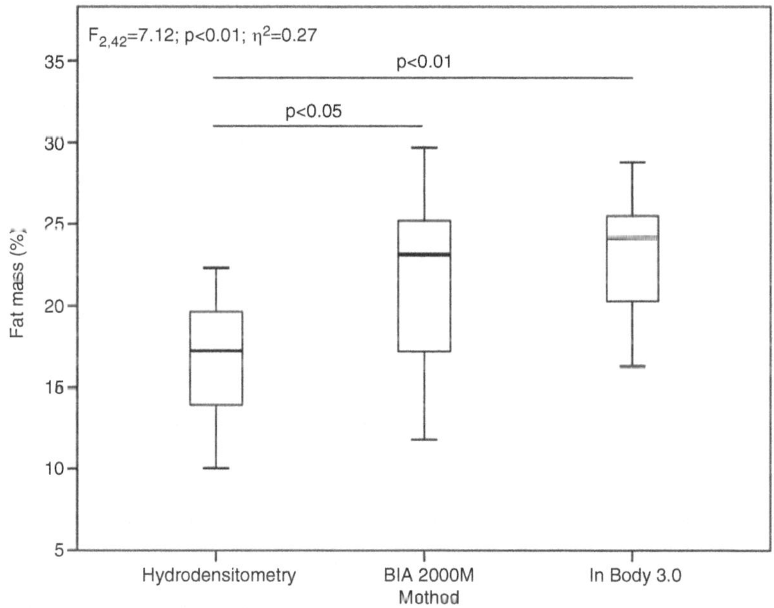

Fig. 12.1 Comparison of fat mass determined using different methods (Mala et al. 2016)

team level) female soccer players from the Czech Republic. BC was measured using three methods: two methods of BIA (InBody 3.0 and BIA 2000M) and hydrodensitometry (HD).

The results of the BIA 2000M and InBody 3.0 methods were strongly correlated ($r = 0.90$, $p < 0.01$). The BIA and HD methods were also significantly correlated ($p < 0.05$), but the correlation coefficient was only 0.63–0.64. Shakeryan et al. (2013) reported that FM detected using HD and BIA analysis (Olympia 3.3 device) in male wrestlers showed a correlation of $r = 0.74$ and a coefficient of determination of $R^2 = 0.72$. Mala et al. (2016) reported a lower correlation (BIA 2000M vs. HD, $r = 0.74$; InBody 3.0 vs. HD, $r = 0.77$). The correlation coefficient does not indicate agreement but merely the strength of the relationship between the two variables, and even data points which are in poor agreement can show a strong correlation (Williams and Bale 1998).

No equation is available for calculation of FM based on BIA for elite female soccer players (national team level). Fornetti et al. (1999) presented an equation for prediction of FFM for female athletes (basketball, crew, cross-country runners, field hockey, golf, gymnastics, soccer, softball, swimming and diving, tennis, track and field and volleyball athletes from Michigan State University) based on BIA, as follows:

$$\text{FFM} = (0.272^* \text{ BH}) + (0.461^* \text{ BM}) - (0.036^* \text{ resistance}) + (0.101^* \, x_c) - 11.567$$

where BH = body height, BM = body mass and x_c = reactance.

The BIA analyser delivered an alternating current of 800 μA at a fixed 50 kHz frequency (Fornetti et al. 1999).

Using of bioelectrical devices BIA 2000M or InBody 3.0 for assessment of FM percentage in elite female soccer players is prerequisite by the use of a prediction equation appropriate for the given sporting population. Analysis of variance showed significant differences between the means of the compared groups, the effect size (Cohen's coefficient of effect size) between BIA and HD was >0.87, the convergent validity was lower than $r < 0.65$, the coefficient of determination was lower than $R^2 < 0.65$, and the standard error of the estimate was higher than the SEM > 2.8 (Mala et al. 2016).

12.3 Characteristics of the Selected Parameters of the Bioimpedance Method in Sports

12.3.1 The Use of Directly Measurable Parameters in Sports

Measurement equipment based on BIA principle directly measures physical parameters (resistance, reactance, or PhA). Consequently, the BIA uses these physical parameters in predictive equations and deduces body composition components (LBM, BCM, MM) as indirect result of it.

Directly measurable parameters also enable monitoring of changes in body composition over a short period (such as a short-term camp for targeted weight reduction in combat athletes with the aim to fit into the best weight category).

Mala et al. (2014) reported that changes in directly measurable parameters (resistance, reactance and PhA) and in total body water (TBW) and its constituents can be used for monitoring changes in BC under nonconstant hydration in elite Czech judo athletes. Measurements were performed in two stages (pre and postreduction period) 6 days apart, i.e. short-term intervention under changes in hydration. The authors reported changes in TBW before and after approximately 10% BW reduction (mild to moderate dehydration) at the expense of both compartments; ECW was predominantly reduced (Table 12.1). A significant change was found not only in TBW but also in ICW; however, even in a state of dehydration, ICW retained an optimal value (average of 62%).

In assessing the effects of hydration changes on athletes' performance, we can only measure changes in TBW and BM, and our understanding is limited by our inability to measure both intra and extracellular components of bodily water (Sawka et al. 2007). Silva et al. (2010) reported a clear relationship between changes in ICW and upper body power output in elite male judo athletes ($n = 27$, 23.2 ± 2.8 years). According to the authors, athletes who showed a greater decrease in upper body power also showed a greater reduction in TBW, even controlling for changes observed in body weight and lean soft tissue of the arms. They built on the theory proposed by Haussinger et al. (1993) that cellular volume is a key signal for

Table 12.1 Significant differences between pre and posttest values

	Pretest		Posttest			
Parameter	*x*	SD	*x*	SD	*t*	*d*
Resistance (Ω)	373.6	34.94	480.6	47.17	−8.44**	0.96
Phase angle (°)	7.54	0.91	7.24	1.05	5.03**	1.26
Reactance (Ω)	50.9	6.52	59.6	6.33	−9.11**	1.05
LBM (kg)	72	8.06	66.76	9.92	3.69**	1.31
TBW (l)	52.87	5.59	47.73	6.17	8.2**	1.37
ICW (l)	31.09	1.83	29.18	2.28	6.3**	1.39
ECW (l)	21.71	3.89	18.56	4.02	6.02**	0.8

LBM lean body mass, *TBW* total body water, *ICW* intracellular water, *ECW* extracellular water, ** significant differences between pretest vs. posttest with probability $p < 0.01$

orientation of cell metabolism and that cellular swelling leads to anabolism, whereas cellular shrinkage promotes catabolism.

To assess other indirectly measurable variables, it is necessary to take resistances obtained using bioimpedance measurements (resistance, reactance, PhA) into account. The values of directly measurable variables influence both the calculation over the predictive equation of derived parameters. Only the mutual change of directly measurable variables in one direction can verify the relevance of the indirectly measured variables (eventual loss of body cell mass must be in the short-term intervention together with simultaneous reduction of phase angle, reactance and cell quote).

Resistance, from the perspective of physics, is the pure ohmic resistance of a conductor to alternating current, and therefore, it could be inversely proportional to TBW. A high proportion of TBW and electrolytes in muscles facilitates conduction of electricity. A material with low resistance conducts well, while a material with high resistance conducts poorly (Cerit et al. 2009). Fluctuations in the proportion of TBW influence the resistance value. Mala et al. (2014) reported significant decreases in reactance and phase angle after a short-term BM reduction in elite judo athletes, due to dehydration. The authors also reported significantly lower values of PhA that point to decreased cellular integrity, poorer membrane condition and poorer distribution of liquids after a radical BM reduction. PhA is directly measurable and is determined by resistance and reactance (amount and functionality of a cell membrane) without need for use of equations, and it directly reflects BCM values. PhA was described by Selberg and Selberg (2002) as the angle of the impedance vector in relation to the ratio between resistance and reactance of electrical voltage. Available literature presented a formula for direct calculation of resistance and reactance, as follows:

$$\text{Phase angle} = \text{reactance}/\text{resistance} \times 180^{\circ}/\pi$$

Despite known prediction error, changes in indirectly measurable parameters have been presented in several studies. Reljic et al. (2013) found that after 5 days of dehydration, no relationship was observed between the average power volume (−8%) and boxing-related tasks in collegiate boxers who had reduced their BM by 3–4% within a few hours by excessive sweat loss. Calvo Rico et al. (2015) reported that MM was reduced after dehydration in judo athletes. The authors reported significant differences in average muscle strength, average muscle power and muscle strength production ($p < 0.05$). Silva et al. (2010), in a study of judo athletes, reported that reductions in TBW and particularly in ICW were observed in athletes who lost more than 2% of their upper body power and forearm maximal strength, thus indicating that small changes in the ICW compartment may interfere with performance. Coufalova et al. (2014) reported significant changes in the proportion of LBM and TBW ($p < 0.01$) in elite Czech judo athletes. The values of cell quote and ECM/BCM index are influenced by changes in resistance, reactance and PhA in terms of calculation of indirectly measurable parameters. A real change (undesired loss) of BCM after a BM reduction is accompanied by a change in PhA, cell quote (CQ) and reactance in one direction in the bioimpedance measurement (loss of BCM and simultaneous decrease of PhA, reactance and CQ).

Matias et al. (2016) presented prediction models for athletes:

$$\mathrm{TBW} = 0.286 + 0.195^{*}S^{2}/R + 0.385^{*}\mathrm{BM} + 5.086^{*}\mathrm{Sex}\ \ (\mathrm{SEE} = 2.42\ \mathrm{kg})$$

$$\mathrm{ECW} = 1.579 + 0.055^{*}S^{2}/R + 0.127^{*}\mathrm{BM} + 0.006^{*}S^{2}/x_{\mathrm{c}} + 0.932^{*}\mathrm{Sex}\quad (\mathrm{SEE} = 1.33\ \mathrm{kg})$$

where sex is 0 if female or 1 if male, BM is body mass in kg, S is stature in cm, R is resistance and x_c is reactance in ohm (Ω).

Assessment of directly measurable parameters (resistance, reactance, PhA) and the proportion of TBW as well as changes in the inter and extracellular components may in some cases be a better choice. O'Brien et al. (2002) reported that the multifrequency BIA method is not sufficiently accurate to assess TBW under conditions of hydration change, which is, however, normal after high-intensity sport performance. This is because it is difficult to assess alterations in electrolytes and the ratio of ICW and ECW, which subsequently affects the conversion of estimated parameters for identifying BC, which may introduce error. Similarly, conversion of LBM and its constituents depends on a consistent level of hydration (lean body mass = total bodily water/0.73) (Sillanpaa et al. 2014). Resistance and reactance, together with PhA, appear to be suitable parameters for monitoring short-term changes after radical loss of bodily water and could be an eventual replacement for indirectly calculated parameters dependent on the abovementioned conditions.

12.3.2 The Use of Indirectly Measurable Parameters for Assessment of Performance

LBM is a variable component of BC. In athletes, the degree of alteration in BC and the amount of LBM depend on the mode of exercise, as well as the frequency, intensity and duration of training. LBM, including all body tissues except for depot fat, is considered a major precondition for good performance. Higher values of LBM provide a better predisposition for sports performance in various disciplines that require explosive activity. Individual differences among players in LBM and LBM proportion have been reported to differ between players in different playing positions (Mala et al. 2014). LBM is an indicator of a high level of aerobic capacity, and better aerobic fitness contributes to a faster recovery between matches, which can positively influence combat performance. In combat sports, in which strength is important, LBM is a predictor of muscle efficiency and overall performance.

LBM contributes to the production of power during high-intensity activities and provides greater absolute strength for resistance to high dynamic and static loads. LBM strongly contributes to strength and power performance (Milanese et al. 2015) and is considered a major precondition for a good performance in sporting games as well as for optimal individual performance. Sedano et al. (2009) found a significant correlation between LBM and kick speed in elite female soccer players and demonstrated a greater dependence between these variables in elite players than in nonelite players. Mala et al. (2014) observed significant differences between sports. Female soccer players showed significantly lower LBM values than volleyball, basketball and handball players. A significant difference ($p < 0.01$) was also found between softball and basketball players. In a study of female pentathletes ($n = 65$), Claessens et al. (1994) compared female pentathletes with female fencers and swimmers and found that the pentathletes showed greater height and BMI than female fencers and swimmers, which in the case of a lower proportion of FM indicates a higher proportion of lean body mass. In pentathlon, LBM is important for the active components of the individual component disciplines: for explosive strength and coordination (fencing), static strength (horse riding), correct muscle synchronisation for maintaining balance (shooting) and muscle work in different modes (complex preparedness in terms of energy systems).

Table 12.2 presents the body composition variables in elite athletes of various sports. All values were recorded using a multifrequency BIA method and were calculated using prediction equations presented in the literature or in the software of the specific method used.

In multifrequency BIA measurement, we also assess other parameters that would characterise the quality of BC more precisely, in terms of monitoring changes at the cellular level. BCM as a component of LBM is defined as a predictor of muscle efficiency for sport performance (Andreoli et al. 2003). BIA measurements of healthy athletes BCM are always higher than the ECM. In intraindividual assessment, it is appropriate to recalculate the BCM to BM; it is also frequently expressed directly by bioimpedance software as the percentage of body cells,

Table 12.2 Body composition parameters in elite athletes (values are expressed as mean ± st. deviation and range (minimal-maximal value))

Sport	Characteristic	LBM (kg)	FM (kg)	FM (%)	BCM (kg)	ECM (kg)	ECM/BCM	BCMrel (%)	PhA (°)	Method/equation
Volleyball	F, n = 18, age = 27.4 ± 4.1 years, BH = 184 3 ± 4.2 cm, BM = 72.1 ± 6.5 kg. Participant of the European Cup Championship from Czech Republic	57.81 ± 4.53 (48.57–65.97)	14.33 ± 2.32 (10.43–18.28)	19.77 ± 1.77 (17.13 ± 22.56)	30.53 ± 2.77 (25.4–37)	23.96–1.87 (21.3–27)	0.79 ± 0.07 (0.7–0.92)	42.39–2.19 (38.35–46.14)	6.91 ± 0.48 (6.0–7.6)	BIA 2000M/Fornetti et al. (1999)
Softball	F, n = 14, age = 23.6 ± 4.9 years, BH = 169.9 ± 7.1 cm, BM = 67.9 ± 9.9 kg. Czech Republic national team	53.35 ± 5.74 (45.24–62.33)	14.76 ± 4.52 (9.98–26.5)	21.35 ± 3.67 (17.51–30.8)	30.94 ± 3.69 (25.2–37)	22.47–2.41 (19.6–26.8)	0.73 ± 0.05 (0.62–0.80)	45.82 ± 3.66 (38.37–51.9)	7.36 ± 0.46 (6.8–8.4)	BIA 2000M/Fornetti et al. (1999)
Basketball	F, n = 14, age = 25.9 ± 4.2 years, BH = 185.3 ± 9 cm, BM = 76.6 ± 7.8 kg. Second place in world championship competition Czech Republic national team	60.3 ± 5.42 (50.12–69.5)	16.34 ± 2.72 (12.36–22.52)	21.22 ± 1.66 (18.72–25.6)	30.61 ± 2.71 (25.2–34.9)	24.54 ± 2.03 (21.8–27.2)	0.8 ± 0.06 (0.7–0.9)	40.2 ± 4.09 (28.33–45.22)	6.81 ± 0.43 (6.1–7.6)	BIA 2000M/Fornetti et al. (1999)
Soccer	F, n = 18, age = 23.3 ± 4.2 years, BH = 167.3 ± 6.8 cm, BM = 61.3 ± 5.5 kg. Czech Republic national team	49.4 ± 4.31 (41.99–55.92)	12.01 ± 2.1 (7.65–16.21)	19.53 ± 2.59 (14.66–24.3)	27.39 ± 2.16 (23.2–30.9)	21.07 ± 1.68 (8.1–24.6)	0.77 ± 0.06 (0.68–0.9)	44.82 ± 3.28 (40.03–53.26)	7.02 ± 0.45 (6.1–7.8)	BIA 2000M/Fornetti et al. (1999)
Handball	F, n = 16, age = 24 ± 3.5 years, BH = 176 ± 6.5 cm, BM = 69.8 ± 9.1 kg. Czech Republic national team	56.95 ± 5.34 (46.84–64.95]	15.59 ± 3.27 (9.88–22.65)	21.43 ± 2.48 (16.44–25.86)	30.59 ± 3.33 (24.3–35.4)	23.89 ± 2.52 (20–27.7)	0.79 ± 0.1 (0.63–0.97)	42.32 ± 3.22 (37.41–47.21)	6.96 ± 0.76 (5.8–8.3)	BIA 2000M/Fornetti et al. (1999)
Modern pen-tathlon	M, n = 10, age = 16.8 ± 1.03 years, BH = 180.63 ± 6.49 cm, BM = 67.3[illegible] ± 5.93 kg. Slovak Republic junior national team	60.75 ± 5.53		8.98 ± 1.22 (7.08–10.9)	35.19 ± 4.58	24.58 ± 2.54	0.7 ± 0.07	57.8 ± 3.41	7.52 ± 0.8	BIA 2000M/software equation
Fencers	M, age = 1[illegible].88 ± 2.17 years, BH = 182.48 ± 7.4 cm, BM = 72.4[illegible] ± 4.33 kg. Slovak Republic junior national team	62.01 ± 4.34		10.78 ± 2.93 (7.31–13.90)	33.9 ± 4.19	28.11 ± 1.73	0.84 ± 0.13	54.52 ± 3.66	6.61 ± 0.74	Inbody 3.0/software equation
Judo	M, n = 10, age = 18.1 ± 1.1 years, BH = 182.2 ± 7.9 cm, BM = 78.6 ± 13.3 kg. Czech Republic junior national team	71.61 ± 9.61		11.34 ± 3.64 (6.01–16.21)	41.01 ± 5.71	30.61 ± 4.51	0.75 ± 0.08	–	7.24 ± 0.7	Inbody 3.0/software equation

F female, *M* male, *BH* body height, *BM* body mass, *LBM* lean body mass, *FM* fat mass, *BCM* body cell mass, *ECM* extracellular mass, *ECM/BCM* extracellular-intracellular mass ratio, *BCMrel* relative body cell mass, *PhA* phase angle, *BIA* bioelectrical impedance analysis

defined as an indicator of athletes' physical fitness and nutritional state. It is usually approximately 60% of the LBM in men and 50% of the LBM in women. The ECM/BCM ratio is a parameter that is predominantly genetically determined, and it indicates the predisposition for implementation of physical activities. This parameter usually does not vary with changes in BC, i.e. it changes only minimally. A negative correlation between ECM/BCM and FM (Mala et al. 2014) indicates the predisposition of an individual to sports. In adult athletes, it ranges from 0.5 to 0.8. The value of ECM/BCM in well-trained athletes is lower and is associated with a high value of PhA (6–10°). The highest value we recorded (10°) was in a judo athlete, whose ECM/BCM was 0.4; together, these values indicate the excellent musculature and BC of the judo athlete, who was a participant in the Olympic Games (Mala et al. 2014).

Several studies have attempted to identify values of LBM that are optimum for specific field positions in various sports (volleyball, handball and soccer). Football players who play different positions clearly have different physical and physiological profiles (Sporis et al. 2009). Researchers reported the absolute values for LBM in elite soccer players, and they found that those of goalkeepers were significantly higher than those of outfield players (Milanese et al. 2015).

Mala et al. (2014) reported that relative values indicate individual differences between elite female handball players; however, it was not possible to identify a clear trend because the highest approximately comparable values for relative LBM (0.87) were recorded in different playing positions (wing, goalkeeper and backcourt players). Mala et al. (2014) reported individual differences of relative values among players in a national female basketball team. However, no clear trend can be identified in LBM proportion in individual playing positions. The highest relative values of LBM (0.81) were recorded in a point guard and a pivot, and the lowest value was detected in a wing player (0.74).

12.4 Fat Mass

Fat mass is a major factor in inter and intraindividual variability of BC, and it can be estimated using indirect calculation in BIA. The proportion of FM is lower in athletes than in inactive individuals. A lower fat mass proportion is desirable for most sporting disciplines. There is an indirect relationship between FM and performance (Claessens et al. 1994). Excess adipose tissue acts as dead weight in activities during which the body mass must be repeatedly lifted against gravity during locomotion and jumping (Reilly 1996); this in turn decreases performance and increases the energy demands of the activity. In sports, players who have more FM than is appropriate will become tired much faster during a game than players with an optimal amount of FM; moreover, they will run slower during the second period of the game (Sporis et al. 2010). The authors also confirmed a strong negative correlation between FM and maximal running speed ($r = -0.68$, $p < 0.01$) in elite

Croatian handball players ($n = 167$, age $= 26.4 \pm 3.8$ years, BH $= 192.1 \pm 8.2$ cm, BM $= 96.0 \pm 8.3$ kg).

To reduce FM, continuous control of BC and regular diagnostics during training periodisation are important. For such continuous monitoring of FM proportion in athletes, a multifrequency method is appropriate, rapid and comfortable. If other conditions for BIA measurement are kept constant and other factors are eliminated (type of bioimpedance analyser, equation for calculation), it provides a relatively accurate recording of changes in FM values in athletes. FM reduction can be accomplished by omitting fat in nutrition and controlling energy intake and consumption (Astrup et al. 2000). Table 12.2 presents the absolute and relative values of FM proportion in various sports.

In a study of soccer field positions in three elite youth soccer teams in the U19 category (participants in the highest national league, $n = 73$, age $= 18.7 \pm 0.3$ years), Mala et al. (2014) reported that midfielders are the shortest players, with the smallest amount of FM.

Lago-Penas et al. (2011) showed differences in the percentage of FM between goalkeepers and midfielders and attackers. In comparison to attackers and defenders, midfielders cover far greater distances (Sporis et al. 2009). No significant difference in FM proportion was found, although other reports have stated that defenders have significantly more FM ($p < 0.05$) than attackers or midfielders (Sporis et al. 2009). Wittich et al. (2001) reported that midfielders show a significantly higher proportion of FM than backs or forwards ($p < 0.05$). In a study of elite female athletes (handball, volleyball, soccer, softball and basketball), Mala et al. (2014) stated that it is not possible to clearly define a trend in FM proportion in individual playing positions.

12.5 Measurement of Morphological Asymmetries Using BIA

Maladaptive morphological asymmetries may influence some components of physical fitness in youths and adults, resulting in unilateral uncompensated physical loading. A high level of asymmetry has been associated with a higher risk of injury in youth and adult athletes (Bak and Magnusson 1997). Asymmetries and shortened muscles may have severe physical consequences, increasing the risk of injury and negative affected sport performance (Maly et al. 2016).

For instance, joint mobility is important for proper execution of specific techniques (kicks, punches) over the whole range, and joint mobility is influenced by muscle shortening or strength asymmetries between agonist and antagonist. In many sporting disciplines, athletes show a clear preference for the left or right limb since childhood (tennis, ice hockey, soccer and other sports), which may cause morphological and functional asymmetries during regular and uncompensated physical loading. Calbet et al. (1998) reported that professional tennis players show arm asymmetry, with approximately 20% greater bone mineral content and muscle mass in the dominant arm.

Multifrequency BIA can be used to reveal morphological asymmetries in the estimated MM and fluid distribution, FM and PhA in individual body segments of athletes. Studies have reported that players in different positions show various asymmetries. Mala et al. (2016) tested three youth elite soccer teams in the U19 category (participants of the highest national league, $n = 73$, age $= 18.7 \pm 0.3$ years) and compared regional MM in all monitored segments; they found that goalkeepers showed a significantly greater proportion of asymmetry than players in other positions. Silvestre et al. (2006) reported that fullbacks and goalkeepers showed higher values of leg lean tissue than players in other positions. Milsom et al. (2014) reported a significantly greater proportion of LBM in all monitored segments (left arm, right arm, trunk, left leg and right leg) ($p < 0.05$)) in the leading team in the English Premier Soccer League in comparison to players at the U18 level ($n = 75$). A significant difference was also reported between the dominant lower limbs of U18 and U21 players ($p < 0.05$). Significant differences in MM proportion between the dominant and non-dominant legs in all field positions indicate lower limb preference in activities such as passing and shooting during the training process (Mala et al. 2016).

In basketball and handball, one of the upper limbs is usually clearly preferred for ball manipulation. In a study of the national female basketball team ($n = 14$, age $= 25.9 \pm 4.2$ years, BH $= 198.0 \pm 9.0$ cm, BM $= 76.6 \pm 7.8$ kg), Mala et al. (2016) reported a significant difference in the fluid volume of the upper limbs ($p < 0.01$) in favour of the preferred arm. Mala et al. (2014) also observed a significant difference in fluid distribution in the arms ($p < 0.01$) of female national level handball players ($n = 16$, age $= 24.0 \pm 3.5$ years, BH $= 176.0 \pm 6.5$ cm, BM $= 72.5 \pm 8.3$ kg); however, players' legs showed equal proportions of fluid volume ($p > 0.05$).

Conversely, in a study of ten active judo athletes, members of the top Czech team, included in the wider national men's team in 2008 (age $= 21.2 \pm 1.2$ years, BH $= 186.0 \pm 1.7$ cm, BM $= 87.2 \pm 4.0$ kg), Mala et al. (2014) reported no lateral differences in the upper and lower parts of the body ($p > 0.05$). Fluid volume in limbs was symmetric, despite the limb dominance of individual athletes. We can conclude that judo promotes symmetrical strength development likely owing to implementation of bilateral techniques.

Maladaptive processes caused by asymmetries can mean a high risk of injury for adolescent athletes later in their sporting careers. The multifrequency BIA method appears to be a suitable means for recording and continuous monitoring of morphological asymmetries in athletes.

Various studies have demonstrated differences between anthropometric indicators and BC parameters in athletes, based on their field positions. Team position is of extreme importance in the interpretation of morphological data because specific field positions present specific demands. No athlete has optimal body morphology and BC for all field positions. Maladaptive effects stemming from different segmental MM proportion between the preferred and non-preferred limbs may increase a player's risk of injury. Asymmetries should be systematically monitored and compensated for, using specific exercises. The results of bioimpedance recording may be beneficial for fitness coaches, physiotherapists, nutrition specialists, doctors and other clinical staff of professional sporting teams.

The use of BIA in sport is a noninvasive, rapid, easily available and comfortable way of continuous monitoring of BC, with the aim of determining the effects of short-term interventions on fitness training or recovery after injury. However, the conditions for bioimpedance measurement must be standardised, and the individual characteristics of each athlete (density, hydration, correct interpretation of data and use of appropriate parameters in specific situations) must be taken into account.

Acknowledgement Project was supported by GACR16-21791S.

References

Andreoli, A., Melchiorri, G., Brozzi, M., Di Marco, A., Volpe, S. L., Garofano, P., et al. (2003). Effect of different sports on body cell mass in highly trained athletes. *Acta Diabetologica, 40*, 122–125.

Astrup, A., Grunwald, G. K., Melanson, E. L., Saris, W. H. M., & Hill, J. O. (2000). The role of low-fat diets in body weight control: A meta-analysis of ad libitum dietary intervention studies. *International Journal of Obesity, 24*(12), 1545–1552.

Bak, K., & Magnusson, S. P. (1997). Shoulder strength and range of motion in symptomatic and pain-free elite swimmers. *American Journal of Sports Medicine, 25*(4), 454–459.

Calbet, J. A. L., Moysi, J. S., Dorado, C., & Rodriguez, L. P. (1998). Bone mineral content and density in professional tennis players. *Calcified Tissue International, 62*(6), 491–496.

Calvo Rico, B., Garcia Garcia, J. M., Monteiro, L. F., & Rioja Collado, N. (2015). Kinematic indicators in combat sports athletes in a pre-competitive dehydrated status. *Archives of Budo Science of Martial Arts and Extreme Sports, 11*, 181–188.

Cerit, S., Akdag, M. Z., Dasdag, S., & Celik, M. (2009). Alteration in body composition of elite professional female players in a Premier League volleyball bout. *Journal of International Dental and Medical Research, 2*(1), 33–36.

Claessens, A. L., Hlatky, S., Lefevre, J., & Holdhaus, H. (1994). The role of anthropometric characteristics in modern pentathlon performance in female athletes. *Journal of Sports Sciences, 12*(4), 391–401.

Coufalova, K., Cochrane, D. J., Maly, T., & Heller, J. (2014). Changes in body composition, anthropometric indicators and maximal strength due to weight reduction in judo. *Archives of Budo, 10*(1), 161–168.

Esco, M. R., Snarr, R. L., Leatherwood, M. D., Chamberlain, N. A., Redding, M. L., Flatt, A. A., et al. (2015). Comparison of total and segmental body composition using DXA and multifrequency bioimpedance in collegiate female athletes. *Journal of Strength and Conditioning Research, 29*(4), 918–925.

Fornetti, W. C., Pivarnik, J. M., Foley, J. M., & Fiechtner, J. J. (1999). Reliability and validity of body composition measures in female athletes. *Journal of Applied Physiology, 87*(3), 1114–1122.

Haussinger, D., Roth, E., Lang, F., & Gerok, W. (1993). Cellular hydration state: An important determinant of protein catabolism in health and disease. *Lancet, 341*(8856), 1330–1332.

Heymsfield, S. B., Lohmen, T. G., Wang, Z., & Going, S. B. (2005). *Human body composition* (536 p). Champaign: Human Kinetics.

Hortobagyi, T., Israel, R. G., Houmard, J. A., O'brien, K. F., Johns, R. A., & Wells, J. M. (1992). Comparison of four methods to assess body composition in black and white athletes. *International Journal of Sport Nutrition, 2*(1), 60–74.

Kyle, U. G., Bosaeus, I., De Lorenzo, A. D., Deurenberg, P., Elia, M., Gomez, J. M., et al. (2004). Bioelectrical impedance analysis principles and methods. *Clinical Nutrition, 23*(5), 1226–1243.

Lago-Penas, C., Lago-Ballesteros, J., & Rey, E. (2011). Differences in performance indicators between winning and losing teams in the UEFA Champions League. *Journal of Human Kinetics, 27*, 137–148.

Mala, L., Maly, T., Zahalka, F., & Bunc, V. (2014). *Fitness assessment: Body composition*. Prague: Karolinum Press.

Mala, L., Maly, T., Zahalka, F., & Cabell, L. (2016). Field position in soccer influencing maladaptive effect in morphological and body composition variables. *Medicine and Science in Sports and Exercise, 48*(Suppl 1), 990

Mala, L., Maly, T., Zahalka, F., Cabell, L., & Bonacin, D. (2016). A comparison of methods for evaluating body composition in elite female soccer players. *Anthropologist, 24*(2), 642–651.

Maly, T., Zahalka, F., & Mala, L. (2016). Unilateral and ipsilateral strength asymmetries in elite youth soccer players with respect to muscle group and limb dominance. *International Journal of Morphology, 34*(4), 1339–1344.

Matias, C. N., Santos, D. A., Júdice, P. B., Magalhães, J. P., Minderico, C. S., Fields, D. A., et al. (2016). Estimation of total bod water and extracellular water with bioimpedance in athletes: A need for athlete – Specific prediction models. *Clinical Nutrition, 35*, 468–474.

Milanese, C., Cavedon, V., Corradini, G., De Vita, F., & Zancanaro, C. (2015). Seasonal DXA-measured body composition changes in professional male soccer players. *Journal of Sports Sciences, 33*(12), 1219–1228.

Milsom, J., Barreira, P., Burgess, D. J., Iqbal, Z., & Morton, J. P. (2014). Case study: Muscle atrophy and hypertrophy in a Premier league soccer player during rehabilitation from ACL injury. *International Journal of Sport Nutrition and Exercise Metabolism, 24*(5), 543–552.

Moon, J. R. (2013). Body composition in athletes and sport nutrition: An examination of the bioimpedance analysis technique. *European Journal of Clinical Nutrition, 67*, 54–59.

O'Brien, C., Young, A. J., & Sawka, M. N. (2002). Bioelectrical impedance to estimate changes in hydration status. *International Journal of Sports Medicine, 23*(5), 361–366.

Pichard, C., Kyle, U. G., Janssens, J. P., Burdet, L., Rochat, T., Slosman, D. O., et al. (1997). Body composition by X-ray absorptiometry and bioelectrical impedance in chronic respiratory insufficiency patients. *Nutrition, 13*(11–12), 952–958.

Reilly, T. (1996). Fitness assessment. In T. Reilly & A. M. Williams (Eds.), *Science and soccer* (pp. 25–50). London: E. & F. Spon.

Reljic, D., Hassler, E., Jost, J., & Friedmann-Bette, B. (2013). Rapid weight loss and the body fluid balance and hemoglobin mass of elite amateur boxers. *Journal of Athletic Training, 48*(1), 109–117.

Sawka, M. N., Burke, L. M., Eichner, E. R., Maughan, R. J., Montain, S. J., & Stachenfeld, N. S. (2007). Exercise and fluid replacement. *Medicine and Science in Sports and Exercise, 39*(2), 377–390.

Sedano, S., Vaeyens, R., Philippaerts, R. M., Redondo, J. C., de Benito, A. M., & Cuadrado, G. (2009). Effects of lower-limb plyometric training on body composition, explosive strength, and kicking speed in female soccer players. *Journal of Strength and Conditioning Research, 23*(6), 1714–1722.

Selberg, O., & Selberg, D. (2002). Norms and correlates of bioimpedance phase angle in healthy human subjects, hospitalized patients, and patients with liver cirrhosis. *European Journal of Applied Physiology, 86*(6), 509–516.

Shakeryan, S., Nikbakht, M., & Kashkoli, H. B. (2013). Validation of percent body fat using skinfold-thickness, bioelectrical impedance analysis and standard hydrostatic method in male wrestler. *Journal of Public Health and Epidemiology, 5*(1), 15–19.

Sillanpaa, E., Cheng, S. L., Hakkinen, K., Finni, T., Walker, S., Pesola, A., et al. (2014). Body composition in 18-to 88-year-old adultscomparison of multifrequency bioimpedance and dual-energy X-ray absorptiometry. *Obesity, 22*(1), 101–109.

Silva, A. M., Fields, D. A., Heymsfield, S. B., & Sardinha, L. B. (2010). Body composition and power changes in elite judo athletes. *International Journal of Sports Medicine, 31*(10), 737–741.

Silvestre, R., West, C., Maresh, C. M., & Kraemer, W. J. (2006). Body composition and physical performance in men's soccer: A study of a National Collegiate Athletic Association Division I team. *Journal of Strength and Conditioning Research, 20*(1), 177–183.

Sporis, G., Jukic, I., Ostojic, S. M., & Milanovic, D. (2009). Fitness profiling in soccer: Physical and physiologic characteristic of elite players. *Journal of Strength and Conditioning Research, 23*(7), 1947–1953.

Sporis, G., Vuleta, D., Vuleta, D. J., & Milanovic, D. (2010). Fitness profiling in handball: Physical and physiological characteristics of elite players. *Collegium Antropologicum, 34*(3), 1009–1014.

Swartz, A. M., Evans, M. J., King, G. A., & Thompson, D. L. (2002). Evaluation of a foot-to-foot bioelectrical impedance analyser in highly active, moderately active and less active young men. *British Journal of Nutrition, 88*(2), 205–210.

Warner, E. R., Fornetti, W. C., Jallo, J. J., & Pivarnik, J. M. (2004). A skinfold model to predict fat-free mass in female athletes. *Journal of Athletic Training, 39*(3), 259–262.

Williams, C. A., & Bale, P. (1998). Bias and limits of agreement between hydrodensitometry, bioelectrical impedance and skinfold calipers measures of percentage body fat. *European Journal of Applied Physiology and Occupational Physiology, 77*(3), 271–277.

Wittich, A., Oliveri, M. B., Rotemberg, E., & Mautalen, C. (2001). Body composition of professional football (Soccer) players determined by dual X-ray absorptiometry. *Journal of Clinical Densitometry, 4*(1), 51–55.

Yannakoulia, M., Keramopoulos, A., Tsakalakos, N., & Matalas, A. L. (2000). Body composition in dancers: The bioelectrical impedance method. *Medicine and Science in Sports and Exercise, 32*(1), 228–234.

Chapter 13
Wavelet Analysis in Impedance Rheocardiography

Rodion Stepanov, Andrey Dumler, Sergey Podtaev, and Peter Frick

13.1 Impedance Rheocardiography

Cardiovascular diseases, primarily chronic cardiac insufficiency, coronary disease (myocardial infarction, stenocardia), and arterial hypertension, are a major public health issue facing developed countries, which, according to predictions, will remain a challenge for the nearest 50 years. Nowadays, in the general structure of mortality, they comprise almost 57% of all events. All these diseases initiate the process of remodeling of the cardiovascular system. The term "remodeling" denotes the whole complex of structural-functional multilevel phenomena associated with the notion of "hypertonic heart," target organ damage, morphological and functional vascular changes manifesting themselves in vasospastic response, and thickening and higher rigidity of vascular walls (Devereux et al. 2004). The cardiovascular disease continuum suggests in turn a continuous chain of interactions of its components, of which there are risk factors associated with changes in the cardiovascular system and the overlay of the atherosclerosis process being initially asymptomatic and then causing nonfatal and fatal complications (Kirchhof et al. 2014). Timely detection of the remodeling process may have a strong effect on the quality of health aid and reduce the death rates from cardiovascular diseases.

Modern medicine makes use of many methods to diagnose cardiovascular diseases, but some techniques have turned out to be invasive and dangerous for

R. Stepanov (✉) · P. Frick
Institute of Continuous Media Mechanics, Perm, Russia
e-mail: rodion@icmm.ru

A. Dumler
E.A. Vagner Perm State Medical University, Perm, Russia

S. Podtaev
Science Technology Center "FM Diagnostics", Perm, Russia

F. Simini, P. Bertemes-Filho (eds.), *Bioimpedance in Biomedical Applications and Research*, https://doi.org/10.1007/978-3-319-74388-2_13

patients, and others—highly complicated and extremely expensive, which makes them inaccessible to most patients. Before the "ultrasonic era," the impedance rheocardiography was the key noninvasive method used in cardiology to study central hemodynamics, but with an advent of ultrasonic techniques, the interest in rheography reduced significantly. A fresh approach to the noninvasive study of hemodynamics with rheography techniques includes applying the advanced mathematical methods and information technologies to the analysis of rheographic signals. The information inherent in the shape of the rheocardiogram curves is considerably valuable, but it necessitates adequate assessment (e.g., (Ebrahim et al. 2016)). In addition, impedance rheocardiography techniques yield relatively accurate, easily reproducible results and are less expensive than ultrasonic techniques.

Impedance cardiography (ICG) can be used to determine one of the key hemodynamic parameters, the stroke volume (SV), and to assess the cardiac output (CO), systemic peripheral vascular resistance, the time interval of left ventricular ejection, and other parameters (Chen et al. 2014). The principle underlying the rheography method is based on registration of changes in the electrical resistance of biological tissues as the high-frequency alternating current travels through them. It is worth noting that the method can be used not only in clinical conditions but also in cardiovascular parameter monitoring, including functional diagnostics (Morris et al. 2014; Mallam and Bhushana Rao 2016). Unlike the standard electrocardiography (ECG), different structures of the ICG signal can be readily recognized and, what is more, the procedure itself takes much less time and requires less experienced personnel compared to echocardiography (Bour and Kellett 2008; Van Eijnatten et al. 2014).

13.2 Measuring Technique and Data Analysis Algorithms

The ICG signal includes a set of characteristic points and waves (Cybulski 2011), denoted in Fig. 13.1 as B, X points and E, O waves. These features are associated with different physiological events in the cardiac cycle.

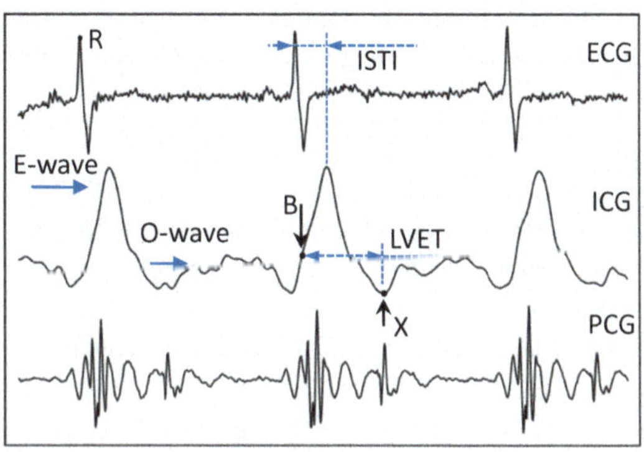

Fig. 13.1 Typical signals of polyrheocardiography: the electrocardiogram (ECG), the differential impedance cardiogram (ICG), and the phonocardiogram (PCG)

Point B is synchronized with the first heart sound that occurs on closing the atrioventricular valves at the very beginning of the ventricular systole. Identification of the localization of point B is sometimes questionable because the ICG signal shape in this area is not always pronounced. Point E shows a maximum on the differential rheogram. It depicts the maximum rate of impedance change, which in turn is related to the maximum ejection rate measured using ultrasonic methods (Cybulski 2011).

Point X corresponds to the closure of semilunar valves at the end of the ventricular systole (the second heart sound). For ICG, the first maximum (E-wave) is related to the systolic cardiac cycle phase, and the second broader maximum with a smaller amplitude (O-wave) is related to the diastolic phase. The E-wave amplitude is proportional to the stroke volume, and the O-wave amplitude correlates with the change in the left atrium volume during the diastolic phase. In some cases, the O-wave amplitude is an important diagnostic parameter (Pickett and Buell 1993). According to Kubicek's formula (Kubicek et al. 1966), the stroke volume is proportional to the time interval of ejection and the maximum value of the first impedance derivative for this cycle. The left ventricular ejection time (LVET) is defined as a time interval between the opening of an aortic valve and its subsequent closure. In some works, e.g., Ermishkin et al. (2007), it has been shown that ICG cannot ensure the identification of LVET with necessary accuracy. Nevertheless, impedance cardiography techniques, in particular the tetrapolar thoracic ICG method, were still rather attractive to specialists due to their simplicity, low cost, and probability of obtaining sufficiently accurate and reproducible express-results (Morris et al. 2016; Bernstein et al. 2015; Payseur et al. 2016). The purpose of the present work is to demonstrate the potentialities of wavelet analysis of rheocardiogram for identification of the myocardial remodeling of patients with cardiovascular diseases.

As a measuring instrument, we used a multifunctional device `iCor` (produced by STC "FM Diagnostics," Perm) that allows one to register rheograms using a tetrapolar method, phonocardiograms (PCG) and electrocardiograms (ECG) (Fig. 13.1). We have developed a program module having user's interface elements and implemented a code for Mathematica package. The code includes data uploading, filtration of artifacts and noises, wavelet transform, definition of hemodynamic parameters, graphical representation of wavelet spectrogram of cardiocycles, and data export using the standard table protocol and the form of diagrams.

In some cases, random errors in estimation of the sample mean can be reduced by increasing the size of the sample, but the contribution of the noise and artifacts persists in statistical characteristics such as variance or correlation. However, the variability of the hemodynamic parameters (e.g., the respiratory variation in SV) opens considerable scope for diagnosis of pathologies (Holme et al. 2016). Studying the variability, one should distinguish between the regular variability, which actually affects each cycle (due to the respiration or some functional tests), and the variability, which occurs during extraordinary and rare cycles that differ significantly from most cardiac cycles.

Selection of cardiocycles appropriate for analysis proceeds in two steps: preliminary assessment of the noise level and further selection of cardiocycles using the similarity criterion. At the first step, cardiocycles are rejected if signal exceeds a level of three times of the standard deviation from the sliding average. At the next step, we keep for further analysis only cardiocycles for which the cross-correlation coefficient is not less than 0.75. This approach makes it possible to significantly reduce errors in hemodynamic parameters caused by the occurrence of artifacts and noises during the measurement procedure (Dumler et al. 2010).

13.3 Wavelets

Advanced signal processing techniques provide more elaborated approaches to the analysis and interpretation. In our study, we carry on developing the use of continuous wavelet transform. We believe that its higher computational complexity as compared with the Fourier transformation is compensated by convenient assessment of amplitude, scale, and phase of signal oscillations in complicated nonstationary signals. Wavelet transform allows us to isolate a given structure in time and frequency space.

We define the wavelet transform of the analyzed function $F(t)$ as

$$w_F(a,b) = \frac{1}{|a|} \int_{-\infty}^{\infty} F(t')\psi^* \left(\frac{t'-b}{a} \right) dt', \tag{13.1}$$

where $\psi(t)$ is the analyzing wavelet, a defines the scale (inverse to the frequency), and b defines the location of the wavelet in time. Then the transform w_F shows the contribution of corresponding structure into the function F.

The function F can be reconstructed using the inverse transform:

$$F(t) = \frac{1}{C_\psi} \int_{-\infty}^{\infty} \int_{-\infty}^{\infty} \psi \left(\frac{t'-t}{a} \right) w_F \left(a, t'\right) \frac{da\, dt'}{a^2}. \tag{13.2}$$

The reconstruction formula (13.2) exists under condition that

$$C_\psi = \frac{1}{2} \int_{-\infty}^{\infty} \frac{|\hat{\psi}(\omega)|^2}{|\omega|} d\omega < \infty. \tag{13.3}$$

Here $\hat{\psi}(\omega) = \int \psi(t) e^{-\imath \omega t} dt$ is the Fourier transform of the analyzing wavelet $\psi(t)$. The recontraction can be used in denoising techniques for suppressing the respiratory and motion artifacts in ICG (e.g., Chabchoub et al. 2016). Estimation of root-mean-square characteristic of oscillations in specific frequency bands does not require an inverse transform (Podtaev et al. 2015).

If the derivative $G(t)$ of the measured function $F(t)$ is required for further analysis, the explicit differentiation of experimental signal can be avoided as follows. The wavelet transform of the function $G(t)$ is

$$w_G(a,b) = \frac{1}{|a|} \int_{-\infty}^{\infty} G(t')\psi^* \left(\frac{t'-b}{a}\right) dt'. \tag{13.4}$$

Applying the differentiation by part to Eq. (13.4), one gets

$$w_G(a,b) = \frac{1}{|a|} \int_{-\infty}^{\infty} F(t')\xi^* \left(\frac{t'-b}{a}\right) dt', \tag{13.5}$$

where $\xi(t) = -\psi'(t)$ is the differentiating wavelet. Thus the wavelet transform of the derivative function $w_G(a,b)$ is directly obtained by the wavelet transform of original signal using as the analyzing wavelet the function $\xi(t)$. This algorithm of wavelet differentiation allows us to combine the filtering and differentiating procedures under the rheocardiogram processing. We use the so-called Mexican hat wavelet $\psi(t) = (1-t^2)\exp(-t^2/2)$ for higher resolution to separate characteristic points in time (see Fig. 13.1). The Morlet wavelet $\psi(t) = \exp(-\frac{t^2}{2} + 2i\pi t)$ was used if better spectral resolution was required.

We illustrate the potentialities of wavelets using the example shown in Fig. 13.2, where wavelet spectrograms of the same signal obtained with different analyzing wavelets are shown. The signal is a superposition of low-frequency harmonics and quasiperiodic bursts with weakly varying characteristics (amplitude and periodicity)—a schematic analog of the cardiac rhythm against the respiratory wave. Analysis of the wavelet spectrogram provided clears up the structure of the signal and allows one to trace the variations of both components. The Mexican hat is able, in this case, to identify better each burst, making it possible to distinguish between inspiration and expiration, whereas the Morlet wavelet (which is a longer wave) is able to display better the periodic component. It should be noted that wavelets have found a wide application in the analysis of medical signals, but their use is mainly confined by filtration carried out using the discrete wavelet transform. The potentials of the wavelet analysis are appreciably greater, and we use the continuous wavelet transform technique in the present study not only to filter rheocardiograph signals but also to regularize the signal processing procedures, to investigate the correlation level of signals within a given frequency band, and to determine quantitative functional characteristics of the cardiac cycle such as the stroke volume.

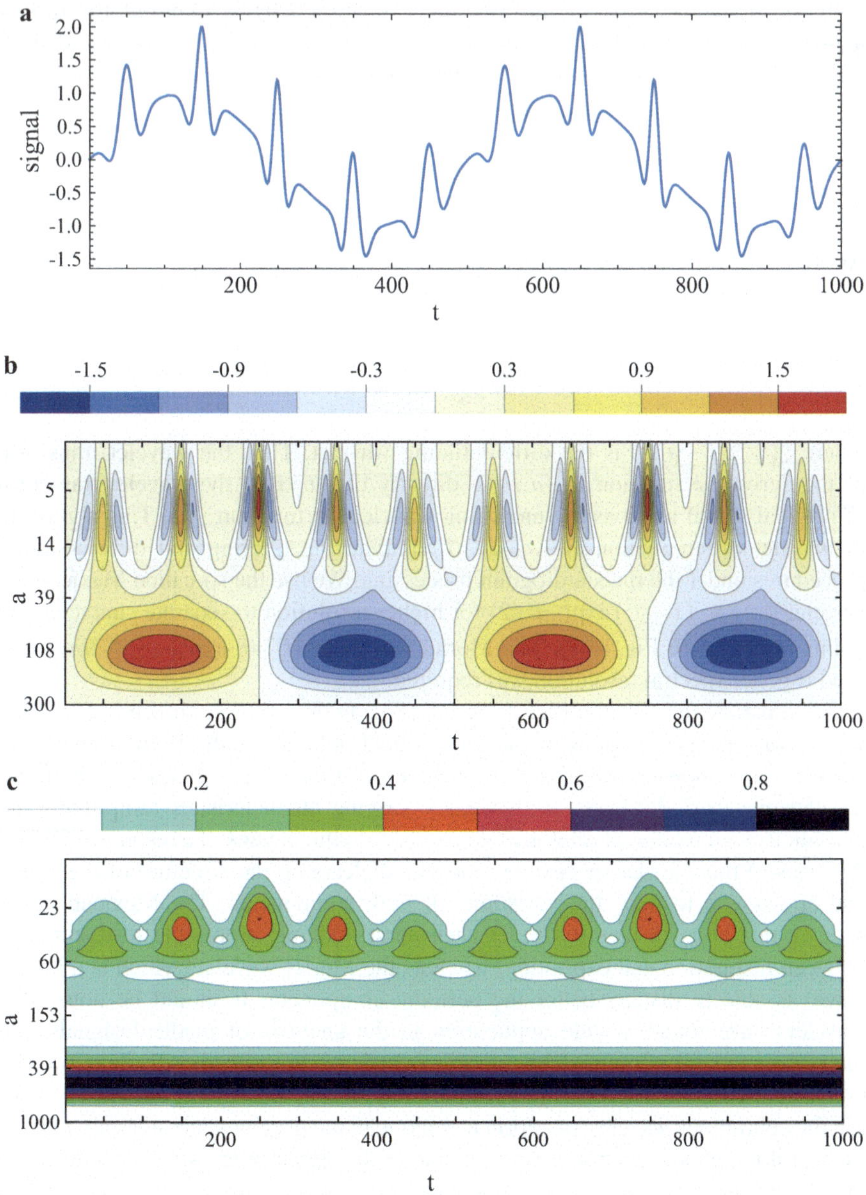

Fig. 13.2 An artificial signal (top) and its wavelet spectrogram obtained using the Mexican hat wavelet (middle) and the Morlet wavelet (bottom)

13.4 Assessment of Beat-to-Beat Cardiovascular Hemodynamic Parameters

We propose here a method to evaluate the functional state of the cardiovascular system on the basis of the analysis of two-dimensional time-frequency wavelet spectrograms of differential rheogram. Specific features of the systolic and diastolic phases of the cardiac cycle can be directly recognized by naked eye analysis of the wavelet spectrogram. Figure 13.3 (top) shows wavelet spectrograms of differential rheogram (made by Mexican hat wavelet) for three sequential cardiocycles. Positive bumps (yellow-red areas) at $a < 1$ correspond to E- and O-waves in each cardiac cycle. It is possible to define the time t_E, the scale a_E, and the amplitude w_E for each extremum point. We assume w_E and a_w to be proportional to the E-wave amplitude and LVET, correspondingly. Note that this approach does not require filtering and deionizing of ICG signal for latter identification of the characteristic points as it is suggested to do with the help of the discrete wavelet transform (Hu et al. 2014). Extremes produced by high-frequency noise or low-frequency respiration wave are naturally separated from cardiac waves on the wavelet plane (Stepanov et al. 2016).

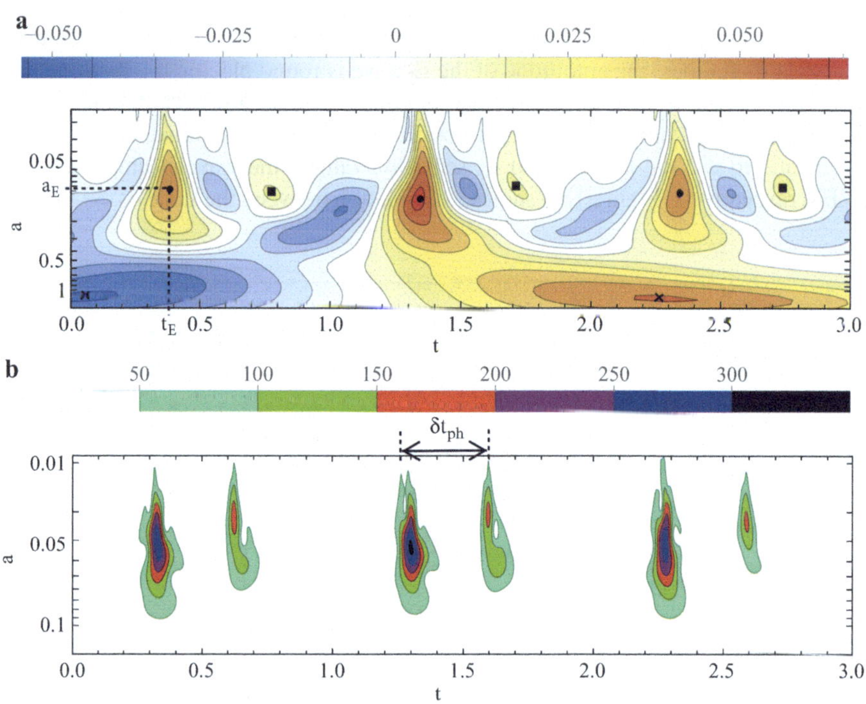

Fig. 13.3 Wavelet spectrograms of the ICG (top) and PCG (bottom) for three sequential cardiocycles: E-wave peaks (points), O-wave peaks (squares), respiration wave extreme (crosses). δt_{ph} is LVET estimated from PCG

The left ventricular ejection time is defined as the time interval of the left ventricular ejection, which occurs between the opening of the aortic valve and its subsequent closure. There are several technologies for detection of opening and closing times of the aortic valve. The most accurate measurement results with respect to the clinical gold standard, the echocardiography, can be achieved using the phonogram (Carvalho et al. 2010). Usually, the ejection time is determined by the time difference between two maxima in PCG signal (like in Fig. 13.1), which is supposed to correspond to the first and second sounds of the cardiac cycle in the phonocardiogram signal. This method is not sufficiently accurate to provide a robust estimation of cardiovascular activity. Its application may lead to an additional error in the estimates of SV. We calculate the ejection time δt_{ph} using the spectrogram (made by the Morlet wavelet) of the PCG data. The wavelet spectrogram of the phonogram provides information to recognize characteristic points of the cardiac cycle. Each cardiac sound produces an elongated vertical structure in the wavelet spectrogram (dashed lines in Fig. 13.3, bottom). Maximal spectral power (spots at $a \approx 0.03$ s) is provided by the blood flow, while the ascending tails correspond to valve vibrations (Meziani et al. 2012).

SV is usually calculated from the ICG signal as the product of the maximum of the first derivative (the E-peak in ICG) by the ejection time, corrected by a factor, which depends on blood resistivity, distance between receiving electrodes, and basic impedance of the body segment limited by receiving electrodes. According to standard methods, the amplitude of the E-wave is found placing the rheogram on the isoline, i.e., compensating the breathing and motional artifacts. Application of the wavelet-based approach makes it possible to separate the pulse and breathing waves and to use the E-wave and its scale to evaluate the wavelet amplitude and LVET, respectively. Figure 13.4 shows a correlation between the scale of the E-wave

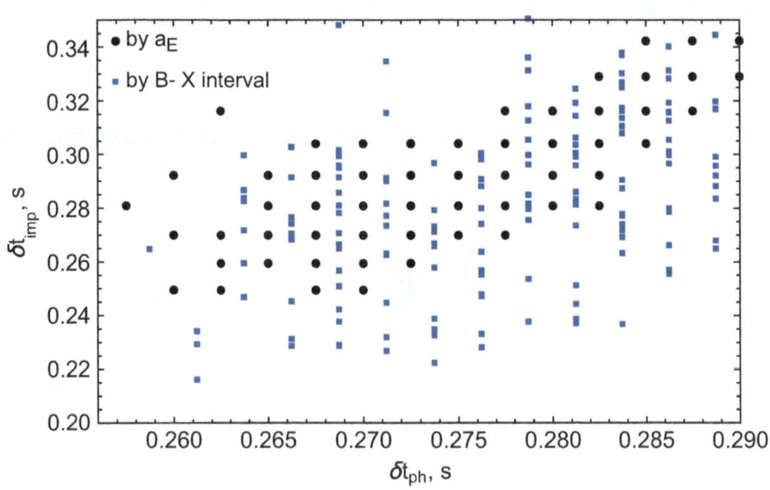

Fig. 13.4 Relation of the ejection time defined from the differential rheogram (δt_{imp}) and PCG (δt_{ph}). For comparison δt_{imp} is calculated using wavelets (points) and X-B points (squares)

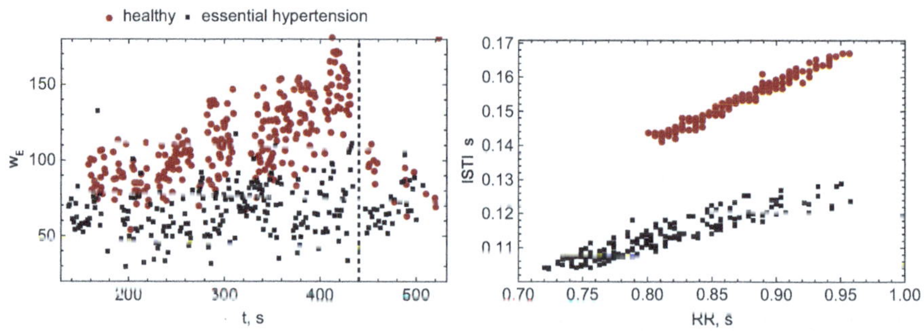

Fig. 13.5 The loading isometric test for healthy subjects (points) and the patient with hypertension (squares): time dependence of E-wave amplitude (left), ISTI vs R-R interval (right)

and the time interval between the opening of the aortic valve and its further closure obtained from PCG. The approach ensures higher accuracy in comparison with the method of determining the ejection time by analyzing the characteristic points (X and B) of the differential rheogram. With this approach, the initial systolic time interval (ISTI) is calculated as the time interval between the R-pick of ECG and the maximum of the E-wave in the wavelet spectrogram instead of the maximum of the E-wave in ICG signal (Van Eijnatten et al. 2014).

The method proposed was tested on 12 healthy men aged 20–25 years (the control group) and 14 patients with essential hypertension. During polyrheocardiogram registration, we carried out the loading isometric test—raising the legs from the horizontal position and keeping them at an angle of 30–45°. For healthy men, the linear growth of the E- and O-wave amplitudes was observed. On completion of loading (the legs returned to the horizontal position), the values of amplitudes of the E- and O-waves come back to their initial values. A change in the stroke volume during this test is attributed to the effect of the Frank-Starling mechanism, and its linear growth under loading corresponds to normal physiological response (Stepanov et al. 2016). For patients with essential hypertension, we observed the weakly pronounced change in the amplitudes of E- and O-waves (the E-wave amplitudes are shown in Fig. 13.5, left) and the shortening of the ISTI interval during the test, which is indicative of serious disorders in vicarious mechanisms. A relationship between the lengths of ISTI and R-R intervals for healthy men and patients is presented in Fig. 13.5, right.

13.5 Correlation Analysis of Respiration, R-R Interval, and Stroke Volume

Our approach can be used to perform a simultaneous time-frequency analysis of both the pulse and respiratory components of the rheography signal. The main cause of impedance variations at the intake of breath is the filling of lung vesicles with air,

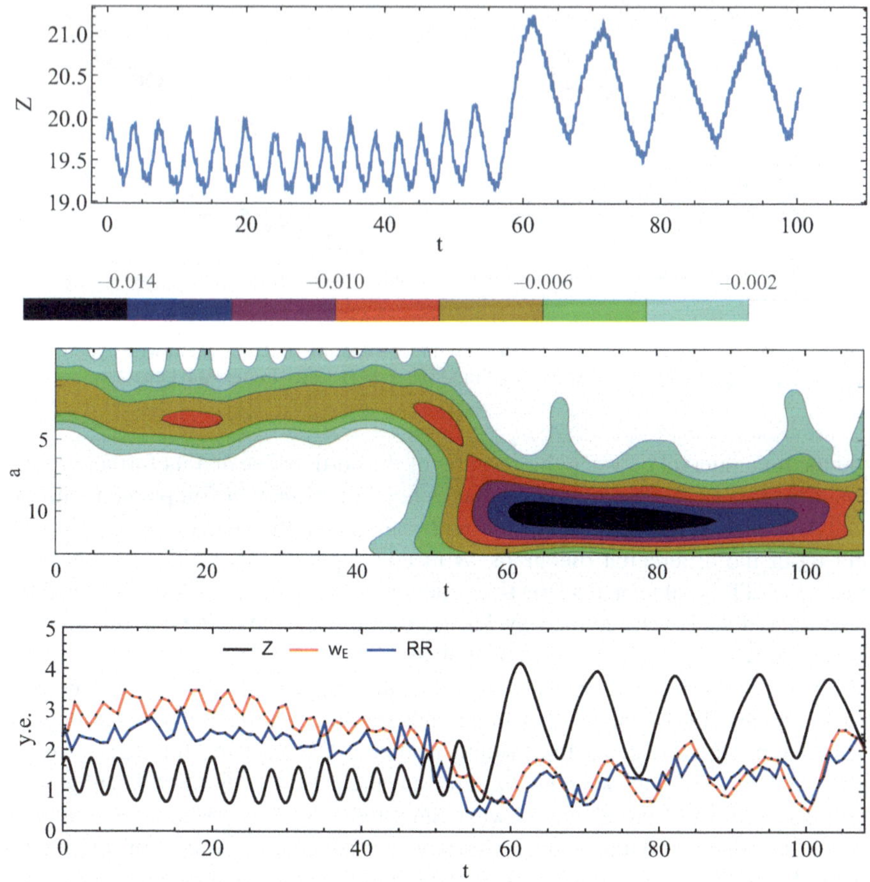

Fig. 13.6 Top: the original ICG signal. Middle: ICG wavelet spectrogram. Bottom: the low-pass filtered ICG (black), the E-wave amplitude (red), and the R-R interval (blue) at normal ($t < 55$) and deep ($t > 55$) breathing

which increases the impedance. The R-R interval length and cardiac expulsion value change concurrently. Figure 13.6 presents the original rheogram, its wavelet image (by the Morlet wavelet) and the ICG signal after low-pass filtering, as well as the E-wave amplitude and the R-R interval variations at normal and deep breathing. The wavelet spectrogram clearly demonstrates the respiratory wave, whose frequency and intensity change markedly at deep breathing (the pulse wave goes beyond the upper boundary of the figure). Figure 13.7 illustrates the distribution of R-R intervals and the stroke volume versus the respiratory cycle phase defined on the basis of a low-frequency (respiratory) variation. The example shows how modulations of the R-R interval and stroke volume change depending on the respiration depth. The pressure in the thoracic cavity during respiration is lower than normal, and therefore the vessels of thoracic cavity organs extend, the amount of blood ejected

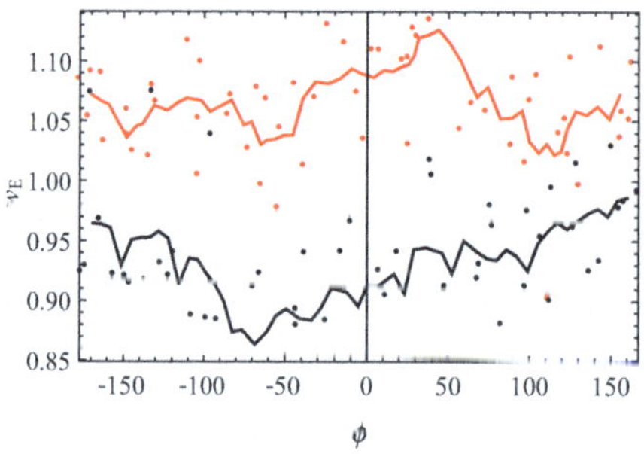

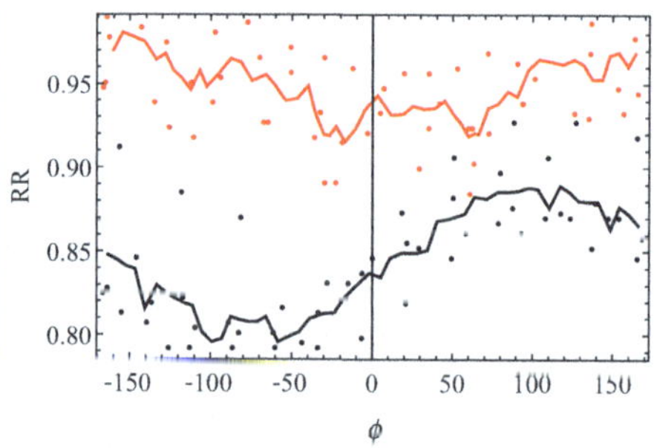

Fig. 13.7 Dependence of E-wave amplitude (left) and R-R interval (right) on the respiratory cycle phase under normal (red points) and deep (black points) breathing. $\phi > 0$ means inhalation and $\phi < 0$ means exhalation

into the left heart reduces, and the cardiac output decreases (Hall and Guyton 2006). Such interpretation of mutual functioning and interactions between the vascular and respiratory systems can serve as a basis for the development of complex functional tests based on respiration.

13.6 Final Remarks

Series-produced rheocardiographs are not provided by the software, which enables automatic data processing. Therefore, signal processing is strongly affected by subjective estimates and requires, in the long run, further training for medical personnel. Besides, the measurements and simulations carried out with these devices yield a set of hemodynamic characteristics obtained while assessing differential rheograms and calculating the stroke volume with the aid of Kubicek's formula. The opinion of top cardiologists on the efficiency of the described approach varies, and some studies provide results, which do not support this method of calculating the stroke volume. Our approach offers opportunities to obtain the characteristics proper of the cardiac cycle diastolic phase and to refine the definition of stroke volume. Data processing has been carried out in automatic mode. With these advantages, a mobile cardiograph for screening diagnostics can be developed. The portable device is intended for use not only in the centers offering specialized cardiac care but also in ordinary polyclinics for primary inspection.

Acknowledgements The work is supported by the Russian Science Foundation under project 14-15-00809.

References

Bour, J., & Kellett, J. (2008). Impedance cardiography – A rapid and cost-effective screening tool for cardiac disease. *European Journal of Internal Medicine, 19*(6), 399–405.

Bernstein, D. P., Henry, I. C., Lemmens, H. J., Chaltas, J. L., DeMaria, A. N., Moon, J. B., Kahn, A. M. (2015). Validation of stroke volume and cardiac output by electrical interrogation of the brachial artery in normals: Assessment of strengths, limitations, and sources of error. *Journal of Clinical Monitoring and Computing, 29*(6), 789–800.

Carvalho, P., Paiva, R., Couceiro, R., Henriques, J., Antunes, M., Quintal, I., Muehlsteff, J., Aubert, X. (2010). Comparison of systolic time interval measurement modalities for portable devices. *Conference proceedings of Annual International Conference of the IEEE Engineering in Medicine and Biology Society* (pp. 606–609). IEEE Engineering in Medicine and Biology Society.

Chabchoub, S., Mansouri, S., & Salah, R. (2016). Impedance cardiography signal denoising using discrete wavelet transform. *Australasian Physical and Engineering Sciences in Medicine, 39*(3), 655–663.

Chen, S. J., Gong, Z., & Duan, Q. L. (2014). Evaluation of heart function with impedance cardiography in acute myocardial infarction patients. *International Journal of Clinical and Experimental Medicine, 7*(3), 719–727.

Cybulski, G. (2011). *Ambulatory impedance cardiography: The systems and their applications.* Lecture Notes in Electrical Engineering. Berlin: Springer.

Devereux, R., Wachtell, K., Gerdts, E., Boman, K., Nieminen, M., Papademetriou, V., Rokkedal, J., Harris, K., Aurup, P., & Dahlof, B. (2004). Prognostic significance of left ventricular mass change during treatment of hypertension. *Journal of the American Medical Association, 292*(19), 2350–2356.

Dumler, A., Zubarev, M., Muraviev, N., Mamatova, A., Salnikova, N., Podtaev, S., Stepanov, R., & Frick, P. (2010). Wavelet analysis of bioimpendancometric data. *Journal of Physics: Conference Series, 224*(1), 012108.

Ebrahim, M., Hegde, S., Printz, B., Abcede, M., Proudfoot, J. A., & Davis, C. (2016). Evaluation of impedance cardiography for measurement of stroke volume in congenital heart disease. *Pediatric Cardiology, 37*(8), 1453–1457.

Ermishkin, V., Lukoshkova, E., Bersenev, E., Saidova, M., Shitov, V. Vinogradova, O., & Khayutin, V. (2007). Beat-by-beat changes in pre-ejection period during functional tests evaluated by impedance aortography: A step to a left ventricular contractility monitoring. In *IFMBE Proceedings: 13th International Conference on Electrical Bioimpedance and the 8th Conference on Electrical Impedance Tomography* (vol. 17, pp. 655–658).

Hall, J. E., & Guyton, A. C. (2006). *Textbook of Medical Physiology* (11th ed.) New York: Elsevier.

Holme, N., Rein, E., & Elstad, M. (2016). Cardiac stroke volume variability measured non-invasively by three methods for detection of central hypovolemia in healthy humans. *European Journal of Applied Physiology, 116*, 2187–2196.

Hu, X., Chen, S., Ren, R., Zhou, B., Qian, Y., Li, H., Xia, S. (2014). Adaptive filtering and characteristics extraction for Impedance Cardiography. *Journal of Fiber Bioengineering and Informatics, 7*(1), 81–90.

Kirchhof, P., Sipido, K.R., Cowie, M.R., Eschenhagen, T., Fox, K.A.A., Katus, H., Schroeder, S., Schunkert, H., Priori, S., & E.C.R.E.A. Work (2014). The continuum of personalized cardiovascular medicine: A position paper of the European Society of Cardiology. *European Heart Journal* **35**(46), 3250–3257.

Kubicek, W., Karnegis, J., Patterson, R., Witsoe, D., Mattson, R. (1966). Development and evaluation of an impedance cardiac output system. *Aerospace Medicine, 37*(12), 1208–1212.

Mallam, M., & Bhushana Rao, K. C. (2016). Efficient reference-free adaptive artifact cancellers for impedance cardiography based remote health care monitoring systems. *SpringerPlus, 5*(1), 1078.

Meziani, F., Debbal, S., & Atbi, A. (2012). Analysis of phonocardiogram signals using wavelet transform. *Journal of Medical Engineering and Technology, 36*(6), 283–302.

Morris, R., Sunesara, I., Darby, M., Novotny, S., Kiprono, L., Bautista, L., Sawardecker, S., Bofill, J., Anderson, B., & Martin, J. N. (2016). Impedance cardiography assessed treatment of acute severe pregnancy hypertension: A randomized trial. *The Journal of Maternal-Fetal & Neonatal Medicine, 29*(2), 171–176.

Morris, R., Sunesara, I., Rush, L., Anderson, B., Blake, P. G., Darby, M., Sawardecker, S., Novotny, S., Bofill, J. A., & Martin Jr., J. N. (2014). Maternal hemodynamics by thoracic impedance cardiography for normal pregnancy and the postpartum period. *Obstetrics and Gynecology, 123*(2, 1), 318–324.

Payseur, J., Rigney, J., Turner, S., Wu, X., Murphy, D., & Rossman, E. (2016). Evaluation of a method utilizing PhysioFlow, a novel signal morphology-based form of impedance cardiography, to measure cardiac output in the conscious beagle. *Journal of Pharmacological and Toxicological Methods, 81*, 115–119.

Pickett, B. R., & Buell, J. C. (1993). Usefulness of the impedance cardiogram to reflect left-ventricular diastolic function. *American Journal of Cardiology, 71*(12), 1099–1103.

Podtaev, S., Stepanov, R., Smirnova, E., & Loran, E. (2015). Wavelet-analysis of skin temperature oscillations during local heating for revealing endothelial dysfunction. *Microvascular Research, 97*, 109–114.

Stepanov, R., Podtaev, S., Dumler, A., & Chugainov, S. (2016a). Assessment of cardiac time intervals by wavelet transform of the impedance cardiogram. *Technology and Health Care, 24*, S803–S809.

Stepanov, R., Podtaev, S., Dumler, A., & Chugainov, S. (2016b). Assessment of systolic heart function by wavelet analysis of the impedance cardiogram. In F. Simini & P. Bertemes-Filho (Eds.), *IFMBE Proceedings: II Latin American Conference on Bioimpedance: 2nd CLABIO, Montevideo, September 30 - October 02, 2015* (pp. 32–35). Singapore: Springer.

Van Eijnatten, M., Van Rijssel, M., Peters, R., Verdaasdonk, R., & Meijer, J. (2014). Comparison of cardiac time intervals between echocardiography and impedance cardiography at various heart rates. *Journal of Electrical Bioimpedance, 5*, 2–8.

Index

A

Absolute imaging, 31, 35
Active electrodes, 19
Admittivity, 119, 122
Alternating current (AC), 204, 221
Amyotrophic lateral sclerosis (ALS), 197
Ani3D package, 121, 123
Ansoft software, 112
Archie's law, 96
Athletes, BIA of body composition, *see* Sports, BIA of body composition

B

Back project algorithm, 32
Barrett's esophagus, 197, 198
Battery-less implantable sensor, 205
BCRL, *see* Breast cancer-related lymphedema (BCRL)
Bellows-type spirometer (Vitalograph), 163, 165
BIA, *see* Bioimpedance analysis (BIA)
Binary medicine, 210
Binatron, 210
Bioelectrical impedance spectroscopy (BIS), 91, 97, 117, 118, 226
  See also Electrical bioimpedance spectroscopy (EBIS)
Bioelectrical impedance vector analysis (BIVA)
  body composition, 232
  dengue, 70
  respiratory system, 201–202
  tolerance ellipses, 226–227
Bio-hybrid tactile sensor array, 95
Bioimpedance analysis (BIA)
  advantages, 244
  body composition assessment (*see* Body composition (BC) assessment)
  definition, 221
  homogeneous isotropic cylindrical conductor, 117
  malnutrition determination, 201–202
  total intra-and extracellular water calculation, 194
Bioimpedance measurements, numerical analysis
  computational technology
    Ani3D package, 121, 123
    conformal tetrahedral mesh, 120, 123
    Delaunay triangulation algorithm, 122
    finite element method, 120
    GMRES based iterative linear solver, 121
    hierarchical meshes, convergence study on, 121
    high-resolution human body geometrical model, 121
    human body tissues, admittivity of, 122
    human torso, simplified geometrical model of, 121, 122
    initial mesh generation, 123
    ITK-SNAP software, 122, 123
    post-processing algorithms, 122
    segmented torso model, 123
    uniform refinement, 120
    unstructured tetrahedral meshes, 120, 123
    VHP man, segmented whole body model of, 124, 132

F. Simini, P. Bertemes-Filho (eds.), *Bioimpedance in Biomedical Applications and Research*, https://doi.org/10.1007/978-3-319-74388-2

Bioimpedance measurements, numerical analysis (*cont.*)
VHP man torso, 121
current conduction
alpha (α) dispersion, 105
anisotropic properties, 103
dipole moments, 104
direction-dependent conductivity, 103
electric flux density, 104, 105
frequency-dependent conductivity, 102, 103
Maxwell-Wagner/beta (β) dispersion, 105
mobility, 102
Ohm's law, 102
relative dielectric constant, 104
specific conductivity, 102
digital bioimpedance models, 132
EIT, 118
electrical impedance tomography, 132
electrical parameters changes, 101
electrode-tissue interface, 107
equations
electrode interface, 109
induced current electrical impedance tomography, 111
magnetic effects, influence of, 109–111
quasi-static case, 108
time retardation effects, 108
FEM-based computational technology, 118
FEMM, 111–115
finite element discretizations, 132
geometrical changes, 101
homogeneous isotropic cylindrical conductor, 117
mathematical model, 118–119
online numerical simulator, 127–128
sensitivity analysis, 124–127
Biopac MP35, 91
Bipolar impedance technique, 11, 12
BIS, *see* Bioelectrical impedance spectroscopy (BIS)
BIVA, *see* Bioelectrical impedance vector analysis (BIVA)
Body cell mass (BCM), 249–251
Body composition (BC) assessment
BIA applications
advantages, 224
affecting factors, 230
bioelectrical spectroscopy, 226
BIVA, 226–227
clinical applications, 231–235
differential impedance analysis, 227
foot-to-foot/leg-to-leg bioimpedance measurements, 228
hand to foot bioimpedance, 228
HH-BIA measurements, 229
limitations, 224
mathematical models and formulas, 231
MF-BIA, 225–226
precision and reproducibility, 223–224
prediction equations, 223
principles and assumptions, 221–223
raw data, use of, 231
SBIA, 229
SF-BIA, 225
SLBIA, 230
in sports (*see* Sports, BIA of body composition)
cadaver analysis, 220
complexity levels, 219
components, 219–220
EBI, 68–70, 79
field techniques, 221
four component models, 220
in vivo methods, 220
laboratory tests, 220–221
practical uses of, 220
two-compartment model, 219–220
Bone examination, EBIS, 206
Brain imaging, 39–41
Breast cancer-related lymphedema (BCRL), 68
Breast imaging, 29, 31, 35–36, 39, 40

C

Cancer screening
BCRL, 68
EBI
breast cancer, 75
colorectal cancer, 68
prostate cancer, 67–68
uterine cervix cancer, 68
and EBIS, 207–208
EIT, prostate cancer imaging, 29
Cardiovascular diseases
remodeling process, 257
wavelet analysis, impedance rheocardiography (*see* Wavelet analysis)
Cardiovascular system
ICG, 202
IPG, 202–203
TBEV, 203
CE clinical EIT system, 38
Chest imaging, 29–30

Chronic diseases and EBIS, 209–210
Chronic lung disease of infancy (CLDI), 38
Chronic obstructive pulmonary disease (COPD), 75
Clinical applications
  body composition assessment, BIA, 231–235
  EBI/EBIS applications
    bones, 206
    cancer, 207–208
    cardiovascular system, 202–203
    cells, 193
    chronic diseases, 209–210
    Debye and Cole model, 191–192
    digestive system (*see* Digestive system, EBI/EBIS applications)
    extracellular space, 192–193
    genitourinary system, 203
    human body systems and functions, 189, 190
    human clinical studies, 188–189
    intracellular space, 194
    muscle, 206–207
    respiratory system, 201–202
    skin, 203–205
  human bioelectricity, study of, 187–188
Cole model, 98, 191–192
Common-mode rejection ratio (CMRR), 19, 21–23, 31
Common-mode (CM) voltage, 18, 20, 24
Comsol software, 112
COMSOL simulation, 150, 153, 154, 159, 176, 177
Conventional wrist-to-ankle 4-electrode scheme, 125, 128
Current carrying (CC) electrodes, 119, 124, 127, 128

D
Dartmouth breast imaging system, 35–36, 39, 40
D-bar methods, 35
Debye model, 191–192
Delaunay triangulation algorithm, 122
Dental caries, 195–196
Dermatron, 205
Dielectric spectroscopy, 89–90
Difference imaging, 31
Differential impedance analysis (DIA), 227
Digestive system, EBI/EBIS applications
  dental canal length, determination of, 196
  dental caries, 195–196
  esophagus, 197–199
  large intestine (colon and rectum), 200–201
  mouth (oral) mucosa, 194
  small intestine, 199–200
  stomach, 198–199
  teeth, 194–195
  tongue, 197
Direct current (DC) method, 204
Dispersion, 9

E
EBI, *see* Electrical bioimpedance (EBI)
EIDORS, 33, 55, 56
EIT, *see* Electrical impedance tomography (EIT)
Electrical bioimpedance (EBI)
  clinical applications (*see* Clinical applications)
  imaging characterization
    breast cancer detection and monitoring, 75
    ECG *vs.* ICG, 76
    EIT, 73–74
    myocardial electrical impedance, 76–77
    pneumotachometer signal *vs.* EBI signal, 75
    proportionality coefficient, 75
    total impedance index (TII), 74
  non-imaging characterization
    BCRL, 68
    BIVA, 70
    body composition evaluation, 69
    colorectal cancer, 68
    dermal evaluation, 67
    fat-free mass (FFM), 69
    gastric motility, 73
    gastrointestinal system, 72
    ischemia and vascularity damage, 71
    knee injuries, 71–72
    Nevisense device, 67
    plethysmography, 71
    prostate cancer, 67–68
    reflux episodes, 72
    single channel multifrequency assessment, 66
    uterine cervix cancer, 68
Electrical bioimpedance spectroscopy (EBIS)
  bones, 206
  cancer, 207–208
  cardiovascular system, 202–203
  cells, 193
  chronic diseases, 209–210
  Debye and Cole model, 191–192

Electrical bioimpedance spectroscopy (EBIS) (*cont.*)
  digestive system (*see* Digestive system, EBI/EBIS applications)
  extracellular space, 192–193
  genitourinary system, 203
  human body systems and functions, 189, 190
  human clinical studies, 188–189
  intracellular space, 194
  muscle, 206–207
  respiratory system, 201–202
  skin, 203–205
  See also Bioelectrical impedance spectroscopy (BIS)
Electrical caries monitor (ECM), 196
Electrical cell model, 9–10
Electrical impedance myography (EIM), 197, 206
Electrical impedance signal analysis, *see* Electrical bioimpedance (EBI)
Electrical impedance spectroscopy (EIS)
  binary multifrequency signals, 6–7
  chirp signal, 6–7
  current injection technique, 7
  definition, 5
  electromagnetic spectrum, 5–6
  noise signal, 6
  sinusoidal signal, 6
  step function, 6
  tissue impedance
    anisotropy, 9
    complex admittance, 8
    complex capacitance, 8
    complex relative permittivity, 10
    conductivity, 7, 8
    dimensionless distribution parameter, 10
    electrical cell model & equivalent circuit, 9–10
    permittivity *vs.* frequency, 10
    relative permittivity, 7, 8
    slab of tissue & equivalent circuit, 8
    theoretical impedance diagram, 9
Electrical impedance tomographic spectroscopy (EITS), 29
Electrical impedance tomography (EIT)
  clinical applications
    brain imaging, 39–41
    breast imaging, 39, 40
    gastrointestinal function, 41
    hyperthermia, 41
    pulmonary function, 36–39
  DX 1800 system, 51
  gastric emptying study, 169–170
  hardware
    calibration, validation and sources of error, 31
    injection protocols, 30
  for human thorax imaging, 50
  imaging
    absolute imaging, 31
    back project algorithm, 32
    breast imaging system, 35–36
    D-bar methods, 35
    difference imaging, 31
    EIDORS software, 33
    forward problem, 34
    gold standard algorithm, 32
    GREIT algorithm, 32
    image reconstruction, 33–34
    Maxwell equations model, 33
    Swisstom AG EIT system, 31–32
  MAX038 function generator, 50
  PC-LPM-16PnP board, 49
  pulmonary oedema (*see* Pulmonary oedema, EIT)
  PulmoVista 500, 51
  RC phantom, 51
  reliability and usability, 50
  Sheffield Mk 1 system, 29–30
  tidal volume determination, 73–74
  tissue engineering applications, 95
  TLV320AIC23 board, 50
  TMS320F206 DSP, 50
  2D image, 139
  variations
    MIT, 42
    MREIT, 42
Electric Cell–substrate Impedance Sensing (ECIS), 91
Electroacupuncture according to Voll (EAV), 205
Electrocardiography (ECG), 76, 258, 259
Electrodermal activity (EDA), 203
Electrodermal potentials (EDP), 204
Electrodermal response (EDR), 204
Electrodermal screening test (EDT), 205
Electrogastrography (EGG), 198, 199
Electrostimulation, 210
Elite Czech judo athletes, 246–248
EndoFlip®, 198, 199
Esophagus, 197–199
Evaluation module (EVM) OMAP-L137 model, 58–60
Extra and intracellular mass (ECM/BCM) ratio, 250, 251

Extracellular mass (ECM), 244, 249, 250
Extracellular space (ES), 192–193
Extracellular water (ECW), 225, 246–248

F
Fat-free mass (FFM)
  body composition, 219–220
  IIH BIA, 229
  prediction error, 224
  prediction for female athletes, 245
  SEE, 223
  SF-BIA, 225
Fat mass (FM)
  body composition, 219, 243
  in elite athletes, 249–251
  HD and BIA analysis, 244–246
  proportion in athletes, 251–252
FEMM, *see* Finite element method magnetics (FEMM)
Field cancerization, 207–208
Finite element method (FEM), 118, 120, 132, 147, 150, 159, 178
Finite element method magnetics (FEMM)
  (bio)impedance, calculation of, 115
  Dirichlet boundary condition, 113
  DXF format, 113
  equipotential lines, 114
  graphical interface of, 112
  LUA, 112
  Neumann boundary condition, 113
  octave interface with, 113
  Octave/Matlab, 112, 113
  simplified simulation structure, 114
  skin bioimpedance, 114
Focused impedance method (FIM)
  advantages and disadvantages, different versions, 161–162
  analytical and numerical methods, sensitivity distribution in
    COMSOL simulation, 150
    electrode plane, 2D sensitivities parallel to, 151
    FEM, 147–150
    impedance variations, 150
    lead fields, 145–147
    normalised impedance, FIM-4, 153
    transfer impedance, 150
    2D positive and negative sensitivity, in two orthogonal planes, 152
  autodiagnosis, pneumonia detection in babies, 167–168
  configuration and visualisation
    FIM-4 (Four-Electrode Version), 143
    FIM-6 (Six-Electrode Version), 142–143
    FIM-8 (Eight-Electrode Version), 140–142
  equipotentials/point sensitivity, 162–163
  experimental methods, sensitivity distribution in
    COMSOL, 153, 154
    phantom measurements, 154
    2D sensitivity, 154–157
    3D sensitivity, 159–161
    3D, visualisation of sensitivity, 157–159
  gastric (stomach) emptying study, 169–171
  instrumentation, 143–145
  lungs ventilation measurement
    localised measurement, 163–165
    upper thorax, mapping of, 165–167
  measurement challenges
    body movements, effect of, 180
    person-to-person variation, 181
    repeatability and reproducibility, 179–180
    skin-electrode movements, 180
  negative sensitivity, 139
  non-invasive breast tumour characterisation, 174–179
  subcutaneous abdominal fat thickness, 171–174
  3D sensitivity of, 139
  TPIM
    basic linear configuration of, 138
    basic square configuration of, 137, 138, 140
    EIT, 139
    equipotential lines in 2D, 138
    linear TPIM sensitive zone equipotential visualisation of, 138, 139
    square TPIM, sensitive zone equipotential visualisation of, 138, 139
  transfer impedance, 137
  weighted back projection technique, 139
Forward problem-solving, 34, 55
Frank-Starling mechanism, 265
Fricke–Morse model, 227

G
Galvanic skin response (GSR), 203
Gastric motility, 199, 200
Gastroesophageal reflux disease (GERD), 197
Gastroesophageal reflux episodes, 72
Gastrointestinal function monitoring, 41

Genitourinary system, EBI/EBIS applications, 203
Geselowitz's lead fields, 146
Graz consensus reconstruction algorithm for EIT (GREIT), 32, 56–57, 95, 96
GREIT software, *see* Graz consensus reconstruction algorithm for EIT (GREIT)

H
Hand-to-hand bioimpedance (HH-BIA), 229
Helmholtz-Gouy-Chapman-Stern electrical lumped model, 107
HP4192A impedance analyzer, 97–98
Hydrodensitometry (HD), 244–246
Hyperthermia, 41
Hypotonic meridian, 210

I
ICG, *see* Impedance cardiography (ICG)
Image reconstruction
  EIT, 33–34
  GREIT, 56–57
  IMPETOM, 56–58
  NOSER, 57–58
  phantom and volunteer experimental data, 56
Impedance cardiography (ICG)
  in cardiovascular assessment, 202
  vs. ECG, 76
  wavelet analysis
    cardiocycles, selection of, 259–260
    cardiovascular parameter monitoring, 258
    continuous wavelet transform technique, 260–262
    E-and O-wave peaks, respiration wave extreme, 263
    vs. echocardiography, 258
    hemodynamic parameters, variability of, 259
    iCor, 259
    loading isometric test, 265
    LVET, 259, 264
    respiration, R-R interval and stroke volume, 265–267
    signal, characteristic points and waves, 258–259
    tetrapolar thoracic ICG method, 259
Impedance flow cytometer, 92, 93
Impedance index, 222
Impedance planimetry, 198
Impedance plethysmography (IPG), 117, 202–203
IMPETOM reconstruction software, 56–58, 61–62
Initial systolic time interval (ISTI), 265
Interdigitated microsensor electrodes (IME), 94
Intracellular water (ICW), 225, 246–248
Inverse problem, EIT
  forward problem-solving, 55
  mathematical background, 54
  regularization, 55
ITK-SNAP software, 123

J
JavaScript 3D library, 127
Johnson thermal noise, 22

K
Kubicek's formula, 259, 267

L
Laplace equation, 112
Lean body mass (LBM)
  elite athletes, 249–251
  short-term intervention, hydration changes, 246–248
Left ventricular ejection time (LVET), 259, 263, 264
Loading isometric test, 265
Lung function imaging, 38

M
Magnetic induction tomography (MIT), 42
Magnetic resonance electrical impedance tomography (MREIT), 36, 37, 42
MALTRON Bioscan 920-II equipment, 175
Mass diffusion process, 107
Maxwell equations model, 33
Medical diagnosis, definition, 79
Mesenchymal stem cells (MSCs), 92, 97
Mesh adaptivity, 133
Mexican hat wavelet, 261–263
MIT, *see* Magnetic induction tomography (MIT)
Morlet wavelet, 261, 262, 264
Morphological asymmetries, 244, 252–254
MREIT system, 36, 37
Multifrequency BIA method, 244, 249, 253

Multiple frequency bioimpedance analysis (MF-BIA), 223, 225–226
Muscle mass (MM), 244, 248, 253, 254
Musculoskeletal system, EBI/EBIS applications
  bones, 206
  muscle, 206–207

N
Negative impedance circuit, 19
Nevisense, 205
NOSER tomographic reconstruction, 57–58
Numerical analysis, bioimpedance measurements, *see* Bioimpedance measurements, numerical analysis

O
OMAP-L137/TMS320C6747 Floating-Point Starter Kit, 53
Oral mucosa, 194

P
pH and multichannel intraluminal impedance (pH-MII) technique, 197–198
Phase angle (PhA), 246–248
Phonocardiogram (PCG)
  LVET, determination of, 263–265
  signal of, 258, 259
Pick-up (PU) electrodes, 119, 124, 127, 128
Pigeon hole imaging (PHI), 177, 182
Plethysmography, 71, 202
Pneumotachometer, 73, 75
Poisson equation, 112, 116
Polyrheocardiography, 258
Prostate cancer imaging, 29
Pulmonary function monitoring, 36–39
Pulmonary oedema, EIT
  circuit designs, 47
  colour-coded representation of normal lungs, 47
  current leakage to adjacent chest sections, 62
  design options
    discrete components with digital processing, 52–53
    discrete components with no digital processing, 51–52
    OMAP-L137/TMS320C6747 Floating-Point Starter Kit, 53
    standard evaluation board, 52
  electrical phantom for testing
    CODEC, 61
    2016 IMPETOM software, 61
    RC phantom and relative error, 61
    saline-filled tank, 59, 62
  fluid occupation, 46
  hardware architecture
    control and data storage block, 49
    current source, 48
    differential amplifier, 48–49
    multiplexers, 49
    synchronous voltmeter, 49
    waveform synthesis, 48
  inverse problem
    forward problem-solving, 55
    mathematical background, 54
    regularization, 55
  standard board, 58–60
  tomographic reconstruction
    GREIT, 56–57
    IMPETOM, 56–58
    NOSER, 57–58
    phantom and volunteer experimental data, 56

R
Rectal EBIS (REBIS), 201
Relative permittivity, 89–90, 104
Relative values, LBM, 249–251
Respiratory system, EBI/EBIS applications, 201–202

S
Sarcopenia diagnosis, 232
Screen driving technique, 18–19
SEE, *see* Standard errors of the estimate (SEE)
Segmental bioimpedance analysis (SBIA), 229, 232
Segment localized electrical bioimpedance analysis (SLBIA), 230
Sensitivity analysis
  CC electrodes, 124, 127
  conventional 4-electrode scheme, 128
  conventional tetrapolar wrist-to-ankle measurement configuration, 125
  density vector field, 124
  homogeneous cylinder, tetrapolar scheme on, 125
  PU electrodes, 124, 127
  reciprocal lead field, 124
  10-electrode segmental BIA scheme, 125, 129–131

Sensitivity analysis (*cont.*)
  tetrapolar scheme measurements, 124, 127
  transfer impedance, 126, 132
  volume impedance density, 126
Sheffield Mk 1 system, 29–31
Single frequency bioimpedance analysis (SF-BIA), 225
Skeletal mass (SM), 232
Skin, EBI/EBIS applications, 203–205
Solartron 1260/1294 impedance gain-phase analyzer, 200
Sports, BIA of body composition
  body cell mass, 243, 249–251
  ECM/BCM ratio, 243, 250, 251
  extracellular mass, 244, 249–251
  fat mass (*see* Fat mass (FM))
  FFM prediction for female athletes, 245–246
  ICW and ECW, 243, 246–248
  lean body mass, 243, 246–251
  morphological asymmetries, 244, 252–254
  multifrequency BIA method, 244, 249, 253
  resistance, reactance and PhA, 246–248
  short-term intervention, hydration changes, 246–248
  sport nutrition, 244
  total body water, 246–248
Standard errors of the estimate (SEE), 223
Stomach, EBI/EBIS applications, 198–199
Swisstom AG EIT system, 31–32

T

TBW, *see* Total body water (TBW)
Tetrapolar impedance measurement (TPIM), 137
  basic linear configuration of, 138
  basic square configuration of, 137, 138, 140
  EIS, 13–14
  EIT, 139
  equipotential lines in 2D, 138
  equipotential surfaces, 158
  linear TPIM sensitive zone equipotential visualisation of, 138, 139
  multiple electrode separations of, 171
  square TPIM, sensitive zone equipotential visualisation of, 138, 139
Thorax imaging, 29
Tissue engineering
  bioelectrical impedance, applications of
    advantages, 98
    bio-hybrid tactile sensor array, 95
    cell number and viability, evaluation of, 91–92
    dielectric spectroscopy, 89–90
    disadvantage, 99
    ECIS, 91
    EIT, 95
    four-electrode interdigitated microsensor, 98
    GREIT, 95, 96
    HP4192A impedance analyzer, 97–98
    impedance flow cytometer, 92, 93
    impedance spectroscopy, 92
    interdigitated microsensor electrodes, 94
    micro-EIT system, 95
    morphological cell characteristics, monitoring of, 97–98
    scaffold characterization, 95–96
    3D cell cultures, monitoring of, 96–97
    xCELLigence system RTCA, 92–94
  challenges, 87–88
  concept model, 89
  definition, 87
Tissue impedance, EIS
  anisotropy, 9
  complex admittance, 8
  complex capacitance, 8
  complex relative permittivity, 10
  conductivity, 7, 8
  dimensionless distribution parameter, 10
  electrical cell model & equivalent circuit, 9–10
  hardware specifications, 23
  measurement
    bipolar technique, 11, 12
    hardware, 13–19
    tetrapolar technique, 12–13 (*see also* Tetrapolar impedance measurement (TPIM))
    transfer impedance, 11
  permittivity *vs.* frequency, 10
  relative permittivity, 7, 8
  slab of tissue & equivalent circuit, 8
  theoretical impedance diagram, 9
  voltage measurement
    artefacts, 19
    capacitances, 22–23
    differential amplifier, 20
    instrumentation amplifier, 20–21
    interference, 19
    Johnson thermal noise, 22
    multiplexers, 22

noise, 19, 22
receive circuit model, 21
Tolerance ellipses, BIVA, 226–227
Tongue, 197
Tongue electrical impedance, 197
Total body water (TBW)
bioelectrical spectroscopy, 226
MF-BIA, 225
pre and posttest values, 246–248
prediction error, 224
SEE, 223
SF-BIA, 225
Total impedance index (TII), 74
Total knee arthroplasty (TKA), 72
Traditional Chinese medicine (TCM), 210
Transbrachial electrical bioimpedance velocimetry (TBEV), 203
Transfer impedance, 11, 119

U
Upper body power, 246, 248

V
Visible Human Project (VHP), 118, 121, 123, 124, 132
Voltage-controlled current source (VCCS)
active electrodes, 19
common-mode voltage, 18
current mirror design, 15
Howland current generator, 16
ideal model, 14
isolated negative-feedback current source, 15
negative impedance circuit, 19
output impedance, 17
screen driving, 18–19
single-end four-electrode system, 17

W
Warburg impedance, 107
Wavelet analysis
impedance cardiography
cardiocycles, selection of, 259–260
cardiovascular parameter monitoring, 258
continuous wavelet transform technique, 260–262
E-and O-wave peaks, respiration wave extreme, 263
vs. echocardiography, 258
hemodynamic parameters, variability of, 259
iCor, 259
loading isometric test, 265
LVET, 259, 264
respiration, R-R interval and stroke volume, 265–267
signal, characteristic points and waves, 258–259
tetrapolar thoracic ICG method, 259
phonocardiogram, 259
LVET, determination of, 263–265
signal of, 258
WebGL technology, 127
Weighted back projection technique, 139
Whole-body electrical impedance, 228
Wound healing, 204

X
xCELLigence system Real-Time Cell Analizer (RTCA), 92–94

Z
Zedscan®, 203

The manufacturer's authorised representative in the EU is Springer Nature Customer Service Centre GmbH, Europaplatz 3, 69115 Heidelberg, Germany. If you have any concerns regarding our products, please contact ProductSafety@springernature.com

Printed and bound by CPI Group (UK) Ltd, Croydon, CR0 4YY

25/09/2026

02222891-0006